Advances in Neuroethics

Advances in neuroscience research are bringing to the forefront major benefits and ethical challenges for medicine and society. The ethical concerns related to patients with mental health and neurological conditions, as well as emerging social and philosophical problems created by advances in neuroscience, neurology and neurotechnology are addressed by a specialized and interdisciplinary field called neuroethics.

As neuroscience rapidly evolves, there is a need to define how society ought to move forward with respect to an ever growing range of issues. The ethical, legal and social ramifications of neuroscience, neurotechnology and neurology for research, patient care, and public health are diverse and far-reaching — and are only beginning to be understood.

In this context, the book series "Advances in Neuroethics" addresses how advances in brain sciences can be attended to for the benefit of patients and society at large.

Veljko Dubljević • Jonathan R. Young
Editors

TMS and Neuroethics

Springer

Editors
Veljko Dubljević
Department of Philosophy and Religious Studies
North Carolina State University
Raleigh, NC, USA

Jonathan R. Young
Department of Psychiatry & Behavioral Sciences
Duke University School of Medicine
Durham, NC, USA

ISSN 2522-5677 ISSN 2522-5685 (electronic)
Advances in Neuroethics
ISBN 978-3-031-92400-2 ISBN 978-3-031-92401-9 (eBook)
https://doi.org/10.1007/978-3-031-92401-9

This Springer imprint is published by the registered company Springer Nature Switzerland AG
The registered company address is: Gewerbestrasse 11, 6330 Cham, Switzerland

Contents

An Introduction to TMS and Neuroethics 1
Ariana D'Alessandro, Jonathan R. Young, and Veljko Dubljević

Historical and Theoretical Reflections Regarding TMS

A History of Transcranial Magnetic Stimulation 11
Sean D. Reeves, Zhi-De Deng, and Jonathan R. Young

Racial Disparities in TMS Research Participation and Clinical Access in the United States 29
Christi Sullivan, Dawson Cooper, Aaron McCright, Dawn Hackman, Alik Widge, Saydra Wilson, and Laura Y. Cabrera

Family and Caregiver Perspectives on TMS Treatment of Refractory Conditions: A Pilot Investigation 47
Ariana D'Alessandro, Iris Coates McCall, and Veljko Dubljević

Transcranial Magnetic Stimulation and Cognitive Enhancement: Present State and Possibilities for the Near Future 65
Bruce Luber and Eva Wiener

Unraveling Ethical Complexities: Transcranial Magnetic Stimulation (TMS) in Non-clinical Settings and the Dilemmas of Neuroenhancement 91
Dov Greenbaum

TMS for Social Neuroscience: New Applications and Ethical Issues 121
Michela Balconi

Bridging Network Neuroscience and Neuroethics: Can TMS and tACS Solve the Creativity Crisis? 139
Daniel Moseley and Flavio Fröhlich

New Directions and Ethical Issues in Clinical TMS Research

Subgrouping "Treatment Responders": Ethical Implications for those Seeking Participation in TMS Treatment and/or Research 153
Stephanie M. Hare and Daniel Roche

TMS for Mood Disorders: An Ethical Perspective 165
Mohsen Poorganji, Emma C. Boyd, Deepinder S. Nagra, and Lawrence G. Appelbaum

"The Fewer Pills the Better": Attitudes Toward Transcranial Magnetic Stimulation for Depressed Adolescents 181
Patricia Henegan, Paul E. Croarkin, Manisha Kodavatiganti, and Laura Y. Cabrera

Transcranial Magnetic Stimulation in Older Adults 199
Tajudeen O. Basiru, Joan A. Camprodon, and Michael D. Kritzer

The Ethics of TMS in the Treatment of Suicidal Ideation 213
Alyssa Bernanke, Claudia Roldan, and Brandon Hamm

Transcranial Magnetic Stimulation for Substance Use Disorders: Ethical Considerations 231
Dhvani D. Mehta, Melissa Hiebert, Victor M. Tang, Heather B. Ward, and Tony P. George

Regulatory Perspectives on Transcranial Magnetic Stimulation 247
Michael Pflanzer and Veljko Dubljević

An Introduction to TMS and Neuroethics

Ariana D'Alessandro, Jonathan R. Young,
and Veljko Dubljević

As transcranial magnetic stimulation (TMS) has evolved from a neuroscientific research tool to a widely expanding clinical treatment, numerous uncertainties remain, both necessitating and complicating ethical evaluation [1, 2]. Collectively, much of the existing TMS literature is limited by sufficient sample sizes [3] and lay audiences may develop misunderstandings about TMS therapy due to the conflation of findings from small studies with anecdotal patient experiences widely shared on social media [4]. There are legitimate, empirically supported concerns about potential interactions between TMS and psychiatric medications and other concurrent treatments [5], uncertainties about the presence of compromised tissue in targeted brain areas [6], and broader ethical questions about whether altering brain function through TMS-induced changes in oscillatory brain states could impact aspects of personhood [7].

There is also a clear need to improve understanding of the ethical concerns, perceived invasiveness, and overall acceptability of TMS across various clinical applications, particularly when considering specific treatment parameters such as stimulation intensity, coil location, and duration of applied frequencies, all of which can influence perceptions of invasiveness and seizure risk [8]. Although the risk of seizure from TMS is frequently described as "very low," which is primarily based

A. D'Alessandro
North Carolina State University, Raleigh, NC, USA
e-mail: amdaless@alumni.ncsu.edu

J. R. Young
Department of Psychiatry & Behavioral Sciences, Duke University School of Medicine, Durham, NC, USA
e-mail: jonathan.young@duke.edu

V. Dubljević (✉)
Department of Philosophy and Religious Studies, North Carolina State University, Raleigh, NC, USA
e-mail: veljko_dubljevic@ncsu.edu

V. Dubljević, J. R. Young (eds.), *TMS and Neuroethics*, Advances in Neuroethics,
https://doi.org/10.1007/978-3-031-92401-9_1

on data from clinical trials and post-marketing surveillance, with limited prospective evaluation in real-world clinical populations across diverse treatment settings [8]. A comprehensive review reported a seizure rate of approximately 0.1% per treatment course, primarily in individuals with predisposing factors such as a history of epilepsy or concurrent use of medications that lower the seizure threshold. Notably, the type of coil used can influence this risk; for instance, the H-coil has been associated with a higher seizure rate of 5.56 per 1000 patients, compared to 0.14 per 1000 patients with figure-8 coil devices [9]. Furthermore, side effects associated with TMS therapy are under-reported in the scientific literature and to the U.S. Food and Drug Administration (FDA), while simultaneously over-emphasized in patient discussion groups and social media communities [10]. This discrepancy contributes to polarization in publicly available information, which patients and care partners often rely upon when making treatment decisions. In addition, there are ongoing uncertainties regarding the prevalence and ethical implications of off-label uses of TMS, including its potential applications for cognitive or performance enhancement [11, 12].

In this volume, we examine the evolving ethical landscape surrounding TMS technology in both clinical practice and research settings. Prior neuroethics scholarship related to neurostimulation therapies has largely focused on anticipating future applications, clarifying foundational concepts such as invasiveness, and evaluating whether emerging uses of these devices align with or challenge societal values. While there are a small number of empirical ethics studies involving TMS patients [13], the perspectives of other key stakeholders, such as care partners, clinicians, and industry representatives, remain largely absent from the existing neuroethics literature.

This volume brings together contributions from investigators developing novel TMS devices and researchers exploring its wide-ranging clinical applications, including addiction, trauma, neurocognition, emotional dysregulation, suicidality, and many more. We aim for this volume to serve as a valuable resource for scientists, engineers, and clinicians working in brain stimulation research and care, providing deeper insight into the ethical complexities surrounding their work. Our goal is to offer a rich and comprehensive view of TMS ethics, shaped by diverse perspectives across disciplines and professional backgrounds, spanning basic science, technology development, clinical practice, and patient experience—from bench to bedside.

1 Overview of Part I: "Historical and Theoretical Reflections Regarding TMS"

In Part I, "Historical and Theoretical Reflections Regarding TMS", we aim to provide both a comprehensive analysis of the historical development of TMS technology and novel insights into its theoretical applications, with a particular emphasis on ethical considerations.

In the chapter "A History of Transcranial Magnetic Stimulation", Sean Reeves, Zhi-De Deng, and Jonathan Young trace the historical evolution of TMS, from its origins in electromagnetic induction to modern clinical applications, highlighting key scientific breakthroughs, regulatory milestones, ethical challenges, and emerging innovations that continue to shape its role in neuromodulation and psychiatry.

The chapter "Racial Disparities in TMS Research Participation and Clinical Access in the United States" shifts focus to racial disparities in both clinical access to TMS and participation in TMS research. Building on prior research into attitudinal barriers to neuromodulation treatments [14], Laura Cabrera and colleagues review the literature to identify barriers to TMS access amongst Black, Indigenous, and People of Color (BIPOC) populations, ultimately finding significant underrepresentation of racialized minorities in TMS research.

In the chapter "Family and Caregiver Perspectives on TMS Treatment of Refractory Conditions: A Pilot Investigation", Ariana D'Alessandro, Iris Coates-McCall, and Veljko Dubljevic synthesize findings from existing literature on stakeholder perspectives toward electroceutical treatments alongside new empirical data gathered from semi-structured interviews with TMS technicians and caregivers of TMS patients. Their analysis offers further insights into the benefits, challenges, and ethical considerations associated with TMS treatment.

In the chapter "Transcranial Magnetic Stimulation and Cognitive Enhancement: Present State and Possibilities for the Near Future", Bruce Luber and Eva Wiener begin exploring TMS as a tool for neuroenhancement, drawing on prior research to demonstrate the potential for TMS to enhance cognitive and functional abilities. With their extensive body of work on TMS and neuroenhancement [15–19], the authors argue that the enhancement potential of TMS will likely grow, particularly when combined with nanotechnological and neural modeling advancements.

Continuing the conversation on TMS-based neuroenhancement in the chapter "Unravelling Ethical Complexities: Transcranial Magnetic Stimulation (TMS) in Non-clinical Settings and the Dilemmas of Neuroenhancement", Dov Greenbaum draws on his prior research into the legal and societal implications of neuroenhancement [20–23] to explore the ethical and regulatory challenges of TMS use beyond the clinic, particularly in military or at-home uses.

In the chapter "TMS for Social Neuroscience: New Applications and Ethical Issues", Michaela Balconi examines the application of TMS in social neuroscience, focusing on how TMS clarifies the neural mechanisms underlying executive function, emotion processing, memory, and social behavior. Drawing on her extensive prior work [24–28], Balconi expands the understanding of how TMS can map and modulate these neural systems.

Finally, in the chapter "Bridging Network Neuroscience and Neuroethics: Can TMS and tACS Solve the Creativity Crisis?", Daniel Moseley and Flavio Frohlich address a specific dimension of neuroenhancement, considering TMS as a potential tool for enhancing creativity. They present evidence that TMS could enhance creative abilities beyond baseline potential, drawing from their expertise in neural oscillations and prior work [29–31] to explore the distinction between therapeutic applications and enhancement efforts, offering a comprehensive perspective on TMS and creativity enhancement.

2 Overview of Part II: "New Directions and Ethical Issues in Clinical TMS Research"

In Part II, "New Directions and Ethical Issues in Clinical TMS Research", we explore the future of clinical TMS, focusing on emerging research investigating its potential to treat a broader range of patient populations and psychiatric conditions. While this section centers on clinical applications, we present a balanced perspective that incorporates careful ethical analysis.

In the chapter "Subgrouping 'Treatment Responders': Ethical Implications for Those Seeking Participation in TMS Treatment and/or Research", Stephanie Hare and Danel Roche open the discussion by examining who is currently receiving TMS treatment and arguing for better-defined criteria for treatment-resistant depression (TRD), as inconsistent definitions may limit access for patients who could benefit from treatment.

In the chapter "TMS for Mood Disorders: An Ethical Perspective", Mohsen Poorganji and colleagues examine the ethical considerations of using TMS for mood disorders, focusing on how TMS-induced neuroplasticity may influence both symptom relief and aspects of personhood. They also address inequities in access to TMS, highlighting how recent advancements in technology and protocols could help reduce disparities and promote more equitable mental health care [32–34].

Patricia Henegan, Paul Croarkin and Laura Cabrera contribute chapter "'The Fewer Pills the Better': Attitudes towards Transcranial Magnetic Stimulation for Depressed Adolescents", which presents empirical data from interviews with adolescents receiving TMS and their parents. Building on previous qualitative work examining perspectives on TMS for adult depression [12, 35, 36], this chapter highlights the ethical importance of studying TMS in younger, more vulnerable populations.

Continuing the focus on vulnerable populations, chapter "Transcranial Magnetic Stimulation in Older Adults" by Tajudeen Basiru, Joan Camprodon, and Michael Kritzer addresses neuropsychiatric conditions in adults over 65, including depression, obsessive compulsive disorder (OCD), Alzheimer's disease, and smoking cessation. While the authors emphasize the need for more research on TMS for older adults, they also present preliminary evidence [37, 38] demonstrating TMS safety and efficacy in this population.

In the chapter "The Ethics of TMS in the Treatment of Suicidal Ideation", Alyssa Bernanke and colleagues explore the use of TMS in treating suicidal ideation, an area of TMS research that remains underexplored. This chapter highlights the ethical complexities of treating suicidality, especially given its presence across multiple psychiatric disorders, which makes it difficult to address with a single treatment modality. However, TMS shows promise, given its efficacy in treating conditions linked to suicidality.

In the chapter "Transcranial Magnetic Stimulation for Substance Use Disorders: Ethical Considerations", Dhvani Mehta and colleagues consider the use of TMS for substance use disorders (SUDs). Drawing from both their own prior research [39–41] and other relevant studies, they examine how to balance the promise of

TMS as a novel treatment with ethical considerations related to efficacy, safety, and the importance of establishing disorder-specific treatment protocols.

Finally, in the chapter "Regulatory Perspectives on Transcranial Magnetic Stimulation", Michael Pflanzer and Veljko Dubljević focus on FDA regulation of TMS devices, including ethical challenges posed by the growing popularity of do-it-yourself (DIY) at-home TMS devices. Building on their previous research on TMS regulation and policy [12, 42, 43], they offer recommendations for strengthening regulatory oversight to ensure safe, effective, and ethically responsible use of TMS technology.

3 Conclusion

As discussed throughout this volume, we address key issues at the intersection of TMS technology and neuroethics. Drawing from an established and growing body of literature, contributing authors apply their expertise to offer new insights and critical perspectives on this evolving field. From theoretical and philosophical discussions on the potential of TMS as a cognitive enhancement tool to practical clinical recommendations for improving and expanding treatment protocols, this volume provides a comprehensive overview of the current state of TMS and its future directions. Overall, we hope to stimulate further dialogue among scholars, clinicians, and policymakers, while contributing valuable knowledge to the broader neuroethics community.

References

1. Illes J, Gallo M, Kirschen MP. An ethics perspective on transcranial magnetic stimulation (TMS) and human neuromodulation. Behav Neurol. 2006;17:149–57.
2. Horvath JC, Perez JM, Forrow L, Fregni F, Pascual-Leone A. Transcranial magnetic stimulation: a historical evaluation and future prognosis of therapeutically relevant ethical concerns. J Med Ethics. 2011;37(3):137–43.
3. Thompson L. Treating major depression and comorbid disorders with transcranial magnetic stimulation. J Affect Disord. 2020;276:453–60.
4. Scheper A, Rosenfeld C, Dubljević V. The public impact of academic and print media portrayals of TMS: shining a spotlight on discrepancies in the literature. BMC Med Ethics. 2022;23(1):25.
5. Chalah MA, Ayache SS. Noninvasive brain stimulation and psychotherapy in anxiety and depressive disorders: a viewpoint. Brain Sci. 2019;9(4):82.
6. Kim TD, Hong G, Kim J, Yoon S. Cognitive enhancement in neurological and psychiatric disorders using transcranial magnetic stimulation (TMS): a review of modalities, potential mechanisms and future implications. Exp Neurobiol. 2019;28(1):1–16.
7. Shivakumar V, Dinakaran D, Narayanaswamy J, Venkatasubramanian G. Noninvasive brain stimulation in obsessive–compulsive disorder. Indian J Psychiatry. 2019;61(7):66.
8. Lerner AJ, Wassermann EM, Tamir DI. Seizures from transcranial magnetic stimulation 2012–2016: results of a survey of active laboratories and clinics. Clin Neurophysiol. 2019;130(8):1409–16.

9. McClintock SM, Reti IM, Carpenter LL, McDonald WM, Dubin M, Taylor SF, Cook IA, O'Reardon J, Husain MM, Wall C, Krystal AD, Sampson SM, Morales O, Nelson BG, Latoussakis V, George MS, Lisanby SH. Consensus recommendations for the clinical application of repetitive transcranial magnetic stimulation (Rtms) in the treatment of depression. J Clin Psychiatry. 2018;79(1):35.
10. Maizey L, Allen CPG, Dervinis M, Verbruggen F, Varnava A, Kozlov M, Adams RC, Stokes M, Klemen J, Bungert A, Hounsell CA, Chambers CD. Comparative incidence rates of mild adverse effects to transcranial magnetic stimulation. Clin Neurophysiol. 2013;124(3):536–44.
11. Wexler A, Nagappan A, Kopyto D, Santarnecchi E, Pascual-Leone A. Off-label promotion of transcranial magnetic stimulation on provider websites. Brain Stimul. 2021;14(3):723–4.
12. Dubljević V, Coates McCall I, Illes J. Neuroenhancement at work: addressing the ethical, legal and social implications. In: Trempe-Martineau J, Racine E, editors. Organizational neuroethics. Cham: Springer; 2019. p. 87–103.
13. Bluhm R, Cortright M, Achtyes ED, Cabrera LY. "They are invasive in different ways": stakeholders' perceptions of the invasiveness of psychiatric electroceutical interventions. AJOB Neurosci. 2021;14(1):1–12.
14. Cabrera LY, Miller MMC, Achtyes ED, McCright AM, Bluhm R. Jumping through the hoops: barriers and other ethical concerns regarding the use of psychiatric electroceutical interventions. Psychiatry Res. 2022;313:114612. https://doi.org/10.1016/j.psychres.2022.114612.
15. Luber B, Beynel L, Deng ZD, Appelbaum LG, Jones T, Harrison A, Murphy DLK, Lo E, McKinley RA, Lisanby SH. Site- and frequency-specific enhancement of visual search performance with online individual alpha frequency (IAF) repetitive Transcranial Magnetic Stimulation (rTMS) to the inferior frontal junction. Cereb Cortex. 2024;34(9):371.
16. Luber B, Davis SW, Deng ZD, Murphy D, Martella A, Peterchev AV, Lisanby SH. Using diffusion tensor imaging to effectively target TMS to deep brain structures. NeuroImage. 2022;249:118863.
17. Luber B, Kinnunen LH, Rakitin BC, Ellsasser R, Stern Y, Lisanby SH. Facilitation of performance in a working memory task with rTMS stimulation of the precuneus: frequency- and time-dependent effects. Brain Res. 2007a;1128(1):120–9.
18. Luber B, Balsam P, Nguyen T, Gross M, Lisanby SH. Classical conditioned learning using transcranial magnetic stimulation. Exp Brain Res. 2007b;183(3):361–9.
19. Luber B, Stanford AD, Bulow P, Nguyen T, Rakitin BC, Habeck C, Basner R, Stern Y, Lisanby SH. Remediation of sleep-deprivation-induced working memory impairment with fMRI-guided transcranial magnetic stimulation. Cerebral Cortex. 2008;18(9):2077–85.
20. Greenbaum D. Is criminal law both redundant and inconsistent? crime and consciousness in light of developments in neuroscience. AJOB Neurosci. 2018;9(1):51–2.
21. Liv N, Greenbaum D. Cyberneurosecurity. In: Policy, identity, and neurotechnology: the neuroethics of brain-computer interfaces. Cham: Springer; 2023. p. 233–51.
22. Greenbaum D, Gerstein M. The control and privacy of your most intimate information: navigating the battle for your brain. AJOB Neurosci. 2023;14(3):1–4.
23. Nakar S, Weinberger S, Greenbaum D. Legal and social implications of predictive brain machine interfaces: duty of care, negligence, and criminal responsibility. AJOB Neurosci. 2015;6(4):40–2.
24. Balconi M. Dorsolateral prefrontal cortex, working memory and episodic memory processes: insight through transcranial magnetic stimulation techniques. Neurosci Bull. 2013;29:381–9.
25. Balconi M, Bortolotti A. Emotional face recognition, empathic trait (BEES), and cortical contribution in response to positive and negative cues. The effect of rTMS on dorsal medial prefrontal cortex. Cogn Neurodyn. 2013;7:13–21.
26. Balconi M, Canavesio Y. High-frequency rTMS improves facial mimicry and detection responses in an empathic emotional task. Neuroscience. 2013;236:12–20.
27. Balconi M, Ferrari C. Repeated transcranial magnetic stimulation on dorsolateral prefrontal cortex improves performance in emotional memory retrieval as a function of level of anxiety and stimulus valence. Psychiatry Clin Neurosci. 2013;67:210–8.

28. Balconi M, Ferrari C. rTMS stimulation on left DLPFC increases the correct recognition of memories for emotional target and distractor words. Cogn Affect Behav Neurosci. 2012;12:589–98.
29. Juengst E, Moseley D. Human enhancement. In: Zalta EN, Nodelman U, editors. The Stanford encyclopedia of philosophy (Spring 2025). Metaphysics Research Lab, Stanford University; 2025. https://plato.stanford.edu/archives/spr2025/entries/enhancement/.
30. Lustenberger C, Boyle MR, Foulser AA, Mellin JM, Fröhlich F. Functional role of frontal alpha oscillations in creativity. Cortex. 2015;67:74–82.
31. Fröhlich F. Network neuroscience. London: Academic Press (Elsevier); 2016.
32. Neacsiu AD, Beynel L, Powers JP, Szabo ST, Appelbaum LG, Lisanby SH, LaBar KS. Enhancing cognitive restructuring with concurrent repetitive transcranial magnetic stimulation: a transdiagnostic randomized controlled trial. Psychother Psychosom. 2022;91(2):94–106.
33. Appelbaum LG, Shenasa MA, Stolz L, Daskalakis Z. Synaptic plasticity and mental health: methods, challenges and opportunities. Neuropsychopharmacology. 2023;48(1):113–20. https://doi.org/10.1038/s41386-022-01370-w.
34. Nagra DS, Stolz LA, Weissman CR, Appelbaum LG. Access to interventional psychiatric treatments in the United States: disparities and proposed solutions. GHES. 2024;2(1):2456.
35. Bluhm R, Sipahi ED, Achtyes ED, McCright AM, Cabrera LY. Stakeholders' ethical concerns regarding psychiatric Electroceutical interventions: results from a US Nationwide survey. AJOB Empir Bioeth. 2023;15:11–21.
36. Cabrera LY, Achtyes ED, Bluhm R, McCright AM. Views about neuromodulation interventions for depression by stakeholder group, treatment modality, and depression severity. Compr Psychiatry. 2023;122:152365.
37. Cappon D, den Boer T, Jordan C, Yu W, Metzger E, Pascual-Leone A. Transcranial magnetic stimulation (TMS) for geriatric depression. Ageing Res Rev. 2022;74:101531.
38. Ahmed MA, Darwish ES, Khedr EM, El Serogy YM, Ali AM. Effects of low versus high frequencies of repetitive transcranial magnetic stimulation on cognitive function and cortical excitability in Alzheimer's dementia. J Neurol. 2012;259(1):83–92.
39. Mehta DD, Praecht A, Ward HB, Sanches M, Sorkhou M, Tang VM, Steele VR, Hanlon CA, George TP. A systematic review and meta analysis of neuromodulation therapies for substance use disorders. Neuropsychopharmacology. 2024;49:649–80.
40. Coles AS, Kozak K, George TP. A review of brain stimulation methods to treat substance use disorders. Am J Addict. 2018;27:71–91.
41. Sorkhou M, Stogios N, Sayrafizadeh N, Hahn MK, Agarwal SM, George TP. Non-invasive neuromodulation of dorsolateral prefrontal cortex to reduce craving in alcohol use disorder: a meta analysis. Drug Alcohol Depend Rep. 2022;1:100076.
42. Dubljević V, Saigle V, Racine E. The rising tide of transcranial direct current stimulation (tDCS) in the media and academic literature. Neuron. 2014;82:731–6.
43. Dubljević V. Neurostimulation devices for cognitive enhancement: toward a comprehensive regulatory framework. Neuroethics. 2015;8(2):115–26.

Historical and Theoretical Reflections Regarding TMS

A History of Transcranial Magnetic Stimulation

Sean D. Reeves, Zhi-De Deng, and Jonathan R. Young

1 Introduction

The modern version of transcranial magnetic stimulation (TMS) was invented in 1985 by Dr. Anthony Barker and his team at the University of Sheffield. TMS is a non-invasive brain stimulation technique that utilizes electromagnetic induction to modulate the brain. Rapidly fluctuating electromagnetic fields pass through the skin, skull, cerebrospinal fluid, and finally neural tissue. These changing magnetic fields establish brief electric fields within the brain, resulting in neuronal depolarization.

Although modern TMS emerged in the 1980s, the application of electricity or electromagnetism to the human body for therapeutic purposes has occurred since the Greek and Roman times [76]. A key influence was Michael Faraday, whose discovery of the principles of electromagnetic induction in 1831 laid the foundation for TMS [1]. Building upon this work, successive generations of engineers and scientists refined these principles, leading to new hypotheses, experiments, and, ultimately, modern applications of electromagnetic stimulation.

S. D. Reeves
The NeuroComputational Ethics Research Group at NC State,
North Carolina State University, Raleigh, NC, USA
e-mail: sdreeves@ncsu.edu

Z.-D. Deng
Noninvasive Neuromodulation Unit, Experimental Therapeutics and Pathophysiology Branch,
National Institute of Mental Health, National Institutes of Health, Bethesda, MD, USA
e-mail: zhi-de.deng@nih.gov

J. R. Young (✉)
Department of Psychiatry & Behavioral Sciences, Duke University School of Medicine,
Durham, NC, USA
e-mail: jonathan.young@duke.edu

V. Dubljević, J. R. Young (eds.), *TMS and Neuroethics*, Advances in Neuroethics,
https://doi.org/10.1007/978-3-031-92401-9_2

As a medical device, TMS has a variety of approved clinical applications. After several large-scale clinical trials funded by both industry and the National Institutes of Mental Health (NIMH), the FDA approved high-frequency repetitive TMS (rTMS) for clinical use for the treatment of major depressive disorder (MDD) in 2008. Subsequent FDA approvals followed along with sets of safety standards [63, 64] and consensus guidelines [73]. Current U.S. Food and Drug Administration (FDA)-approved applications of TMS also include obsessive-compulsive disorder (OCD), migraine with aura, tobacco smoking cessation, anxious depression, and presurgical cortical mapping [18, 48].

Major technological advancements that have impacted the field of TMS include navigation TMS (nTMS), which uses camera-guidance to stereotactically guide coil placement with precision [69] and patterned forms of TMS such as theta-burst stimulation (TBS) [38]. Most recent advances include functional magnetic resonance imaging (fMRI)-based targeting and accelerated treatment schedules [20].

While a review of biomedical ethics is outside the scope of this chapter, it is critical to consider bioethical principles while studying any historical development, TMS notwithstanding. Bioethical concepts such as autonomy, beneficence, nonmaleficence, and justice informed the development of TMS as a medical intervention throughout each stage of history. These issues are critical for how scholars, clinicians, and technologists think about future developments in TMS, such as improving efficacy, minimizing side effects, elucidating neural mechanisms, expanding clinical indications, developing innovative procedures or biotyping for predicting efficacy, and influencing risk–benefit ratio evaluations and informed consent. We explore the foundational development of the technology and consider the context by which somatic interventions first developed for psychiatric disorders, with the most effective and widely used to date being electroconvulsive therapy (ECT).

We examine the last 150 years of research and technological advances leading up to Barker's invention of modern TMS. We then detail how experimental studies have refined coil design, pulse frequencies, and clinical applications. The following section discusses key regulatory milestones of TMS and its functions that expand treatment options for neurological disorders. We conclude with a consideration of current hurdles that exist such as side effects, clinical practices, and accessibility, which are also further explored in subsequent chapters. In combining the history that predated the advent of modern TMS with its contemporary historical developments, we seek to build a rich narrative that will facilitate a better understanding of TMS and its potential impact on human health.

2 The Origins of TMS: A Historical Foundation

The earliest documented explorations of electricity and its therapeutic application began in ancient Greece and Rome [76]. In first-century AD Rome, Scribonius Largus suggested that the electric discharge of torpedo fish—aquatic creatures that use electricity to stun their prey—could be used to treat headaches and gout [50].

Similarly, the Greek philosopher Plato described their numbing effects as having healing properties [25]. From the Renaissance through the eighteenth century, the concept of applying electricity to the human body persisted despite little understanding of both electricity or the human nervous system.

Scientific advancements in the study of bioelectromagnetism, including Luigi Galvani's discovery of animal electricity, Alessandro Volta's development of the voltaic pile (battery), and Giovanni Aldini's pioneering experiments in galvanic stimulation of humans, laid the foundation for transcranial electrical stimulation as a potential intervention for psychiatric disorders [4]. These discoveries set the stage for Michael Faraday's groundbreaking work in the early nineteenth century in which he discovered electromagnetic induction— now known as Faraday's Law — the core physical principle describing the induction of an electric field due to a changing electromagnetic field, which would later underpin the development of TMS.

The premise of electromagnetic induction was discovered by Faraday in 1831 through a series of experiments that were administered by wrapping a copper coil, the conductor material, around a wooden cylinder in a helix shape. Both ends of the wire were connected to a galvanometer and a battery. The preliminary results showed that electric currents would induce if an active conductor is met with a moving or changing magnetic field [1, 3]. Modern TMS systems continue to be built on this basic premise: large capacitors release high electrical currents through a set of coil sections before returning to the capacitors, meanwhile establishing a rapidly changing magnetic field.

While Faraday laid the groundwork for understanding the basis for electromagnetic induction, it was not until 1865 when James Clerk Maxwell formulated the mathematical framework describing electricity and magnetism — now known as Maxwell's Equations — at the University of Cambridge [59]. Nikola Tesla further explored electromagnetic induction, being one of the first to study high frequency alternating current (AC) and its physiological effects on the body. By 1894, he had patented inventions on high-frequency magnetic fields that later influenced modern TMS systems. The development of Tesla coils, which generate strong oscillating or pulsed magnetic fields through resonant transformer circuits, is a key component of modern TMS (short, powerful magnetic pulses through a coil placed on the scalp induce electric fields in the brain and induce neural activity) [88].

In the late nineteenth century, Jacques-Arsène d'Arsonval, inspired by Tesla and others, became one of the first to systematically explore early versions of electromagnetic stimulation on human physiology [76]. In 1896, d'Arsonval successfully induced magnetically generated phosphenes—the sensation of seeing light without its actual presence—demonstrating that electromagnetism could affect humans [14, 76, 79]. Building on this work, Adrian Pollacsek and Berthold Beer replicated d'Arsonval's experiments—inducing magnetic phosphenes through electromagnetic stimulation—and patented its application in 1902 to treat depression and neuroses via eddy currents: electric currents produced by a conductor with changing magnetic fields [14, 27].

In 1910, Silvanous Thompson validated d'Arsonval's findings that magnetic brain stimulation could induce phosphenes [33, 79]. In his experiments, volunteers exposed to fluctuating magnetic field densities reported seeing flickers of colorless or bluish hues of light, whether in a dark room or with their eyes covered [45]. Despite these fascinating findings, it would be decades until a device powerful enough for focal magnetic brain stimulation would be engineered and clinical applications discovered. This time period instead witnessed the development of direct forms of neuromodulation, most notably ECT, which emerged in the 1930s [81] and used direct electrical currents to induce seizures as a treatment for severe psychiatric conditions.

The development of electroconvulsive therapy (ECT) was influenced by an earlier seizure-based treatment, pharmacoconvulsive therapy, a method of chemically-inducing seizures developed by Ladislas Meduna in 1935. Meduna, a Hungarian neuropathologist and neurospychiatrist, observed high concentration of glia in postmortem brains of patients with epilepsy and a scarcity in individuals with schizophrenia, leading him to propose the induction of seizures in patients with psychosis. This approach had significant drawbacks: painful injections, inconsistent responses, induced fear reactions, and physical bruising from distress during convulsions [26]. Additionally, its inconsistent efficacy raised ethical concerns about exposing patients to a procedure with sporadic success. ECT, developed by Ugo Cerletti and Lucio Bini in 1938, offered a more controlled and safer alternative to chemically-induced seizures. By the 1940s and 1950s, ECT was popularized in the psychiatric communities [81], but ethical concerns regarding informed consent and autonomy, broad usage, adverse cognitive side effects, and utilization in the absence of anesthesia or muscle relaxants would lead to negative public perceptions and stigmatization [39]. Public perception eventually led to legislative actions that reduced access to ECT treatments in some states until it regained popularity in the late 1970s and 1980s during which ECT was itself modernized with unilateral electrode placement and the introduction of novel electrical waveforms [80]. Nonetheless, the ethical considerations highlighted by the history of ECT and public perception would influence TMS research and development decades later.

During the 1950s, Alexander Kolin identified a limitation to Silvanous Thompson's 1910 experiments: the lack of magnetic localization [41]. In response, he constructed a bar with a pyramid-like structure and a wire-loop conductor at the tip for maximum flux density. He then demonstrated empirically that phosphenes appeared when the tip of the structure was positioned at the temporal area of his volunteers, establishing an ability to concentrate a magnetic field and induce focal stimulation. It had yet to stimulate the brain with electromagnetic induction, but it demonstrated peripheral nerve tissues could be excited by eddy currents [40]. It was later confirmed in 1965 when Reginald Bickford and Benjamin Flemming replicated Kolin's muscle excitability experiments [9, 75].

By the 1980s, scientists integrated techniques from peripheral nerve stimulation and transcranial electrical stimulation to develop a novel approach to stimulate the brain. The foundation for this work was laid in 1982 by P.A. Merton and Michael Polson, whose research significantly influenced the development of TMS. Merton demonstrated that human muscles, the motor cortex, the visual cortex, and the spinal cord could be activated using electrical currents applied through surface electrodes. While effective for studying muscle latency and relaxation, this method often caused significant discomfort and pain in research participants, limiting its practical application [6, 49].

Recognizing these limitations, Anthony Barker and Reza Jalinous sought an alternative, non-invasive method to study the nervous system while replicating Merton's findings. In 1985, Barker integrated Merton's cortical stimulation approach with Michael Polson's 1982 research on peripheral nerve stimulation using electromagnetic pulses rather than direct electrical currents. Their breakthrough demonstrated that magnetic stimulation could effectively activate the motor cortex without the pain or discomfort associated with electrical stimulation [6, 7, 60, 67].

This discovery marked a turning point in neuroscience and clinical neuromodulation, leading to the development of TMS as a non-invasive alternative to electrical stimulation. Over the decades, TMS would not only become a powerful research tool but also address long-standing ethical concerns associated with ECT by offering a safer, more targeted method of brain stimulation. From Faraday's foundational discoveries to Barker's groundbreaking innovations, the development of TMS illustrates the profound impact of scientific curiosity and the ongoing reassessment of ethical considerations. As the next section explores, numerous scientists played a pivotal role in transforming TMS from a purely an experimental tool into a clinically validated treatment.

3 The Origins of TMS: The Development of Modern TMS

In 1985, Barker demonstrated that pulsed magnetic fields could non-invasively stimulate the human motor cortex, marking a major advancement in neuromodulation. He described magnetic stimulation as "pain-free, requir[ing] no direct contact with the scalp, non-invasive, and easy to use" [6].

The following decade was pivotal for the development of TMS as researchers began exploring its practical applications. Inspired by the successes of neuromodulation by ECT, scientists sought to overcome the limitations of electrical stimulation by harnessing TMS's non-invasive nature and precise targeting capabilities. Many researchers around the world such as Shinji Ueno and Yumi Matsuda, Masahiko Kujirai, Álvaro Pascual-Leone, Mark George, and Robert Chen laid the groundwork for TMS as both a research tool and a clinical intervention.

One of the first advancements to TMS would come in the 1990s as Barker's original TMS device featured a round, flat coil applicator 100 mm in diameter. The coil design dispersed maximum eddy currents through the annulus—the region between the coil's outer and inner edges—but lacked concentration of flow [6, 66]. In 1992, Ueno and Matsuda introduced a figure-of-eight coil design which localized the dispersion of eddy currents at the junction point of the two coils, increasing the density by 2–3 times. This was due to the two opposing magnetic fields running parallel and tangent to the scalp [74]. This design innovation enabled more precise stimulation of targeted cortical regions.

Two investigators explored TMS' capability of multiple-pulse stimulation in 1993. The first, by Masahiko Kujirai, introduced the paired-pulse protocol, which allowed researchers to study intracortical inhibition and facilitation within the motor cortex.[1] The protocol delivers a conditioning stimulus over the motor cortex, followed by a test stimulus. The protocol demonstrated that specific interstimulus intervals modulated motor evoked potential (MEP)—electrical signals generated by stimulating the motor cortex and measuring the responses in corresponding nerves or muscles [42, 87].

The study, led by Álvaro Pascual-Leone, applied rapid-rate rTMS at varying frequencies and intensities over the motor cortex and other brain regions. The aim was to assess the physiological responses and potential adverse effects of TMS. Participants were monitored for side effects such as scalp discomfort, muscle contractions, and seizure risk, as well as cognitive and neurological functions before and after stimulation. Their findings confirmed that rTMS was generally well tolerated and showed potential cognitive benefits. The significance of this experiment was that it initiated discussions on neuroplasticity and the broader role of TMS on brain function [53, 54]. Pascual-Leone's work later influenced researcher Mark George, who played a key role in developing the evidence of rTMS as a treatment for MDD.

Mark George, a neuroscientist and psychiatrist, investigated TMS as a potential treatment for depression. Inspired by ECT—the most effective treatment for depression to date—he hypothesized in 1995 that TMS could alleviate depression by targeting and modulating dysfunctional regions of the prefrontal cortex. Neuroimaging studies had revealed hypoactivity in the left prefrontal cortex and hyperactivity in the right, leading George to theorize that rTMS could restore balance in activity of these regions [29, 54].

Results from this work showed a reduction in depressive symptoms based on the Hamilton Depression Rating Scale (HDRS) score. However, due to limited data on TMS for psychiatric use, George conducted an open-label pilot study with six treatment-resistant patients applying at least 5 days of daily rTMS over the left prefrontal cortex. This study laid the foundation for future studies demonstrating efficacy and safety of rTMS as a therapeutic intervention for depression [28, 29].

That same year, Luciano Fadiga, investigated whether humans possess neural mechanisms that are activated both during movement and observation [23]. Using a TMS device with a figure-of-eight coil, Fadiga and his team stimulated the left premotor cortex of 12 healthy participants to induce MEP's, hypothesizing that the premotor cortex communicates with the motor cortex and spinal cord through corticospinal connections. The results revealed increased MEPs between the brain and muscles that were excited. These findings provided strong evidence for motor resonance mechanisms in humans. Equally important, they demonstrated that TMS could serve as a tool to study motor control, cognitive brain functions, neuroplasticity, and movement disorders.

[1] Short-interval intracortical inhibition (SICI) occurred at intervals of 1–5 ms, reducing MEP amplitudes, whereas intracortical facilitation (ICF) was observed at 6–20 ms intervals, leading to increased MEP amplitudes [44, 87].

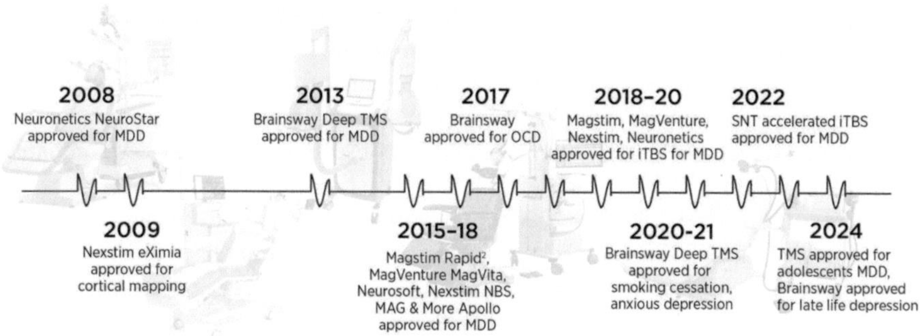

Fig. 1 Key milestones in TMS regulatory approvals, highlighting therapeutic applications for major depressive disorder (MDD), cortical mapping, obsessive-compulsive disorder (OCD), smoking cessation, anxious depression, as well as adolescent and late-life depression

By 1997, researchers were focusing on the potential of rTMS for both therapeutic applications in psychiatry and neurology. Mark George, building on his 1995 pilot study, refined his research parameters for rTMS in depression treatment: targeting the left dorsolateral prefrontal cortex (L-DLPFC) for 4 weeks—5 days a week, administering 20 min of 2 Hz stimulation per session. Using the HDRS, all 12 patients showed a reduction in depression scores, with no reported side effects.

That same year, Robert Chen investigated the effects of frequency-dependent rTMS on cortical excitability and inhibition. He explored the relationship between long-term potentiation (LTP)—which enhances neuronal excitability—and long-term depression (LTD)—which suppresses neuronal excitability. Chen hypothesized that low-frequency (0.9 Hz) rTMS for 15 min would reduce cortical excitability, mimicking LTD. His findings confirmed that frequency-dependent magnetic stimulation could modulate brain activity, sparking interest in rTMS as a potential treatment for neurological disorders such as epilepsy and myoclonus [15, 87].

The demonstration of LTD and LTP was a breakthrough in TMS research. Its discovery deepened the understanding of neuroplasticity and laid the groundwork for its clinical applications in depression. As researchers continued refining stimulation parameters and targeting methods, the next critical step was obtaining regulatory approval to ensure safe and standardized clinical use. The following section explores the history of FDA regulation and the pivotal studies that established TMS as a viable treatment option for psychiatric disorders, neuromapping, migraines, and smoking cessation (see Fig. 1 for key milestones).

4 FDA Approval and the Regulatory Journey of TMS

FDA approval of rTMS treatment for adults with MDD was granted in 2008 for the NeuroStar TMS System, developed by Neuronetics [19, 56]. This approval provided a new option for patients with treatment-refractory depression (TRD)—those who had not responded adequately to traditional pharmacotherapy [37, 46]. The decision was based on two large-scale clinical trials demonstrating the safety and efficacy of rTMS: one conducted by O'Reardon in 2007, sponsored by Neuronetics, and another independent study funded by the National Institute of Mental Health (NIMH) [56]. Both studies followed similar treatment protocols and concluded that patients receiving rTMS exhibited favorable response and remission rates compared to sham stimulation [37].

Following this breakthrough, TMS applications continued to expand. In 2009, the Nexstim eXimia NBS System received FDA approval for cortical mapping [19]. This marked a significant advancement in TMS technology, providing a non-invasive alternative for presurgical neuromapping. Before this approval, Direct Cortical Stimulation (DCS) was the standard for neuromapping. However, comparisons between TMS to DCS using magnetoencephalography (MEG) demonstrated that TMS was comparably effective, particularly in applications such as epilepsy surgery and brain tumor detection [19, 69].

In 2013, the FDA-cleared BrainsWay's Deep Transcranial Magnetic Stimulation (dTMS) for the treatment of MDD in adults. This approval was significant because it demonstrated how coil design improvements could enhance the depth and efficacy of magnetic stimulation. While the figure-of-eight coil improved focality and intensity of conventional rTMS, BrainsWay's newly approved H1-Coil design, housed within a helmet applicator, allowed for deeper and broader stimulation of the DLPFC, offering greater flexibility in treatment delivery [34, 43, 56].

Additional TMS devices received FDA clearance in the following years. These included the Magstim Rapid^2 and the Magventure MagVita (2015), Neurosoft TMS (2016), Nexstim NBT system for MRI-based coil positioning (2017), and Mag & More Apollo TMS system (2018). However, some of the most notable regulatory approvals extended beyond MDD treatment. In 2013, the eNoura Cerena TMS device was approved for the treatment of migraines with aura, and in 2017, the eNeura SpringTMS received additional FDA clearance [19]. Also in 2017, BrainsWay's H7-Coil design was granted regulatory clearance for treating OCD, extending TMS applications into anxiety-related conditions [11, 34, 65].

Further innovations in TMS led to advancements such as theta-burst stimulation (TBS). In 2018, the FDA approved Magventure's MagVita TMS Therapy System with intermittent theta-burst stimulation (iTBS) for treating MDD [19]. This protocol significantly reduced treatment duration, offering a more time-efficient alternative to conventional rTMS [16]. Unlike standard rTMS, TBS applies bursts of high-frequency (50 Hz) rTMS at an overall stimulation rate of 5 Hz, which more closely mimics natural brain rhythms [68]. By 2019, the FDA had extended intermittent TBS (iTBS) approval to Nexstim's NTB 2 and Magstim's Horizon TMS with Navigation, and in 2020, Neuronetics joined the list of FDA-cleared iTBS devices.

In 2020, BrainsWay's Deep TMS system received FDA approval as an adjunct treatment for smoking cessation, utilizing a specialized H4-Coil design to target bilateral insula and prefrontal cortices. This approval marked a major milestone in neuromodulation for addiction treatment. That same year, Deep TMS was also approved for the treatment of anxious depression, expanding its role in addressing comorbid mood and anxiety disorders [19, 35, 55, 86].

In 2022, the FDA granted clearance for Stanford Neuromodulation Therapy (SNT), an accelerated iTBS protocol designed to deliver high-dose iTBS over a condensed timeframe. This protocol administers 10 treatment sessions per day over 5 consecutive days, targeting the left DLPFC based on functional connectivity with the subgenual cingulate cortex. Unlike conventional rTMS, which typically requires daily sessions over 4–6 weeks, SNT enables faster symptom relief while maintaining efficacy [85].

The most recent FDA approvals in 2024 have further expanded TMS accessibility and clinical applications. Notably, the FDA extended NeuroStar's TMS protocol for MDD to adolescents aged 15–21 and approved BrainsWay's dTMS for treating late-life depression in patients aged 68–86 [51, 52]. These regulatory milestones are particularly significant as they broaden TMS treatment eligibility across a wider patient demographic. By extending the age range of approval, these decisions affirm TMS as a safe and effective intervention across different life stages, fostering continued innovation in neuromodulation therapies.

As ongoing clinical research refines stimulation protocols and broadens therapeutic indications, these recent FDA approvals underscore the growing role of TMS in personalized, non-invasive treatments. By offering more efficient protocols, targeting diverse neuropsychiatric conditions, and expanding treatment eligibility, these advancements reflect a commitment by the scientific field, industry and regulatory agencies to maximize the potential therapeutic applications of TMS (See Table 1 for clinical indications for TMS and standard protocols).

5 Integrating the Past with the Future of TMS

The historical development of TMS provides essential context for understanding both its scientific evolution and the ethical challenges surrounding its use in neuromodulation. Over the years, TMS has gained recognition as a viable treatment for psychiatric and neurological disorders, generating both public interest and rigorous scientific investigation. Despite its demonstrated benefits, several limitations remain, including side effects, time commitment, accessibility barriers, targeting precision, and variable efficacy. These challenges have shaped ethical debates surrounding TMS, at times amplifying concerns about its risks while overlooking its therapeutic potential.

5.1 Historical Challenges

One of the most frequently cited concerns regarding TMS is its side effects, which commonly include headaches, scalp discomfort at the stimulation site, and muscle twitching. These effects are generally mild to moderate and tend to subside over time or as patients acclimate to treatment. However, more severe side effects, such

Table 1 Clinical indications for TMS and standard protocols

Condition	Target	Off-label uses	Parameters	Citations
MDD	Figure-8 rTMS, left DLPFC	Low-frequency (1 Hz) right DLPFC	10 Hz, 120% MT, 3000 pulses/session, 36 sessions, (5x/week for 6 weeks)	[5, 10, 31]
	H1-coil dTMS, prefrontal cortex, including bilateral DLPFC		18 Hz, 120% MT, 1980 pulses/session (5 sessions/week for 4 weeks+biweekly maintenance sessions for up to 12 weeks)	[72]
	Left DLPFC		iTBS (50-Hz triplets at 5 Hz), 120% MT, 600 pulses/session, 3-min session, 20 sessions (1 session/day for 4 week)	[10]
	Region in left DLPFC functionally anticorrelated with subgenual cingulate		fMRI-guided, accelerated iTBS, 90% MT, 1800 pulses/session, 50-min intersession interval, 10 sessions/day over 5 days	[20]
OCD	H7-coil dTMS, mPFC and ACC	Traditional rTMS, pre-SMA, bilateral TMS	20 Hz, 100% leg MT, 2000 pulses/session, 29 sessions, (5x/week for 6 weeks) with symptom provocation	[12, 65]
Smoking cessation	H4-coil dTMS, bilateral insula and prefrontal cortices	Alternative coil placements, various frequencies	10 Hz, 120% MT, 1800 pulses/session, 3 weeks, (5x/week)	[86]
Migraine with Aura	Occipital cortex	Multi-pulse protocols for chronic migraine	Single pulses at 0.9 T, self-administered at migraine onset	[44]

M Major Depressive Disorder, *OCD* Obsessive-Compulsive Disorder, *MT* motor threshold, *iTBS* Intermittent Theta-Burst Stimulation, *DLPFC* dorsolateral prefrontal cortex, *SMA* supplementary motor area, *mPFC* medial prefrontal cortex, *ACC* anterior cingulate cortex

as seizures, though rare, require careful screening and monitoring. The estimated seizure risk associated with TMS is approximately 0.1% per full treatment course, with the highest risk occurring in individuals with a history of epilepsy or seizures. By comparison, some oral medications such as bupropion have a higher seizure risk of 0.4%–0.5%, which is more difficult to monitor in a controlled clinical setting [63, 64]. While these risks highlight the need for rigorous safety protocols, they do not outweigh the potential benefits of TMS, particularly when appropriate patient selection and adherence to clinical guidelines are ensured.

Another significant challenge is the financial and time commitment required for standard TMS treatment. Traditional rTMS protocols involve daily sessions for 4–6 weeks, which presents a significant burden for patients with work and family obligations. In addition, out-of-pocket costs per session can be prohibitive, especially for individuals without adequate insurance coverage. Even for insured patients, insurance reimbursement policies vary significantly, with some providers limiting coverage and others requiring multiple failed medication trials before approving TMS [47]. These financial constraints disproportionately impact economically disadvantaged populations, exacerbating inequities in mental health treatment accessibility.

A further complicating factor is the lack of standardized criteria for defining treatment-resistant depression (TRD) across insurance providers and healthcare systems. This inconsistency creates barriers to treatment approval, as some insurance companies require failure of four or more antidepressant medications before authorizing TMS treatment, despite evidence suggesting that TMS is more cost-effective—measured in quality-adjusted life years (QALY)—than continued pharmacological treatment [77, 82]. Many researchers advocate for earlier intervention with TMS, recommending its use after just one failed antidepressant trial [82]. However, rigid insurance policies often delay access, limiting the potential for early and effective neuromodulation interventions.

Beyond financial considerations, treatment efficacy remains a key consideration for patients and clinicians. The cost-benefit ratio of TMS is particularly relevant given that response rates vary among individuals. Although TMS has demonstrated significant efficacy in TRD, some patients experience limited duration of symptom relief, necessitating maintenance treatments or alternative neuromodulation strategies [22]. Precise stimulation of targeted neural circuits is critical to maximizing treatment success, as improper targeting may reduce efficacy, increase side effects, and alter the risk–benefit profile. A meta-analysis of 126 online TMS studies found that fMRI-guided targeting was associated with increased disruptive effects of rTMS on cognitive performance, suggesting that more precise targeting may enhance the efficacy of TMS in modulating neural activity and highlighting the need for careful parameter selection [8]. Optimizing TMS parameters—such as frequency, intensity, and duration—remains a largely trial-and-error process, further complicating treatment standardization and individual response prediction. Emerging research on depression subtypes suggest that treatment response heterogeneity may be influenced by distinct neurobiological markers [83, 84]. This underscores the need for precision-medicine approaches to TMS.

5.2 Considerations for the Future of TMS

Many of the challenges associated with TMS have been met with innovative solutions. TMS has a long-standing history of innovation and scientific collaboration, with ongoing advancements aimed at addressing current limitations, improving accessibility, and enhancing patient outcomes.

While an area of focus is reducing physiological side effects, particularly seizure risk, researchers are now exploring controlled seizure induction with TMS technology as a therapeutic tool rather than a side effect. Magnetic Seizure Therapy (MST), pioneered by Sarah H. Lisanby in the mid-2000s, applies focused magnetic fields to induce seizures, similar to ECT but with fewer cognitive side effects. MST offers greater precision in targeting specific brain regions, providing a more refined and controlled approach to seizure-based interventions [21]. With continued research and careful patient screening, new TMS-based protocols could lead to more personalized and effective neuromodulation strategies.

Another promising advancement is the optimization of TMS efficacy through personalized treatments. Functional neuroimaging, particularly fMRI, is emerging as a tool for biotype-specific patient matching, improving response rates, and reducing treatment resistance [71]. Recent advances in connectivity-based targeting have made it possible to identify personalized TMS targets with millimeter precision, leveraging individual brain network architecture to optimize treatment efficacy, although large-scale trials have yet to determine its true clinical impact and standardize methodologies for real-world application [13].

Technological advancements such as robotic-assisted coil positioning are also gaining momentum. Robotic guidance enhances coil placement accuracy, reduces human error, and minimizes patient fatigue by ensuring optimal positioning throughout treatment sessions [70]. These developments increase precision while reducing variability in treatment outcomes.

Further innovation is seen in Controllable Pulse Parameter TMS (cTMS), co-developed by Angel Peterchev and Sarah H. Lisanby. Unlike conventional TMS, cTMS allows real-time control over waveform shape, pulse width, and intensity using an insulated gate bipolar transistor (IGBT) system for dynamic neuromodulation [57, 58]. This technology enables reproducible dosing, improving both safety and efficacy while paving the way for more customized treatment protocols [64].

To address cost and accessibility barriers, researchers are developing accelerated TMS (aTMS) protocols, which compress treatment schedules into fewer days [62]. This approach could reduce financial burden and time commitment associated with traditional TMS, making treatment more accessible to a wider patient population.

Addressing these challenges is essential for optimizing TMS as a mainstream therapeutic tool. By advancing treatment precision, improving accessibility, and developing more efficient protocols, researchers, clinicians, and policymakers can work toward reducing barriers to care and combating misinformation. The ongoing evolution of TMS highlights its potential for broader clinical applications, reinforcing its role as a transformative intervention in psychiatry and neurology.

Collaboration through organizations such as the Clinical TMS Society (CTMSS) and conferences such as NYC Neuromodulation and the International Brain Stimulation Conference has played a critical role in supporting research, clinical development, and technology advancements in neuromodulation. Federal support has also been instrumental in advancing TMS research, with funding mechanisms such as the National Institute of Health (NIH) Brain Research through Advancing Innovative Neurotechnologies (BRAIN) Initiative driving innovation in TMS technology, precision targeting, and mechanistic understanding. These efforts have enabled major advancements in TMS methodologies, including personalized neuromodulation strategies, integration with functional neuroimaging, and the exploration of novel stimulation parameters. The goal of this chapter is to establish an interdisciplinary framework that integrates historical insights, scientific advancements, and ethical considerations. By analyzing the origins, evolution, and ongoing innovations in TMS, we can identify overlooked patterns and contribute to meaningful progress in both the scientific community and societal perceptions of TMS.

Acknowledgments Thank you to the members of the Neuro-Computational Ethics and Research Group at NC State University for their valuable feedback (in alphabetical order): Ashley Beatty, Savannah Beck, Hunter Bissette, Sean Brantley, Dario Cecchini, Ari D'Alessandro, Veljko Dubljević, Macy Ferrell, Jack Harris, Steven Peppers, Michael Pflanzer, Ishita Pai Raikar, Shaun Respess, and Ryan Sterner. An additional thank you to Tyler Anderson, Alicia Johnson, and Brandy Reeves for their academic feedback and support.

Funding S. Reeves has no financial disclosures. Z.-D. Deng is supported by the National Institute of Mental Health (NIMH) Intramural Research Program (ZIAMH002955). J. Young is supported by Department of Veterans Affairs (VA) Clinical Science Research and Development Service (CSR&D) Career Development Award (CDA-2: 1IK2CX002610-01); VA Mid-Atlantic Mental Illness Research, Education, and Clinical Center (MIRECC), Durham VA Health Care System; Duke University Department of Psychiatry & Behavioral Sciences; and Duke Interdisciplinary Studies Bass Connections Program.

References

1. Al-Khalili J. The birth of the electric machines: a commentary on Faraday (1832) 'Experimental researches in electricity'. Philos Trans Royal Soc A Math Phys Eng Sci. 2015;373(2039):20140208. https://doi.org/10.1098/rsta.2014.0208.
2. Alonzo A, Charvet L. Home-based tDCS: design, feasibility and safety considerations. In: Brunoni A, Nitsche M, Loo C, editors. Transcranial direct current stimulation in neuropsychiatric disorders. Cham: Springer; 2016. p. 351–61. https://doi.org/10.1007/978-3-319-33967-2_23.
3. Anderson R. The referees' assessment of faraday's electromagnetic induction paper of 1831. Notes Rec R Soc Lond. 1993;47(2):243–56. https://doi.org/10.1098/rsnr.1993.0031.
4. Arêas FZ, et al. Giovanni Aldini and his contributions to non-invasive brain stimulation. Arq Neuropsiquiatr. 2020;78(11):733–5. https://doi.org/10.1590/0004-282x20200080.
5. Bares M, et al. Low frequency (1-Hz), right prefrontal repetitive transcranial magnetic stimulation (rTMS) compared with venlafaxine ER in the treatment of resistant depression: a double-blind, single-centre, randomized study. J Affect Disorder. 2009;118(1–3):94–100. https://doi.org/10.1016/j.jad.2009.01.032.
6. Barker AT, Jalinous R, Freeston IL. Non-invasive magnetic stimulation of human motor cortex. Lancet. 1985;325(8437):1106–7. https://doi.org/10.1016/s0140-6736(85)92413-4.
7. Bedder M, Parker L. Magnetic peripheral nerve stimulation (MPNS) for chronic pain. J Pain Res. 2023;16:2365–73. https://doi.org/10.2147/jpr.s409331.
8. Beynel L, et al. Effects of online repetitive transcranial magnetic stimulation (rTMS) on cognition: a meta-analysis and recommendations for future studies. Brain Stimul. 2019;12(2):565. https://doi.org/10.1016/j.brs.2018.12.870.
9. Bickford RB, Flemming BD. Neuronal stimulation by pulsed magnetic fields in animals and man. In: Digest of the 6th international conference on medical electronics and biological engineering; 1965. p. 112.
10. Blumberger DM, et al. Effectiveness of theta burst versus high-frequency repetitive transcranial magnetic stimulation in patients with depression (Three-D): a randomised non-inferiority trial. Lancet. 2018, 1683–1692;391(10131) https://doi.org/10.1016/s0140-6736(18)30295-2.

11. Carmi L, Alyagon U, Barnea-Ygael N, Zohar J, Dar R, Zangen A. Clinical and electrophysiological outcomes of deep TMS over the medial prefrontal and anterior cingulate cortices in OCD patients. Brain Stimul. 2017;11(1):158–65. https://doi.org/10.1016/j.brs.2017.09.004.
12. Carmi L, et al. Efficacy and safety of deep transcranial magnetic stimulation for obsessive-compulsive disorder: a prospective multicenter randomized double-blind placebo-controlled trial. Am J Psychiatry. 2019;176(11):931–8. https://doi.org/10.1176/appi.ajp.2019.18101180.
13. Cash RFH, Zalesky A. Personalized and circuit-based transcranial magnetic stimulation: evidence, controversies, and opportunities. Biol Psych. 2024;95(6):510–22. https://doi.org/10.1016/j.biopsych.2023.11.013.
14. Chae J-H, Nahas Z, Li X, George MS. Transcranial magnetic stimulation in psychiatry: research and therapeutic applications. Int Rev Psychiatry. 2001;13(1):18–23. https://doi.org/10.1080/09540260125143.
15. Chen R, et al. Depression of motor cortex excitability by low-frequency transcranial magnetic stimulation. Neurology. 1997;48(5):1398–403. https://doi.org/10.1212/wnl.48.5.1398.
16. Cheng C-M, Li C-T, Tsai S-J. Current updates on newer forms of transcranial magnetic stimulation in major depression. Adv Exp Med Biol. 2021;1305:333–49. https://doi.org/10.1007/978-981-33-6044-0_18.
17. Chung SW, Hoy KE, Fitzgerald PB. Theta-burst stimulation: a new form of TMS treatment for depression? Depress Anxiety. 2014;32(3):182–92. https://doi.org/10.1002/da.22335.
18. Cleveland Clinic Medical: TMS (transcranial magnetic stimulation): what it is. Cleveland Clinic (2022). https://my.clevelandclinic.org/health/treatments/17827-transcranial-magnetic-stimulation-tms
19. Cohen SL, Bikson M, Badran BW, George MS. A visual and narrative timeline of US FDA milestones for transcranial magnetic stimulation (TMS) devices. Brain Stimul. 2022;15(1):73–5. https://doi.org/10.1016/j.brs.2021.11.010.
20. Cole EJ, et al. Stanford neuromodulation therapy (SNT): a double-blind randomized controlled trial. Am J Psychiatry. 2022;179(2):132–41. https://doi.org/10.1176/appi.ajp.2021.20101429.
21. Deng Z-D, et al. Clinical outcomes of magnetic seizure therapy vs electroconvulsive therapy for major depressive episode. JAMA Psychiatry. 2024;81(3):240. https://doi.org/10.1001/jamapsychiatry.2023.4599.
22. Elsayed OH, Ercis M, Pahwa M, Singh B. Treatment-resistant bipolar depression: therapeutic trends, challenges, and future directions. Neuropsychiatr Dis Treat. 2022;18:2927–43. https://doi.org/10.2147/ndt.s273503.
23. Fadiga L, Fogassi L, Pavesi G, Rizzolatti G. Motor facilitation during action observation: a magnetic stimulation study. J Neurophysiol. 1995;73(6):2608–11. https://doi.org/10.1152/jn.1995.73.6.2608.
24. Farzan F. Single-pulse transcranial magnetic stimulation (TMS) protocols and outcome measures. In: Transcranial magnetic stimulation. New York: Humana Press; 2014. p. 69–115. https://doi.org/10.1007/978-1-4939-0879-0_5.
25. Finger S, Ferguson I. The role of *The Gentleman's Magazine* in the Dissemination of knowledge about electric fish in the eighteenth century. J Hist Neurosci. 2009;18(4):347–65. https://doi.org/10.1080/09647040801895489.
26. Fink M. Electroconvulsive therapy resurrected: its successes and promises after 75 years. Can J Psychiatry. 2011;56(1):3–4. https://doi.org/10.1177/070674371105600102.
27. George MS. Stimulating the brain. Sci Am. 2003;289(3):66–73. JSTOR, https://www.jstor.org/stable/26060008
28. George MS, Taylor JJ. Theoretical basis for transcranial magnetic stimulation. In: A clinical guide to transcranial magnetic stimulation. Oxford: Oxford University Press; 2014. p. 1–7. https://doi.org/10.1093/med/9780199926480.003.0001.
29. George MS, Wassermann EM, et al. Daily repetitive transcranial magnetic stimulation (rTMS) improves mood in depression. Neuroreport. 1995;6(14):1853–6. https://doi.org/10.1097/00001756-199510020-00008.
30. George MS, Lisanby SH, Sackeim HA. Transcranial Magnetic Stimulation. Arch Gen Psychiatry. 1999;56(4):300. https://doi.org/10.1001/archpsyc.56.4.300.

31. George MS, et al. Daily left prefrontal transcranial magnetic stimulation therapy for major depressive disorder. Arch Gen Psychiatry. 2010;67(5):507. https://doi.org/10.1001/archgenpsychiatry.2010.46.
32. Grunhaus L, Dannon PN, Gershon AA. Transcranial magnetic stimulation: a new tool in the fight against depression. Dialogues Clin Neurosci. 2002;4(1):93–103. https://doi.org/10.31887/dcns.2002.4.1/lgrunhaus.
33. Halsey G, Yu W, Guo L. Transcranial magnetic stimulation. In: Neural interface engineering. Cham: Springer; 2020. p. 49–65. https://doi.org/10.1007/978-3-030-41854-0_3.
34. Harmelech T, Roth Y, Tendler A. Deep TMS H7 coil: features, applications & future. Exp Rev Med Dev. 2021;18(12):1133–44. https://doi.org/10.1080/17434440.2021.2013803.
35. Harmelech T, Hanlon CA, Tendler A. Transcranial magnetic stimulation as a tool to promote smoking cessation and decrease drug and alcohol use. Brain Sci. 2023;13(7):1072. https://doi.org/10.3390/brainsci13071072.
36. Hasey G. Transcranial magnetic stimulation in the treatment of mood disorder: a review and comparison with electroconvulsive therapy. Can J Psychiatry. 2001;46(8):720–7. https://doi.org/10.1177/070674370104600804.
37. Horvath JC, et al. Transcranial magnetic stimulation: a historical evaluation and future prognosis of therapeutically relevant ethical concerns. J Med Ethics. 2010;37(3):137–43. https://doi.org/10.1136/jme.2010.039966.
38. Huang Y-Z, et al. Theta burst stimulation of the human motor cortex. Neuron. 2005;45(2):201–6. https://doi.org/10.1016/j.neuron.2004.12.033.
39. Kim H-H. Electroconvulsive therapy: a historical and legal perspective. Am J Psych Res J. 2018;13(3):2–4. https://doi.org/10.1176/appi.ajp-rj.2018.130305.
40. Kolin A. Magnetic fields in biology. Phys Today. 1968;21(11):39–50. https://doi.org/10.1063/1.3034587.
41. Kolin A, Brill NQ, Broberg PJ. Stimulation of irritable tissues by means of an alternating magnetic field. Exp Biol Med. 1959;102(1):251–3. https://doi.org/10.3181/00379727-102-25209.
42. Kujirai T, et al. Corticocortical inhibition in human motor cortex. J Physiol. 1993;471(1):501–19. https://doi.org/10.1113/jphysiol.1993.sp019912.
43. Levkovitz Y, et al. Efficacy and safety of deep transcranial magnetic stimulation for major depression: a prospective multicenter randomized controlled trial. World Psychiatry. 2015;14(1):64–73. https://doi.org/10.1002/wps.20199.
44. Lipton RB, et al. Single-pulse transcranial magnetic stimulation for acute treatment of migraine with aura: a randomised, double-blind, parallel-group, sham-controlled trial. Lancet Neurol. 2010;9(4):373–80. https://doi.org/10.1016/s1474-4422(10)70054-5.
45. Lövsund P, et al. Magnetophosphenes: a quantitative analysis of thresholds. Med Biol Eng Comput. 1980;18(3):326–34. https://doi.org/10.1007/bf02443387.
46. Lucena MF, et al. Top 100 cited noninvasive neuromodulation clinical trials. Expert Rev Med Devices. 2019;16(6):451–66. https://doi.org/10.1080/17434440.2019.1615440.
47. Marder KG, et al. Psychiatric applications of repetitive transcranial magnetic stimulation. Focus. 2022;20(1):8–18. https://doi.org/10.1176/appi.focus.20210021.
48. Mayo Clinic. Transcranial Magnetic Stimulation. Mayo Clinic, 7 April, 2023. https://www.mayoclinic.org/tests-procedures/transcranial-magnetic-stimulation/about/pac-20384625
49. Merton PA, et al. Scope of a technique for electrical stimulation of human brain, spinal cord, and muscle. Lancet. 1982;320(8298):597–600. https://doi.org/10.1016/s0140-6736(82)90670-5.
50. Mould RF, Aronowitz AN. Evolution of the knowledge of electricity and electrotherapeutics with special reference to x-rays & cancer. Nowotwory J Oncol. 2006;56(6):653–22.
51. O'Brien E. FDA Clears NeuroStar TMS for Treatment of MDD in Adolescents. Psychiatric Times, 24 Apr. 2024., www.psychiatrictimes.com/view/fda-clears-neurostar-tms-for-treatment-of-mdd-in-adolescents
52. Park B. Brainsway deep TMS device cleared for older patients with MDD. MPR, 7 June 2024., www.empr.com/home/news/brainsway-deep-tms-device-cleared-for-older-patients-with-mdd/

53. Pascual-Leone A, et al. Safety of rapid-rate transcranial magnetic stimulation in normal volunteers. Electroencephalogr Clin Neurophysiol. 1993;89(2):120–30. https://doi.org/10.1016/0168-5597(93)90094-6.
54. Pascual-Leone A, et al. Responses to rapid-rate transcranial magnetic stimulation of the human motor cortex. Brain. 1994;117(4):847–58. https://doi.org/10.1093/brain/117.4.847.
55. Pell GS, et al. Efficacy of deep TMS with the H1 coil for anxious depression. J Clin Med. 2022;11(4):1015. https://doi.org/10.3390/jcm11041015.
56. Perera T, et al. The clinical TMS Society consensus review and treatment recommendations for TMS therapy for major depressive disorder. Brain Stimul. 2016;9(3):336–46. https://doi.org/10.1016/j.brs.2016.03.010.
57. Peterchev AV, et al. A transcranial magnetic stimulator inducing near-rectangular pulses with controllable pulse width (CTMS). IEEE Trans Biomed Eng. 2008;55(1):257–66. https://doi.org/10.1109/tbme.2007.900540.
58. Peterchev AV, Murphy DL, Lisanby SH. Repetitive transcranial magnetic stimulator with controllable pulse parameters. J Neural Eng. 2011;8(3):Article 036016. https://doi.org/10.1088/1741-2560/8/3/036016.
59. Pisano R. Briefs on symmetries in the history of physics–mathematics: Maxwell (1865–1873), Noether (1915–1918) and Einstein (1905–1926). J Phys Conf Ser. 2024;2877:012101. https://doi.org/10.1088/1742-6596/2877/1/012101.
60. Polson MJ, et al. Stimulation of nerve trunks with time-varying magnetic fields. Med Biol Eng Comput. 1982;20(2):243–4. https://doi.org/10.1007/bf02441362.
61. Roland PE, Hilgetag CC, Deco G. Cortico-cortical communication dynamics. Front Syst Neurosci. 2014;8:5. https://doi.org/10.3389/fnsys.2014.00019.
62. van Rooij SJ, Arulpragasam AR, McDonald WM, Philip NS. Accelerated TMS - moving quickly into the future of depression treatment. Neuropsychopharmacology. 2023;49(1):128–37. https://doi.org/10.1038/s41386-023-01599-z.
63. Rossi S, Hallett M, Rossini PM, Pascual-Leone A. Safety, ethical considerations, and application guidelines for the use of transcranial magnetic stimulation in clinical practice and research. Clin Neurophysiol. 2009;120(12):2008–39. https://doi.org/10.1016/j.clinph.2009.08.016.
64. Rossi S, et al. Safety and recommendations for TMS use in healthy subjects and patient populations, with updates on training, ethical and regulatory issues: expert guidelines. Clin Neurophysiol. 2021;132(1):269–306. https://doi.org/10.1016/j.clinph.2020.10.003.
65. Roth Y, Tendler A, Arikan MK, Vidrine R, Kent D, Muir O, MacMillan C, et al. Real-world efficacy of deep TMS for obsessive-compulsive disorder: post-marketing data collected from twenty-two clinical sites. J Psychiatr Res. 2020;137:667–72. https://doi.org/10.1016/j.jpsychires.2020.11.009.
66. Rothwell J. Transcranial brain stimulation: past and future. Brain Neurosci Adv. 2018;2 https://doi.org/10.1177/2398212818818070.
67. Rothwell J. How TMS came of age. Trans Mag Stimul. 2024;1:100002. https://doi.org/10.1016/j.transm.2024.100002.
68. Rounis E, Huang Y-Z. Theta burst stimulation in humans: a need for better understanding effects of brain stimulation in health and disease. Exp Brain Res. 2020;238(7–8):1707–14. https://doi.org/10.1007/s00221-020-05880-1.
69. Ruohonen J, Karhu J. Navigated transcranial magnetic stimulation. Clin Neurophysiol. 2010;40(1):7–17. https://doi.org/10.1016/j.neucli.2010.01.006.
70. Shin H, et al. Robotic transcranial magnetic stimulation in the treatment of depression: a pilot study. Sci Rep. 2023;13(1):Article 14074. https://doi.org/10.1038/s41598-023-41044-1.
71. Siddiqi SH, Fox MD. Targeting symptom-specific networks with transcranial magnetic stimulation. Biol Psych. 2024;95(6):502–9. https://doi.org/10.1016/j.biopsych.2023.11.011.
72. Tendler A, et al. Deep TMS H1 coil treatment for depression: results from a large post-marketing data analysis. Psychiatry Res. 2023;324:Article 115179. https://doi.org/10.1016/j.psychres.2023.115179.

73. Trapp NT, et al. Consensus review and considerations on TMS to treat depression: a comprehensive update endorsed by the National Network of depression centers, the clinical TMS Society, and the International Federation of Clinical Neurophysiology. Clin Neurophysiol. 2025;170:206–33. https://doi.org/10.1016/j.clinph.2024.12.015.
74. Ueno S, Matsuda T. Magnetic stimulation of the human brain. Ann N Y Acad Sci. 1992;649(1):366–8. https://doi.org/10.1111/j.1749-6632.1992.tb49632.x.
75. Uzair M, et al. Transcranial magnetic stimulation in animal models of neurodegeneration. Neural Regen Res. 2021;17(2):251–65. https://doi.org/10.4103/1673-5374.317962.
76. Vidal-Dourado M, et al. Magnetic fields in noninvasive brain stimulation. Neuroscientist. 2013;20(2):112–21. https://doi.org/10.1177/1073858413491145.
77. Voigt J, et al. Cost effectiveness analysis comparing repetitive transcranial magnetic stimulation to antidepressant medications after a first treatment failure for major depressive disorder in newly diagnosed patients: a lifetime analysis. PLoS One. 2017;12(10):e0186950. https://doi.org/10.1371/journal.pone.0186950.
78. Wali FA. Electromagnetic stimulation blocks isometric twitch contraction in the frog sciatic nerve—gastrocnemius muscle preparation. Life Sci. 1990;46(24):1747–53. https://doi.org/10.1016/0024-3205(90)90138-h.
79. Walsh V, Cowey A. Magnetic stimulation studies of visual cognition. Trends Cogn Sci. 1998;2(3):103–10. https://doi.org/10.1016/s1364-6613(98)01134-6.
80. Weiner RD. ECT and seizure threshold: effects of stimulus wave form and electrode placement. Biol Psychiatry. 1980;15(2):225–41.
81. Weiner RD, et al. Electroconvulsive therapy and other forms of brain stimulation. In: Steffens DC, Zdanys K, editors. The American Psychiatric Association publishing textbook of geriatric psychiatry. 6th ed. American Psychiatric Publishing; 2022. p. 601–42. ISBN 978-1-61537-340-6.
82. Weissman CR, et al. Repetitive transcranial magnetic stimulation for treatment-resistant depression. J Clin Psychiatry. 2023;84(3) https://doi.org/10.4088/jcp.22com14575.
83. Williams LM. Defining biotypes for depression and anxiety based on large-scale circuit dysfunction: a theoretical review of the evidence and future directions for clinical translation. Depress Anxiety. 2017;34(1):9–24. https://doi.org/10.1002/da.22556.
84. Williams N. Stanford neuromodulation therapy (SNT): a double-blinded, randomized, and controlled trial. Brain Stimul. 2021;14(6):1736–7. https://doi.org/10.1016/j.brs.2021.10.494.
85. Williams LM, et al. Identifying response and predictive biomarkers for transcranial magnetic stimulation outcomes: protocol and rationale for a mechanistic study of functional neuroimaging and behavioral biomarkers in veterans with Pharmacoresistant depression. BMC Psychiatry. 2021;21(1):Article 35. https://doi.org/10.1186/s12888-020-03030-z.
86. Zangen A, et al. Repetitive transcranial magnetic stimulation for smoking cessation: a pivotal multicenter double-blind randomized controlled trial. World Psychiatry. 2021;20(3):397–404. https://doi.org/10.1002/wps.20905.
87. Ziemann U. Thirty years of transcranial magnetic stimulation: where do we stand? Exp Brain Res. 2017;235(4):973–84. https://doi.org/10.1007/s00221-016-4865-4.
88. Zohuri B. Transcranial and repetitive transcranial magnetic stimulation driving a noninvasive depression treatment. J Heal Sci. 2020;8(3):87–99. https://doi.org/10.17265/2328-7136/2020.03.003.

Racial Disparities in TMS Research Participation and Clinical Access in the United States

Christi Sullivan, Dawson Cooper, Aaron McCright, Dawn Hackman, Alik Widge, Saydra Wilson, and Laura Y. Cabrera

1 Background

Neuromodulation therapies provide additional, non-medication-based options for treatment-resistant individuals, playing an important role in the psychiatric toolbox. Examples of neuromodulation include transcranial magnetic stimulation (TMS), electroconvulsive therapy (ECT) vagus nerve stimulation (VNS), and deep brain stimulation (DBS). Each of these therapies have unique benefits, disadvantages, and risks. Some of the biggest benefits of neuromodulation is the ability to fine-tune treatment parameters and their use in combination with other therapies like medication and psychotherapy with side effects that are generally quick to resolve. Some neuromodulation treatments are time intensive, often requiring multiple visits on a consistent schedule, and can also have higher costs compared to oral medication.

C. Sullivan (✉) · D. Cooper · A. Widge · S. Wilson
Department of Psychiatry and Behavioral Sciences, University of Minnesota Medical School, Minneapolis, MN, USA
e-mail: perki322@umn.edu; coope862@umn.edu; awidge@umn.edu; wils2548@umn.edu

A. McCright
Department of Sociology, College of Social Science, Michigan State University, East Lansing, MI, USA
e-mail: mccright@msu.edu

D. Hackman
San Jose State University, San Jose, CA, USA
e-mail: dawn.hackman@sjsu.edu

L. Y. Cabrera
School of Public Health and Health Sciences, UMass Amherst, Amherst, MA, USA

Department of Engineering Science and Mechanics, and Rock Ethics Institute, The Pennsylvania State University, University Park, PA, USA

Center for Neural Engineering, The Pennsylvania State University, University Park, PA, USA
e-mail: lcabrera@psu.edu

V. Dubljević, J. R. Young (eds.), *TMS and Neuroethics*, Advances in Neuroethics,
https://doi.org/10.1007/978-3-031-92401-9_3

As the use of neuromodulation expands, it is important to understand who is accessing and receiving treatment, the patterns of potential patients adopting neuromodulation for the treatment of depression, and the barriers faced by those who choose neuromodulation treatments. While neuromodulation is being utilized across the USA, there are emerging patterns of racial treatment disparities in access to most neuromodulation treatments [1–9]. A 2024 review article found that geographic, cultural, provider bias, and financial barriers all play a role in perpetuating treatment disparities in the field of interventional psychiatry [9]. This emergence of treatment disparities is of particular interest in the case of TMS, as the inventors of TMS, Dr. Anthony Baker and colleagues chose not to patent the device in order to allow others to manufacture and use TMS devices, hoping this would enable a quick and effective dissemination of the technology [10].

TMS is a non-invasive treatment that uses magnetic fields to stimulate nerve cells in the brain. This treatment targets areas of the brain associated with mood regulation. During a TMS session, an electromagnetic coil is placed on the scalp near the forehead, delivering painless magnetic pulses. The treatment typically involves daily sessions, 5 days a week, for 4–6 weeks, with each session lasting about 30–40 min. It is used most commonly to treat major depressive disorder (MDD) and is especially useful for individuals who have not responded to traditional treatment like medication or psychotherapy.

Multiple factors impact the dissemination of and access to TMS. Internal and systemic biases in healthcare can significantly affect treatment recommendations, with minoritized patients often facing disparities in access to advanced treatments like TMS. Implicit biases held by healthcare providers, combined with systemic issues such as unequal access to mental health services, lower insurance coverage, and mistrust in the healthcare system, can result in minoritized patients being less likely to be referred for cutting-edge treatments, despite having similar or greater need. This contributes to ongoing disparities in mental health outcomes for these populations. In this chapter, we will use the example of TMS for MDD to highlight racial treatment disparities and discuss potential solutions to emerging disparities.

2 Depression on the Rise

Depression is a leading cause of disability and economic burden [11]. In the USA, 18.5% of adults experience symptoms of depression [12], spanning age, race, and region [13, 14]. Depression also continues to rise across younger age groups, particularly among adolescents and younger adults, which has been further exacerbated by the COVID-19 pandemic [15]. Both effective and accessible treatments are needed to address these trends.

Traditional first-line treatments for depression include medication and psychotherapy. At least 20% of individuals do not experience symptom reduction from these treatments [16]. Individuals who fail at least two treatments are considered to have treatment-resistant depression (TRD). More than 3 million people in the USA have TRD [17]. Neuromodulation is often the next step for individuals with

TRD. This means that access to neuromodulation treatments relies on a robust and effective mental health treatment infrastructure to have provided first-line treatments before they can be deployed.

3 TMS Treatment Barriers

Compared to medications, some neuromodulation treatments are more time intensive, often requiring multiple visits on a consistent schedule, and can also have higher costs compared to oral medication. Lack of access to care for first-line treatment further limits the availability of TMS. In the case of TMS, a course of treatment requires daily clinic visits to achieve benefit with the best studied TMS protocols for MDD requiring 6–9 weeks of treatments. Treatment of this length introduces significant logistical challenges including the need for transportation, child-care, and time away from work or schooling which has a direct impact on the accessibility of TMS across diverse populations. The cost associated with the multiple visits also contributed to delays in adoption decisions by healthcare insurers in the USA. Despite TMS achieving FDA approval for the treatment of MDD in 2008 it took almost a decade before Medicare public insurance had a coverage policy [18].

4 Racial Treatment Barriers

Treatment barriers are not experienced equally across the population. Black, Indigenous, and Other People of Color (BIPOC) groups experience barriers to care at higher rates than their white counterparts [19]. BIPOC groups are often referred to in the literature as racialized, as these groups are often subject to systemic racism and discrimination. The word racialized is specifically chosen in this case because race is an identity that is dependent on the community to which one belongs. We as a society decide who is and is not part of certain racial groups based on skin color and facial features. As a result, the racial categories used often neglect the socio-historical characteristics and experiences of individuals.

5 The Role of Race in TMS Uptake

Despite its role as the second most commonly used neuromodulation treatment for depression [20], TMS has not been previously examined for the presence of racial treatment disparities or barriers to care. Furthermore, understanding who receives TMS for depression and how different groups view neuromodulation therapies may help increase accessibility and subsequent usage. The following section presents a systematic review conducted in the fall of 2021 to establish a baseline of racial uptake of TMS for MDD. The goal of this review was to identify if there were disparities present in TMS participation within the USA.

6 Exploratory Literature Review

To confirm this potential health disparity, a systematic review was performed (https://osf.io/ve6z2/, registered 10/16/21 and revised 10/26/21). A trained medical librarian conducted an electronic search of the following databases: Medline (1946 to October 15, 2021), Embase (1947 to October 15, 2021), and PsycINFO (1987 to October Week 2 2021) via Ovid; Social Services Abstracts (1979-present); Cochrane Library (Issue 10 of 12, October 2021) via Wiley; Campbell Library;and CINAHL via EBSCOhost (1937-present). No databases were excluded as part of this search. The final searches were conducted on October 18, 2021. Per the Cochrane Collaboration's Methodological Expectations of Cochrane Intervention Reviews, a combination of controlled vocabulary and natural language was used to find studies whose primary objective was treating depressive symptoms in adults using TMS in the USA. No filters or hedges were used to limit retrieval based on publication type or location of study. All types of depressions were included to ensure all relevant studies were captured. We excluded studies that included non-clinical endpoints that required participation with additional technologies outside of common depression standard of care, such as functional neuroimaging or electroencephalography, to avoid conflating TMS technology attitudes with attitudes toward other technologies. A full search strategy can be found in Appendix A.

All resulting citations were uploaded into Endnote (https://endnote.com/) for deduplication and then were added to Covidence for the screening process. Three authors [CS, DC, DH] independently reviewed titles and abstracts to screen for eligible studies. Following the initial screening, two authors [CS, DC] independently conducted a full-text review of the remaining papers. Disagreement among the reviewers was resolved by consensus with an additional reviewing author [SW]. Reasons for excluding studies at this stage are reported in the PRISMA flowchart (Fig. 1).

6.1 Data Extraction

A reviewer independently extracted data from the included studies into a predefined table. The following information was extracted: title, author, year published, location, percent of participants identifying as White, and other available race/ethnicity data. If the locations spanned multiple regions, the study was designated as multi-site. Racial data collection and reporting was inconsistent across the included studies. All corresponding authors were contacted to collect information about how race was collected in the study. Responses were received from 10 of 24 authors stating that race was collected through chart review or self-report. To account for the incongruent categorization, study populations were divided into two groups: White non-Hispanic and racialized: non-White and/or Hispanic. This categorization was most common among all studies and appeared to be the most encompassing way to capture race for our analysis. This grouping of race into White and non-White is inherently problematic and harmful; however, it was necessary due to the number of

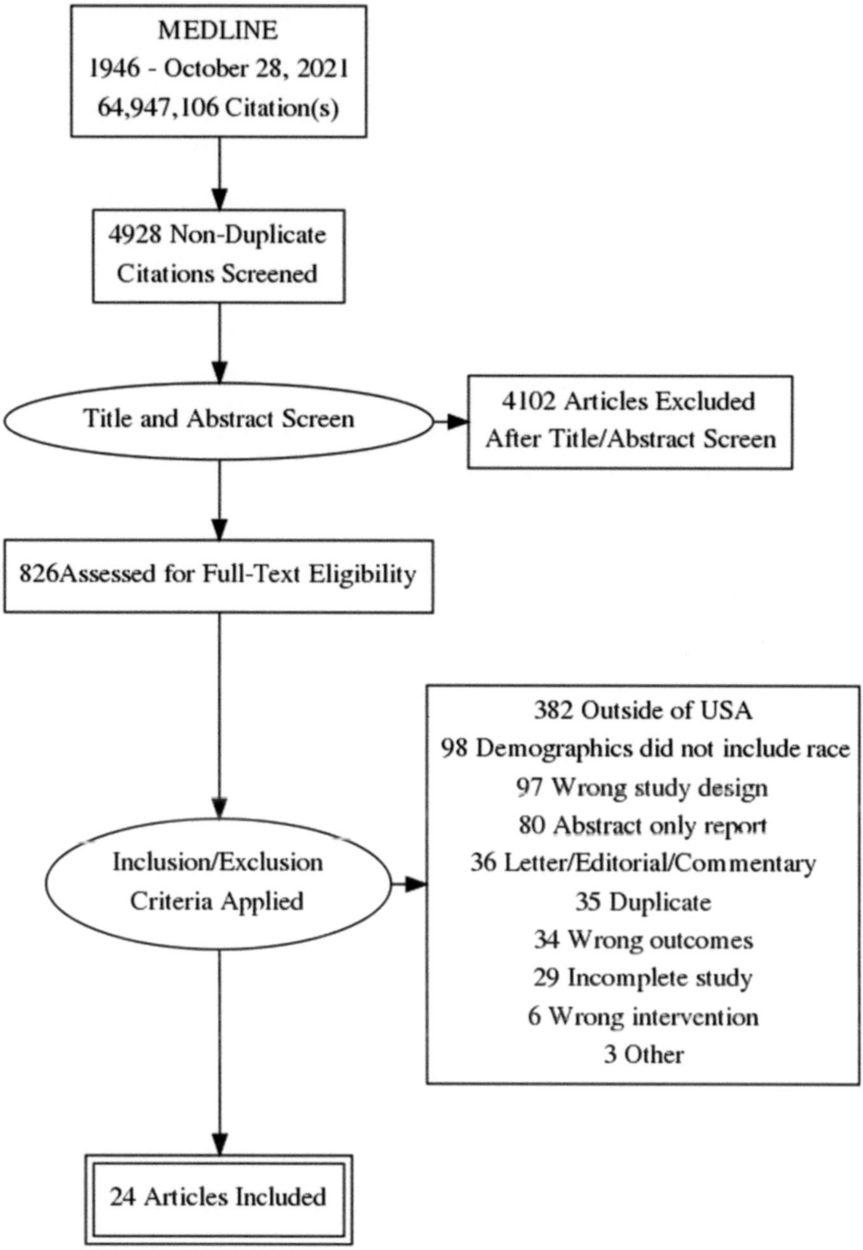

Fig. 1 PRISMA flow chart

studies who did not report more specific racial data. Extracted data are summarized in Appendix B.

7 Literature Review Results

7.1 Participation in TMS Research

A total of 24 studies were identified and included in the final analysis. A total of 98 studies were excluded because they did not report race as a demographic variable but were otherwise eligible.

Eligible studies were categorized by U.S. Census Bureau region including Midwest (n = 5; 334 participants), Northeast (n = 6; 107 participants), South (n = 7; 356 participants), West (n = 1; 77 participants), and multisite (n = 5;353 participants). The purpose of this categorization was to identify if racial disparities existed across all parts of the country or if they were concentrated in sub-regions of the USA. All but one study in the West region were part of multisite studies. White populations were overrepresented in 96% of studies included in our analysis. This overrepresentation was apparent across all major regions on a participant level (Fig. 2). Racialized populations were not equitably engaged in clinical research of TMS.

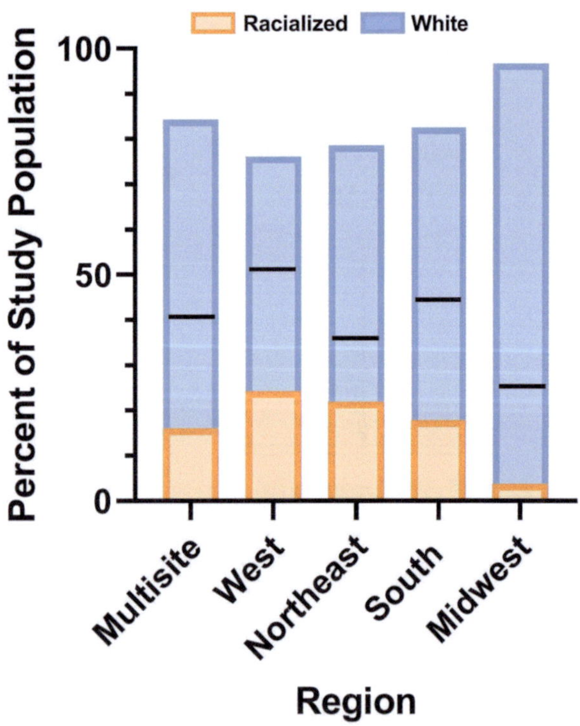

Fig. 2 Racial demographics of participants (n = 1227) who received TMS for depression in the USA by region. The black line in each bar represents the proportion of racialized persons in that region [21]. Multisite studies were compared to the racial demographic makeup of the entire USA (40.7% racialized)

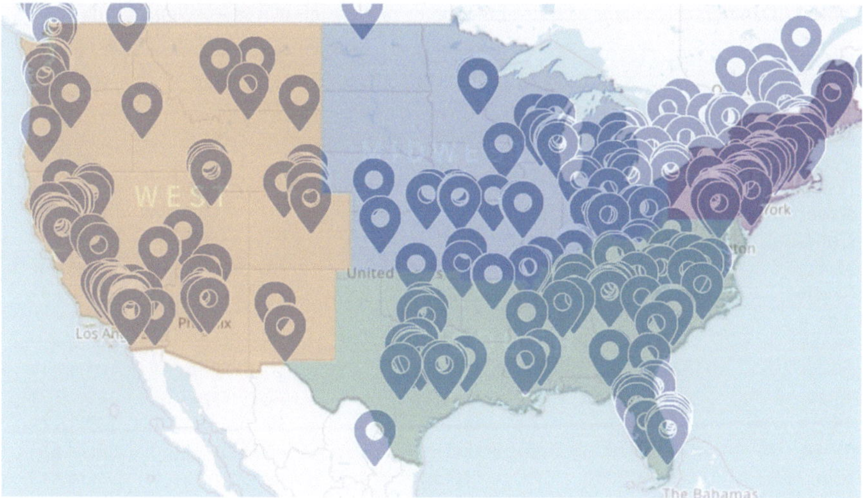

Fig. 3 Map of the U.S. Census Bureau Regions, West (Orange), Midwest (Blue), South (Green), and Northeast (Purple) overlayed onto a map of TMS treatment locations retrieved from the Clinical TMS Society website (https://www.clinicaltmssociety.org/find-a-clinic, Date of Retrieval: October 2022)

The racial disparities by region may be explained in part by geographic availability of TMS. As shown in Fig. 2, there is a consistent gap of 20–30% between percent racialized study participation and the percent of the region who identifies as racialized. The only region for which this is not true is the Northeast. One possible explanation for this difference is that the Northeast region has a higher density of TMS providers compared to the West, South, and Midwest regions. Limited access to a treatment directly impacts who is able to receive a treatment. Figure 3 shows the TMS treatment locations by region.

8 TMS Attitudinal Barriers

In addition to structural treatment barriers, attitudinal barriers also play a role in access to TMS [22]. Research surrounding the role of attitudinal barriers in TMS uptake is limited but growing. The primary source of knowledge on this topic comes from a nationwide online survey conducted in 2022. This survey assessed the barriers and attitudes toward multiple neuromodulation treatments for depression [22]. Individuals were recruited from two groups: non-psychiatrist (general public, caregivers, and depressed participants) and psychiatrists. Each individual viewed one of eight possible video vignettes (four neuromodulation types and two severity levels of disease [moderate or severe]). In this section, we focus on the sub-sample of individuals surveyed about TMS (n = 603). For this secondary analysis, we did not include the caregivers subgroup. Participants were given a series of questions assessing attitudes and concerns, including the willingness to pursue the treatment,

Table 1 Demographic characteristics of the survey participants across stakeholder group

	Psychiatrists (n = 117)	Depressed patients (n = 240)	General public (n = 246)	Total (n = 603)
	n(%)	n(%)	n(%)	n(%)
Gender				
Male	81(69.2)	68(28.3)	127(51.6)	276(45.8)
Female	36(30.8)	172(71.7)	119(48.4)	327(54.2)
Experimental Condition				
Moderate	64(54.7)	122(50.8)	131(53.3)	317(52.6)
Severe	53(45.3)	118(49.2)	115(46.7)	286(47.4)
Age				
18–24	0(0)	58(24.2)	31(12.6)	89(14.8)
25–34	4(3.4)	48(20.0)	51(20.7)	103(17.1)
35–44	24(20.5)	51(21.3)	44(17.9)	119(19.7)
45–54	30(25.6)	44(18.3)	35(14.2)	109(18.1)
55–64	23(19.7)	28(11.7)	37(15.0)	88(14.6)
65 or older	36(30.8)	11(4.6)	48(19.5)	95(15.8)
Race/Ethnicity				
White	100(85.5)	208(86.7)	200(81.3)	508(84.2)
Racialized	17(14.5)	32(13.3)	46(18.7)	95(15.8)
Education				
High diploma or GED		111(46.3)	90(36.6)	201(33.3)
Associate's degree		41(17.1)	41(16.7)	82(13.6)
Bachelor's degree		50(20.8)	82(33.0)	132(21.9)
Graduate degree	117(100)	38(15.8)	33(13.4)	188(31.2)

and eight perceived barriers to the target intervention. Participants were also asked to state how willing they would be to receive TMS treatment from the point of view of the patient in the video vignette they were presented. Demographic characteristics of the survey sample are located in Table 1.

8.1 Perceived Barriers

Individuals ranked their top three perceived barriers to utilizing TMS. Barriers included limited evidence of effectiveness, lack of understanding of TMS, out-of-pocket costs, lack of insurance coverage, low public trust in the mental health system, treatment not being available in specific geographic area, stigma about treatment, and frequency of treatment. A summary of responses among the stakeholder groups can be found in Appendix C and D. The top three barriers for Depressed Patients and General Public were the same, ranking limited evidence of effectiveness as their top barrier, followed by a lack of understanding of TMS and out-of-pocket costs. Psychiatrists ranked lack of insurance coverage as their top barrier, followed by out-of-pocket costs and frequency of treatment.

Table 2 Results from ordinal logistic regression analyses assessing differences among the eight barriers and their relative importance using three predictor variables

Barrier	Predictor White—Racialized	Psychiatrist—Not	Severe Condition—Moderate condition
Limited evidence of TMS' effectiveness	OR = 1.00, p = 0.99	**OR = 0.32, p < 0.001**	OR = 1.09, p = 0.58
Lack of understanding of TMS	OR = 0.95, p = 0.82	**OR = 0.38, p < 0.001**	OR = 0.89, p = 0.45
Out-of-pocket costs	OR = 1.00, p = 0.99	OR = 1.37, p = 0.10	OR = 1.08, p = 0.61
Lack of insurance coverage	OR = 0.96, p = 0.86	**OR = 2.74, p < 0.001**	OR = 0.80, p = 0.15
Low public trust in the mental healthcare system	OR = 0.58, p = 0.03	**OR = 0.11, p < 0.001**	OR = 1.15, p = 0.48
TMS is not available in all geographic areas	OR = 0.72, p = 0.18	OR = 1.96, p = 0.0014	OR = 1.11, p = 0.57
Stigma about TMS	OR = 0.84, p = 0.51	OR = 0.34, p = 0.0013	OR = 1.59, p = 0.02
Frequency of TMS	OR = 1.07, p = 0.76	**OR = 2.36, p < 0.001**	OR = 0.86, p = 0.38

Statistically significant differences are bolded (OR = odds ratio)

Race did not predict the likelihood of selecting any of the eight barriers. Severity of disease similarly had no predictive significance among all barriers. Psychiatrists differed in their ranking of the overall importance of all barriers compared to non-psychiatrists, except for out-of-pocket cost, availability of treatment, and stigma about treatment.

8.2 Attitudes Towards Seeking TMS Treatment

Racialized participants did not report significantly different willingness from White participants to seek TMS treatment. Participants assigned to the severe depression group did not report significantly different willingness to seek TMS in this hypothetical situation compared to those who observed the moderate depression vignette. Psychiatrists were more likely to seek treatment compared to non-psychiatrists (odds ratio = 2.22, $p < 0.00625$) (Table 2).

After determining the differences of participants' rankings among the eight barriers, we next sought to understand which variables predicted participants' single most important barrier. Race and disease severity did not significantly predict any of the eight barriers. Psychiatrists and potential patients ranked the importance of two out of eight barriers differently. Psychiatrists were less likely to select lack of understanding of TMS as their most important barrier compared to non-psychiatrists (OR = 0.29, $p < 0.00625$). In contrast, psychiatrists were more likely to rank lack of insurance coverage as their top barrier (OR = 2.92, $p < 0.00625$) (Table 3).

Table 3 Results from binomial logistic regression analyses assessing differences among participants' most important barrier using three predictor variables

	Predictor		
Barrier	White—Racialized	Psychiatrist—Not	Severe Condition—Moderate condition
Limited evidence of TMS' effectiveness	OR = 0.98, p = 0.93	OR = 0.48, p = 0.01	OR = 1.24, p = 0.25
Lack of understanding of TMS	OR = 1.00, p = 0.99	**OR = 0.29, p < 0.001**	OR = 0.77, p = 0.22
Out-of-pocket costs	OR = 1.09, p = 0.77	OR = 0.87, p = 0.62	OR = 1.02, p = 0.92
Lack of insurance coverage	OR = 1.38, p = 0.35	**OR = 2.92, p < 0.001**	OR = 0.72, p = 0.16
Low public trust in the mental healthcare system	OR = 0.39, p = 0.02	OR = 4.52e-8, p = 0.99	OR = 0.74, p = 0.41
TMS is not available in all geographic areas	OR = 0.86, p = 0.72	OR = 1.97, p = 0.05	OR = 1.14, p = 0.66
Stigma about TMS	OR = 0.59, p = 0.17	OR = 0.34, p = 0.08	OR = 1.47, p = 0.24
Frequency of TMS	OR = 1.31, p = 0.52	OR = 1.77, p = 0.08	OR = 1.32, p = 0.33

Statistically significant differences are bolded (OR = odds ratio)

9 Discussion

Racial and ethnic demographics serve an important role in addressing health disparities. Without accurate racial and ethnic information about study populations, we are susceptible to missing pertinent information about populations who are most in need [23]. We also risk inadequate monitoring of health status and quality of care within racialized communities when racial demographic data is missing from publication. Current NIH requirements for diverse study participant recruitment are open-ended, without concrete expectations if the recruitment outcomes are not met. A report at study completion summarizing racialized recruitment plans should be implemented to identify successes and failures in recruitment. This might be implemented as part of the new emphasis on Plans for Enhancing Diverse Perspectives in Research [24].

This review identified a clear under-participation of racialized non-white individuals in TMS research. It is probable that a corresponding racial disparity in clinical TMS participation exists, and this needs to be further assessed in order to assess what policy changes might be required to address such disparities. Medicare utilization data may be a possible source of confirming clinical TMS proportional participation. Large registry studies are often used to report general TMS outcomes and practice and could collect and report racial data [25, 26].

Combining the findings of the survey with the literature review suggests that racialized non-white populations, when specifically engaged and informed, are willing to consider TMS as a treatment option. We specifically did not find a significant difference in trust in the mental health system as a barrier to TMS use based on race. Distrust has been frequently proposed to explain some of the differential research participation and clinical participation in these groups. Our results suggest that this explanation may be overly simplistic. Poor uptake of TMS within racialized communities may be a by-product of failure to integrate those communities earlier in the research/development process.

10 Recommendations

Editorial boards can and should require racial reporting as part of all clinical trial publications. Some journals have already begun this important work through the addition of new columns, disparity focused publication collections, and guidelines for how to relay racial demographics within a publication [27]. From the federal level, NIH reporting requirements should be strengthened. Since 1993, all NIH-funded projects have required the annual reporting of racial/ethnic categories [21]. In our sample, 42% of studies were NIH supported and conducted in the years after enactment of this requirement. Despite the reporting requirement, this information rarely makes its way into publication. In our review, only 20% of eligible papers reported race. This is higher than previously collected data where only 13.4% of NIH-funded studies reported race [28].

Beyond reporting and education, we must increase efforts to recruit from a diverse population throughout all phases of research. When we do not include a diverse population throughout the clinical trial process, we exclude sections of the population from the newest therapies. Disparate recruitment also leads to potential sampling bias leading to lower generalizability of the results [29]. Racial disparities in treatment participation could be avoided by enrolling diverse populations earlier in the research process and enhancing awareness of the technology across communities. Racially representative recruitment would allow researchers to better identify treatment barriers and cultural considerations early on to improve the treatment and increase treatment accessibility post approval. It is also important to increase clinical outreach of TMS to underserved populations, as this can help enhance TMS racial representation in research settings using clinics.

The struggle to enroll across multiple racial demographics is not unique to TMS. In the 20 years following the 1993 NIH Revitalization Act aimed at increasing racialized enrollment, only 2% of over 10,000 NIH-funded trials reached goals outlined by the Act [26]. While these numbers are disappointing, one group has been able to effectively increase racialized enrollment. The National Cancer Institute created the Minority-Based Community Clinical Oncology program, which specifically funds research at minority serving clinics and hospitals. This program has had positive outcomes, increasing racialized enrollment in oncology trials from 14% in 1999 to 25% 2019 [30]. The program focused on bringing research into clinics already serving racialized groups and offering the same high caliber research care typically available at university centers. This strategy could be employed in mental health research using TMS. In psychiatry research, this might mean setting up new research centers in clinics that primarily serve racialized populations. Many of these may be new sites that have historically been excluded from conducting research. A key feature to this success would be providing ample research resources to these new sites, as there is a high level of institutional knowledge needed to make a clinical trials program sustainable.

Outside of research, efforts to reduce treatment disparities consist of reducing economic barriers through equitable insurance coverage, recruitment of racialized clinicians, and utilization of medical transport to either bring patients to clinic for treatment or deliver treatment through mobile TMS options. For more on this topic see [9].

11 Limitations

A limitation of the survey is that survey responses reflect respondent attitudes and perceived barriers in a hypothetical scenario but may not accurately predict real-world behavior. Direct studies with racialized communities who have been offered TMS or who are undergoing TMS are needed to fully assess actual barriers to TMS. Also, the survey's video vignette presenting TMS technology was limited in length and focused on the procedure and common side effects only. If more efficacy data were discussed, participants might not have reported a perceived lack of scientific evidence for TMS. On another hand, if the vignettes more clearly conveyed some of the complex time and transportation factors associated with a 30+ day TMS treatment course, differences in perceived barriers among racialized and non-racialized participants may have been more evident.

The review likely underestimated the racial research participation disparity by focusing on TMS for depression only. This resulted in exclusion of studies with primary outcomes other than change in clinical depression symptoms. Many TMS studies are seeking biomarkers of target engagement or response and include these additional measures, which would not have been captured by this review. Additionally, we may have missed studies that collected race but did not report the racial demographics in their final publication, despite conducting a comprehensive search. More precise estimates would be possible if racial reporting were standardized and compulsory in publication. Despite these limitations, to our knowledge, this is the first comprehensive systematic review to report demographics of participation in TMS research for depression in the USA.

12 Conclusion

Overall, our results from the literature review and survey analysis suggest that requiring demographic reporting, increasing TMS education and anti-bias training for psychiatric providers, as well as improving TMS education of racialized patients are necessary for addressing barriers and improving access of TMS among diverse populations. These actions are not to be done to simply check a box on a form; these actions are rather the fuel for transformative change in TMS participation and generate public health impact. From early phase research to clinical use, TMS should be accessible and utilized by those who are eligible, regardless of racialized status.

Acknowledgments None.

Funding The survey portion of this project was supported by the National Institute of Health BRAIN Initiative grant [grant number #RF1MH117802].

Appendix A: Search Strategy

Ovid MEDLINE(R) and Epub Ahead of Print, In-Process, In-Data-Review & Other Non-Indexed Citations, Daily and Versions(R) <1946 to October 28, 2021>.

#	Query	Results from 28 Oct 2021
1	Transcranial Magnetic Stimulation/	12,921
2	Magnetics/	25,755
3	(transcranial magnet* stimulat* and (repetitive or repetitious or repeat or repeating or repeated or repeatedly)).tw,kw.	6242
4	TMS.tw,kw,kf.	5098
5	or/1–4	40,637
6	exp Migraine Disorders/	29,123
7	exp Depressive Disorder/	115,036
8	exp Obsessive-Compulsive Disorder/	15,590
9	smoking cessation/	30,670
10	"tobacco use cessation"/	1316
11	"Tobacco Use Cessation Devices"/	2133
12	exp Smoking/and therapy.fs.	8544
13	(ocd or "obsessive compulsive" or "obsessive-compulsive" or unipolar or depression or depressed or depressive or mdd or trd or migraine? or smoking or tobacco).tw,kw,kf.	833,817
14	or/6–13	866,887
15	exp Sensory Thresholds/	57,444
16	rest* motor threshold?.tw,kw.	947
17	(rmt or mt).tw,kw,kf.	41,330
18	((corticomotor or motorcortico or corticalmotor or "cortical motor" or "motor cortical" or "motor cortex") adj3 threshold?).tw,kw.	120
19	((cortico or cortical) and excitability).tw,kw.	7318
20	or/15–19	105,667
21	(variability or variable or variation or varies or varying or drift or chang* or "day-to-day" or daily or weekly or monthly or yearly or hour* or "per day").tw,kw,kf.	5,568,019
22	("intra-individual" or "intra-subject").tw,kw,kf.	7018
23	(between adj3 (subject? or individual?)).tw.	79,697
24	21 or 22 or 23	5,615,492
25	exp animals/	24,739,768
26	humans/	19,833,961

(continued)

#	Query	Results from 28 Oct 2021
27	25 not 26	4,905,807
28	5 and 14 and 20 and 24	306
29	28 not 27	297
30	(exp child/ or exp. adolescent/ or exp. infant/) not exp. adult/	1,989,284
31	29 not 30	290
32	limit 31 to english language	286
33	limit 32 to year = "1985 -current"	286

Appendix B: Breakdown of Studies and Associated N's

Region	Northeast	South	Midwest	West	Multisite	2020 US census
Number of studies	6	7	5	1	5	X
Number of patients	107	356	334	77	353	X
Percentage white patients	78.5	82.6	96.7	66.2	84.4	59.3

Appendix C

Response rates of stakeholder groups regarding whether a barrier was in their top three most important to utilize TMS. Each participant was asked to select three barriers.

Top 3 Barriers (%)				
BARRIER	Psychiatrists (n = 117)	Depressed patients (n = 240)	General public (n = 246)	Total (n = 603)
Limited evidence of treatment's effectiveness	30	61	63	56
Lack of understanding of tms	32	52	55	49
Out-of-pocket cost	61	47	49	50
Lack of insurance coverage	69	45	46	50
Low public trust in mental health system	4	30	27	24
Treatment is not available in all geographic areas	43	23	25	28
Stigma about treatment	9	28	22	22
Frequency of treatment	55	35	33	38

Appendix D

Response rate of participants regarding the most important barrier to utilize TMS.

Participant's Most Important Barrier (%)				
Barrier	Psychiatrists (n = 117)	Depressed patients (n = 240)	General public (n = 246)	Total (n = 603)
Limited evidence of treatment's effectiveness	15	26	32	26
Lack of understanding of tms	8	24	20	19
Out of pocket cost	18	18	17	18
Lack of insurance coverage	32	11	14	16
Low public trust in mental health system	0	6	9	6
Treatment is not available in all geographic areas	12	6	7	8
Stigma about treatment	3	10	6	7
Frequency of treatment	13	10	7	9

References

1. Lad SP, Kalanithi PS, Arrigo RT, et al. A socioeconomic survey of spinal cord stimulation (SCS) surgery. Neuromodul Technol Neural Interf. 2010;13(4):265–9. https://doi.org/10.1111/j.1525-1403.2010.00292.x.
2. Chan AK, McGovern RA, Brown LT, et al. Disparities in access to deep brain stimulation surgery for Parkinson disease: interaction between African American race and medicaid use. JAMA Neurol. 2014;71(3):291–9. https://doi.org/10.1001/jamaneurol.2013.5798.
3. Fox J, Lekoubou Looti AZ, Bishu KG, Ovbiagele B. Recent patterns of vagal nerve stimulator use in the United States: is there a racial disparity? Epilepsia. 2019;60(4):756–63. https://doi.org/10.1111/epi.14695.
4. Jones KC, Salemi JL, Dongarwar D, et al. Racial/ethnic disparities in receipt of electroconvulsive therapy for elderly patients with a principal diagnosis of depression in inpatient settings. Am J Geriatr Psychiatry Off J Am Assoc Geriatr Psychiatry. 2019;27(3):266–78. https://doi.org/10.1016/j.jagp.2018.11.007.
5. Strzalek A. Accessing neuromodulation: Race and ethnicity "should not cloud treatment options" for pain patients. NeuroNews International. February 22, 2021. Accessed April 26, 2022. https://neuronewsinternational.com/accessing-neuromodulation-race-and-ethnicity-should-not-cloud-treatment-options-for-pain-patients/
6. Cramer SW, Do TH, Palzer EF, et al. Persistent racial disparities in deep brain stimulation for Parkinson's disease. Published online April 13, 2022:2022.04.05.22273075. https://doi.org/10.1101/2022.04.05.22273075
7. Frassica M, Kern DS, Afshari M, et al. Racial disparities in access to DBS: results of a real-world U.S. claims data analysis. Front Neurol. 2023;14:1233684. https://doi.org/10.3389/fneur.2023.1233684.
8. Peebles IS, Phillips TO, Hamilton RH. Toward more diverse, inclusive, and equitable neuromodulation. Brain Stimul Basic Transl Clin Res Neuromodulation. 2023;16(3):737–41. https://doi.org/10.1016/j.brs.2023.04.013.

9. Nagra DS, Stolz LA, Weissman CR, Appelbaum LG. Access to interventional psychiatric treatments in the United States: Disparities and proposed solutions. Glob Health Econ Sustain. 2024;2(1):2456. https://doi.org/10.36922/ghes.2456.
10. Cochran C. Meeting Professor Anthony Barker. NeuroScience & TMS Treatment Centers. July 2, 2016. Accessed May 30, 2024. https://hopeforyourbrain.com/uncategorized/meeting-professor-anthony-barker/
11. Greenberg P, Chitnis A, Louie D, et al. The economic burden of adults with major depressive disorder in the United States (2019). Adv Ther. 2023;40(10):4460–79. https://doi.org/10.1007/s12325-023-02622-x.
12. Products—Data Briefs—Number 379—September 2020. September 24, 2020. Accessed April 26, 2022. https://www.cdc.gov/nchs/products/databriefs/db379.htm
13. Bailey RK, Mokonogho J, Kumar A. Racial and ethnic differences in depression: current perspectives. Neuropsychiatr Dis Treat. 2019;15:603–9. https://doi.org/10.2147/NDT.S128584.
14. Jia H. National and State Trends in Anxiety and Depression Severity Scores Among Adults During the COVID-19 Pandemic — United States, 2020–2021. MMWR Morb Mortal Wkly Rep. 2021;70 https://doi.org/10.15585/mmwr.mm7040e3.
15. Ettman CK, Abdalla SM, Cohen GH, Sampson L, Vivier PM, Galea S. Prevalence of Depression Symptoms in US Adults Before and During the COVID-19 Pandemic. JAMA Netw Open. 2020;3(9):e2019686. https://doi.org/10.1001/jamanetworkopen.2020.19686.
16. Sullivan CRP, Olsen S, Widge AS. Deep brain stimulation for psychiatric disorders: From focal brain targets to cognitive networks. NeuroImage. 2021;225:117515. https://doi.org/10.1016/j.neuroimage.2020.117515.
17. The Prevalence and National Burden of Treatment-Resistant Depression and Major Depressive Disorder in the United States. Psychiatrist.com. Accessed April 26, 2022. https://www.psychiatrist.com/jcp/depression/prevalence-national-burden-treatment-resistant-depression-major-depressive-disorder-in-us/
18. LCD—Transcranial Magnetic Stimulation (TMS) (L34641). Accessed May 30, 2024. https://www.cms.gov/medicare-coverage-database/view/lcd.aspx?LCDId=34641
19. Caraballo C, Ndumele CD, Roy B, et al. Trends in racial and ethnic disparities in barriers to timely medical care among adults in the US, 1999 to 2018. JAMA Health Forum. 2022;3(10):e223856. https://doi.org/10.1001/jamahealthforum.2022.3856.
20. Lipsman N, Sankar T, Downar J, Kennedy SH, Lozano AM, Giacobbe P. Neuromodulation for treatment-refractory major depressive disorder. CMAJ Can Med Assoc J. 2014;186(1):33–9. https://doi.org/10.1503/cmaj.121317.
21. U.S. Census Bureau QuickFacts: United States. Accessed June 2, 2022. https://www.census.gov/quickfacts/fact/table/US/PST045221
22. Cabrera LY, Miller MMC, Achtyes ED, McCright AM, Bluhm R. Jumping through the hoops: Barriers and other ethical concerns regarding the use of psychiatric electroceutical interventions. Psychiatry Res. 2022;313:114612. https://doi.org/10.1016/j.psychres.2022.114612.
23. Mehalick ML, Laje G. A pilot assessment of psychiatrists' basic knowledge and perceptions of transcranial magnetic stimulation. Brain Stimul Basic Transl Clin Res Neuromodulation. 2020;13(6):1852–3. https://doi.org/10.1016/j.brs.2020.06.050.
24. Plan for Enhancing Diverse Perspectives (PEDP) | Brain Initiative. Accessed September 7, 2022. https://braininitiative.nih.gov/about/plan-enhancing-diverse-perspectives-pedp
25. Aaronson ST, Carpenter LL, Hutton TM, et al. Comparison of clinical outcomes with left unilateral and sequential bilateral Transcranial Magnetic Stimulation (TMS) treatment of major depressive disorder in a large patient registry. Brain Stimulat. 2022;15(2):326–36. https://doi.org/10.1016/j.brs.2022.01.006.
26. Sackeim HA, Aaronson ST, Carpenter LL, et al. Clinical outcomes in a large registry of patients with major depressive disorder treated with transcranial magnetic stimulation. J Affect Disord. 2020;277:65–74. https://doi.org/10.1016/j.jad.2020.08.005.
27. Dixon LB, Lewis-Fernández R, Isom J, et al. Reporting on and reviewing for race/racialization, ethnicity, and culture. Psychiatr Serv. 2022;73(5):482–3. https://doi.org/10.1176/appi.ps.20220099.

28. Mindlis I, Livert D, Federman AD, Wisnivesky JP, Revenson TA. Racial/ethnic concordance between patients and researchers as a predictor of study attrition. Soc Sci Med. 1982;2020(255):113009. https://doi.org/10.1016/j.socscimed.2020.113009.
29. Chen MS Jr, Lara PN, Dang JHT, Paterniti DA, Kelly K. Twenty years post-NIH Revitalization Act: Enhancing minority participation in clinical trials (EMPaCT): laying the groundwork for improving minority clinical trial accrual. Cancer. 2014;120(S7):1091–6. https://doi.org/10.1002/cncr.28575.
30. Alegría M, Chatterji P, Wells K, et al. Disparity in depression treatment among racial and ethnic minority populations in the United States. Psychiatr Serv Wash DC. 2008;59(11):1264–72. https://doi.org/10.1176/appi.ps.59.11.1264.

Family and Caregiver Perspectives on TMS Treatment of Refractory Conditions: A Pilot Investigation

Ariana D'Alessandro, Iris Coates McCall, and Veljko Dubljević

1 Introduction

Transcranial magnetic stimulation (TMS), a form of non-invasive brain stimulation was first approved by the Food and Drug Administration (FDA) in 2008 for the treatment of major depressive disorder (MDD) in patients who had not achieved significant therapeutic results from other forms of treatments and has since been found to be an evidently effective, safe, and well-tolerated treatment for MDD [10, 14]. However, the process of receiving TMS treatment is relatively burdensome, with regular and frequent trips to the clinic, a scarcity of interventional psychiatrists, and potential need for life-long periodic maintenance TMS [48]. This presents barriers not only to patients who may face challenges navigating these hurdles on their own, but also, for this very reason, their caregivers on whom the onus often falls for getting patients the treatment they need—caregivers have many responsibilities when it comes to helping people access TMS. As such, it is important to consider all stakeholder perspectives on the intervention beyond simply clinical efficacy, and acknowledge any gaps there may be in the experiences of all those who are implicated [7]. This chapter discusses the effects of TMS in comparison with a frequently associated treatment, electroconvulsive therapy (ECT), considers public perceptions of other treatments involving brain stimulation, follows with an

A. D'Alessandro
North Carolina State University, Raleigh, NC, USA
e-mail: amdaless@alumni.ncsu.edu

I. C. McCall
Centre for Addiction and Mental Health, Toronto, ON, Canada
e-mail: iris.mccall@camh.ca

V. Dubljević (✉)
Department of Philosophy and Religious Studies, North Carolina State University, Raleigh, NC, USA
e-mail: veljko_dubljevic@ncsu.edu

V. Dubljević, J. R. Young (eds.), *TMS and Neuroethics*, Advances in Neuroethics,
https://doi.org/10.1007/978-3-031-92401-9_4

evaluation of caregiver and clinician perspectives of other electroceutical treatments, and discusses stakeholder perspectives on TMS. We then examine the findings from our empirical work, specifically these important yet often overlooked caregiver perspectives on the use of TMS for psychiatric conditions.

1.1 Neurophysiological Effects of ECT Versus TMS

An important finding from prior literature on electroceuticals and our current work is that TMS is often compared to or discussed in relation to ECT. Therefore, it is important to clarify relevant similarities and differences in order to conduct a qualitative ethical analysis of TMS treatment. Although they may be perceived similarly, there are substantial differences between ECT and TMS. The main difference is that TMS induces an electrical current in the brain through an alternating electromagnetic field without inducing a seizure or requiring anesthesia [15]. The variation of coil types for TMS indicate the volume and distribution to which the brain is stimulated, with the figure eight coil being one design that stimulates further beneath the cortical surface than other designs [8]. This may impact the way in which stakeholders, including caregivers, view the intervention.

Another consideration is the long-term cognitive side effects of ECT. Despite evidence of its high efficacy for treating MDD and other psychiatric disorders, ECT has been associated with a number of cognitive impairments [1], including but not limited to transient deficits in orientation, attention, speech fluency, executive functions, and both short and long-term memory [11, 13, 25, 30]. However, a recent meta-analysis and systematic review found that memory, attention, language, spatial perception, and orientation showed no significant long-term detriments [19]. Yet anecdotally, between 29% and 55% of respondents believed they experienced long-lasting or permanent memory changes including complete retrograde amnesia [23, 38]. Although there seems to be a discrepancy between data from clinical research and self-reporting of adverse effects with ECT, perceived long-term effects of the treatment are important considerations for qualitative research, as they may play a significant role in recovery.

Conversely, for TMS, while most side effects are equally transient, as long as seizure is avoided there are no clear long-lasting adverse effects of short-term exposure to rTMS at moderate intensities [47]. By both self-reporting and clinician ratings, TMS displayed robust antidepressant effects in treatment-resistant depression with a low side effect and medical risk profile [35]. Nonetheless, TMS often remains reserved as a last-resort treatment for depression when other forms of treatment have been unsuccessful at treating symptoms. Considering these differences between ECT and TMS, TMS may emerge as the preferred choice for relevant stakeholders, including patients considering treatment options for mental health conditions and their caregivers. Although TMS is typically considered after unsuccessful symptom reduction with other methods, and considering the potential positive effects of TMS treatment, TMS may actually be more effective as a first line of treatment for depression [46].

2 Public Perceptions and Media Portrayals of Psychiatric Electroceutical Interventions (PEIs)

Although this chapter is not centered around public perceptions of neurotechnological interventions, we include a section discussing this in order to investigate potential underlying biases and potential barriers to treatment. Comparatively speaking, TMS is a relatively new treatment among other PEIs such as electroconvulsive therapy (ECT), which have been implemented for decades. Electroceutical interventions in general have a complicated history, and patients have mixed opinions and understandings of their safety and efficacy [36]. It is imperative that we take skepticism and criticism of existing psychiatric therapies seriously if we are to fully appreciate the forces that impact the adoption of these novel technologies. Not only have PEIs been discussed in academic and therapeutic settings, but these types of interventions have taken a significant and often controversial role in various forms of media. Therefore, it is important to consider public perceptions in addition to caregiver and professional perspectives on PEIs, as a wide range of sources may influence them, whether these sources be academic, media, or popular culture. Regardless of the type of source, they all contribute to initial ideas or biases toward PEIs amongst stakeholder groups. One study examined the opinions of both patients and members of the general public on PEIs. The majority of the responses in both the patient and public groups had a cautionary attitude, with smaller proportions being strictly positive or negative. Unsurprisingly, negative perceptions of PEIs were largely related to discussion of ECT, with many participants mentioning specific media sources that shaped their perceptions [6]. Another study indicated that there are discrepancies between academic and print media sources regarding TMS treatment and found that the latter had a more optimistic tone when describing TMS, with less concern for adverse effects and more of an emphasis on the benefits. Print media sources also lacked specificity around the details of TMS application, which may indicate that these sources are presenting incomplete or inadequate information to the public [37].

Deep brain stimulation (DBS), which is considered a medically invasive treatment, seems to be associated with mixed attitudes when it comes to public discussion. Gilbert and Ovadia found that DBS may be portrayed in an unrealistically optimistic light in media reports without proper ethical analysis. This puts patients in a potentially dangerous position in terms of informed consent, as they may be relying on incomplete information about the risks of treatment [16]. This is a particularly interesting finding, especially in contrast with attitudes toward other non-invasive treatments such as ECT. However, invasiveness is often judged by subjective factors to the individual, which can play a substantial role in overall attitudes toward any treatment, whether electroceutical or not. For example, some patients may define invasiveness in terms of emotional invasiveness, which may be experienced through psychotherapy, permanence or reversibility of treatment, or according to the medical definition of invasiveness that requires breaking the skin [9].

ECT has been used for eight decades and yet remains one of the most controversial medical treatments for psychiatric illness. Historically, ECT has often played an exclusively negative role in popular media and has been cited as a source of stigma and fear surrounding the treatment. Paired with ECT's long history of involuntary and non-anesthetized administration, this modality often holds negative connotations to the general public, including those who may at some point be recommended for ECT or another PEI treatment [18]. Another study surveyed attitudes about ECT from both medical students and the general public and found several differences between the groups. In particular, participants from the general public discussed having gained much of their knowledge about ECT from popular media and were more hesitant to either receive ECT or have a family member receive the treatment [26].

The observations made about media representations of PEIs highlight a need for further, ethically driven research to balance and inform public perceptions on therapeutic neurotechnologies. Although patients are required to receive information about the procedure before starting treatment, it is important to consider what kinds of information they may already have prior to speaking with a physician, and what kinds of associations they may make between different technologies (e.g., between ECT and TMS). This clarification of assumptions and potential misconceptions may reduce barriers to receiving TMS treatment, in that it may allow patients and their families to view TMS as a viable treatment option and reduce the stigma surrounding it.

3 Caregiver Perspectives on Electroceutical Treatments

Although the focus of this empirical work is on TMS, other research has examined caregiver and other stakeholder perspectives on different treatments involving brain stimulation such as ECT, DBS, and transcranial direct current stimulation (tDCS). As TMS is on the rise as a treatment option for various conditions, more research is needed to gain insight on the effect it has on the lives of patients and care partners. The following section will consider the perspectives of caregivers to patients receiving other PEIs, especially with regard to stigma surrounding treatment, long-term effects of treatment, and sense of self while engaging in treatment. We consider these perspectives to develop the background for our discussion with caregivers to TMS patients, as other treatments within the same category may hold similar beliefs and attitudes from stakeholders.

3.1 Caregiver Perspectives on Electroconvulsive Therapy

Mignone, Zufferey, & De Anstiss conducted a study in which families of patients undergoing ECT were interviewed in order to gain perspective on the social implications of treatment [27]. One major finding was that the stigma surrounding ECT, in addition to the stigma surrounding mental health in general, led family members to feel socially isolated. Family members also discussed a lack of medical support post-treatment, feeling as though the treatment was misunderstood or viewed negatively by others, and feeling a sense of unity with other care partners [27]. Similarly, another study by Bluhm et al. considered stakeholder perspectives including those of patients, members of the public, and psychiatrists. While all groups shared certain ethical concerns, one major difference between non-clinicians and psychiatrists was the former citing concern over evidence of treatment safety, while psychiatrists were more concerned that patients who may benefit from PEIs may not actually receive the treatment [4]. Stakeholder perception of invasiveness can go beyond the technical medical definition of breaking the skin, as invasiveness has also been described in social or psychological sense. For instance, many participants described ECT as invasive, even though the treatment does not meet the medical definition for it [3].

3.2 Caregiver Perspectives on Deep Brain Stimulation

Similar to ECT, DBS is often highly stigmatized. However, unlike ECT, DBS is medically invasive and requires the surgical implantation of electrodes into the brain of the patient to directly stimulate affected areas, as well as a battery pack inserted into the chest and a subcutaneous wire running up the neck connecting the two. DBS may be used to treat conditions such as MDD, obsessive-compulsive disorder (OCD), or Parkinson's disorder (PD) [32]. One study investigated patient perspectives prior to and after receiving DBS for treatment-resistant depression (TRD) including outcome expectations, prior knowledge about risks and potential benefits of DBS, and overall perceived benefit after treatment. One important finding from TRD patients was that while many understood the risks of treatment, they had little concern for what may happen to them in the long term [43, 44]. Because TMS is often used to treat MDD or TRD, it is important to be aware of factors that may be included in the patient's experience of managing depression that may alter the patient's decision-making process or their personal valuations of long-term risk vs reward [20]. We again see the need here for assessment of caregiver perspectives, as they are crucial in making treatment decisions.

Another important consideration raised by Smith and colleagues is whether or to what extent DBS in patients with OCD influences their sense of identity. In this interview study with adolescent patients and their caregivers, participants raised concerns about authenticity in the sense that DBS may make the patient feel as though their thoughts or emotions are not "theirs" or that they are being controlled. All patients and several caregivers believed that DBS would affect the patient's sense of identity in some way. That being said, many participants also viewed OCD as a barrier to the development of a sense of self, and therefore were optimistic that DBS could strengthen the patient's sense of identity [40].

3.3 Caregiver Perspectives on Transcranial Direct Current Stimulation

tDCS, a non-invasive brain stimulation technique that uses weak electrical currents to induce changes in neurological function, is a modality that is gaining popularity amongst pediatric populations. One study aimed to investigate child and caregiver perspectives on tDCS for commercial use, relating to issues such as safety, enhancement, and regulation. Among parents who were interviewed, comparisons were made between tDCS and pharmaceutical treatments for neurological conditions, with many parents believing tDCS to be a safer option of the two [39]. tDCS is also being suggested as potentially beneficial for symptoms of posttraumatic stress. In an interview study aimed to gather perspectives from veteran or active-duty military personnel patients, as well as their caregivers, themes surrounding tDCS treatment included: self-efficacy, coherence, perceived effectiveness, ethicality, and burden [41]. An important finding was that patients and caregivers shared a perspective that they had been provided information about treatment, but felt they had little understanding of how it worked. This identifies a pressing need for the information provided about PEIs to be understandable by individuals without any professional background in such topics.

4 Clinician and Researcher Perspectives on Electroceutical Treatments

4.1 Clinician Perspectives on Electroconvulsive Therapy

In one study, eight clinicians involved in ECT procedures were interviewed regarding their protocols for recommending certain patients for ECT and their opinions on risks and concerns of the treatment [42]. Participants described ECT patients as having experienced severe symptoms of psychiatric illnesses such as depression, and often described ECT as a life-saving intervention for such patients. Interviewees expressed awareness of public concern and stigma surrounding ECT and discussed their methods for addressing these concerns with patients by emphasizing the positive effects as being justified regardless of the potential for negative effects. These clinical perspectives on ECT as a "last-resort" treatment for depression are valuable in that they may be indicative of or comparable to clinical perspectives on TMS, especially with regard to addressing patient and public concerns about PEIs. However, associations are limited given the study's publication prior to FDA approval, when ECT was the only permissible treatment for drug-refractory depression.

4.2 Clinician Perspectives on Deep Brain Stimulation

One significant ethical consideration for any PEI, especially an invasive method such as DBS, is its use in pediatric populations. Pham and colleagues interviewed clinicians who had experience treating childhood and adolescent OCD, and whether or not they believed DBS should be an approved treatment. A notable takeaway was that treatment resistance within the diagnosis of OCD is often poorly defined. Current guidelines recommend DBS after two prior failed randomized controlled trials with other treatments in adults, and opinions from clinicians were mixed on whether or not this guideline should extend to pediatric patients. Some ethical considerations identified in this study related to the extent of differences between adult and pediatric OCD cases including differing remission rates and the influence of biological developmental stages on the definition of treatment resistance [29]. Because TMS is often recommended in instances of treatment resistance, it is essential for consensus and clarity to be established in classifying the disorders, or severity of disorders, that are aimed to be diminished through treatment.

4.3 Clinical Researchers' Perspectives on Transcranial Direct Current Stimulation

To gather perspectives on ethical issues with tDCS application, one study surveyed researchers who had published work within the field of experimental or clinical tDCS use. Overall, participants were more cautious in their judgments of tDCS efficacy, with few describing the technology as either highly effective or ineffective for clinical or enhancement purposes. Additionally, participants raised concerns about the safety of tDCS use in healthy individuals for neuroenhancement and indicated that there was a need for further research, including larger controlled trials on tDCS to more thoroughly investigate the potential for long-term adverse effects. The majority of participants were opposed to tDCS technology being available to the public, but there was little discussion of how to regulate these devices, which indicates further need for research to better inform regulatory protocols for tDCS [33].

Another study investigated best practices for the use of tDCS technology, assessing its use for different mental disorders, efficacy depending on initial brain states, and as individualized treatment guided by a trait-based framework rather than a disorder-centered approach. In reviewing literature on the topic, Dedoncker and colleagues raised several questions regarding clinical application of tDCS technology, including: which brain regions should be stimulated to maximize efficacy, what kinds of stimulation parameters should be implemented, what stimulation duration would be most beneficial, whether other treatments such as psychotherapy should be used sequentially or concurrent with tDCS, and what the optimal between-session interval is [12]. These important questions on the state of the art of tDCS treatment highlight the need for further consensus and neurobiological evidence for treatment, so as to adequately address potential ethical challenges for its clinical use.

5 Stakeholders Perspectives on TMS

5.1 Clinicians

Clinicians and medical staff working in TMS clinics have an important perspective to share about the experience of and any ethical considerations related to TMS treatment. One study interviewed healthcare workers in a TMS clinic (e.g., attending physicians, nurses, receptionists, technicians, and many more) and gained insight about pre-treatment procedures, communication and rapport, environment, aftercare, and staff morale. Overall, healthcare workers indicated that demonstrating the technology to patients before treatment and overall building a supportive and comfortable environment showed promise in providing better treatment outcomes for patients [24]. Another study gathered psychiatrists' opinions about different PEIs and found that most of the participants had a positive attitude toward ECT, particularly its effectiveness, and more hesitance in recommending more invasive PEIs such as DBS. Additionally, participants discussed their impressions of patient views of PEIs, which indicated that patients were often hesitant about such treatments, potentially due to the negative public image of them [5].

5.2 Patients

Patient perspectives on TMS treatment of psychiatric disorders are still largely unexplored in the academic literature on TMS. However, one study asked patients open-ended questions about their experiences with rTMS treatment and revealed themes of frustration and helplessness prior to rTMS as well as beliefs that their expectations about treatment outcomes affected actual outcomes [34]. Another study examined how patients with depression viewed the effect of the disease on their lives as well as how TMS had impacted their lives [49]. One key finding was that while patients often expressed negative worldviews and beliefs about themselves while living with depression, they had generally positive attitudes and expectations toward both TMS and pharmacological treatments. Additionally, many patients spoke about the positive impact TMS had on their mood overall and daily function.

First-person accounts of undergoing TMS offer valuable insight into patients' perspectives on its effectiveness, while also highlighting the discrepancies and tensions between the expectations of family members, clinicians, and patients' actual experiences. In her memoirs, Jenny Lawson reflects that although TMS helped alleviate some of her symptoms, it did not cure her, and she stresses that while the treatment may benefit some, it should not be universally recommended as a first-line option without carefully considering these complexities [21].

6 Caregiver Perspectives on TMS for Psychiatric Conditions—Our Empirical Findings

6.1 Methods

The following study was conducted to gather insight from individuals with first-hand experience with TMS treatment to assess the benefits, harms, and other ethical considerations. There were six total participants in this study, which included three family members of patients receiving TMS treatment, and three healthcare technicians working in TMS clinics. Participants were recruited by snowball sampling, and participants from a larger study with patients receiving TMS were instructed to inform family members of the study. This study is IRB approved (NC State University IRB approval No. 22375), and verbal consent to record responses was obtained at the beginning of each entry. Participants engaged in semi-structured interviews in which they were asked questions about either their family member's or patients' experiences with TMS treatment. The questions, informed by literature on patient's experience with similar technologies [2, 3, 5] generally aimed to gain insight on participants' initial understanding of TMS, comparisons to other electroceutical treatments, informed consent, adverse effects, benefits, as well as costs and non-financial barriers to access of TMS. Interviews typically lasted between 30 min and 1 h, and each participant was compensated with a $60 gift card. Interviewees were renamed using a study code for de-identification and recorded via Zoom and then transcribed and analyzed using an abductive inference framework for qualitative research [45].

6.2 Results

The iterative analytical process described above resulted in the following parent codes, further discussed in this section:

- TMS vs. ECT
- Perceived invasiveness
- Mood, personality, and social changes
- Stigma, skepticism, and informed consent
- Financial and other barriers
- Adverse effects

6.2.1 TMS Compared to ECT

When asked about initial knowledge of TMS, many participants indicated that it was often associated with ECT. For instance, one TMS technician mentioned that many patients that they saw for TMS treatment were initially afraid to receive treatment due to their prior knowledge or perceptions about ECT, thinking that the two treatments were similar or identical, and being concerned that they may experience adverse effects more common with ECT such as memory loss:

> I can't tell you how many times I have spoken with patients and they come in shaking and they are scared to death because they think this is ECT (TMS-AHP001).

> I can't tell you how many people I've talked to who've had relatives or friends that do ECT [...] Sometimes people do get memory loss, and I think that really freaks people out and they think that's going to happen with TMS therapy (TMS-AHP001).

One family member emphasized the difference between TMS and ECT by referring to ECT as a more antiquated technique that was perceived as a less desirable treatment:

> I just kind of think of that as shock therapy in the olden days basically, that [...] maybe it's a little more frightening than using the TMS (TMS-FAM002).

Overall, participants seemed aware of the differences between ECT and TMS and acknowledged that TMS was a safer and more favorable modality.

6.2.2 Perceived Invasiveness

In terms of invasiveness, participants consistently ranked TMS as less invasive than ECT, and in some cases less invasive than both pharmaceutical and psychotherapeutic treatments. One technician discussed patient concerns with identity in relation to initial knowledge about TMS in comparison to ECT:

> ... just having to go under, and then also experiencing memory loss or just not being themselves anymore, that's another big concern. And so they link that with TMS, and they think that's what's going to happen to them ... (TMS-AHP001).

Another technician said that they explain TMS to patients as less invasive than ECT relative to the occurrence of seizures:

> Usually I tell them like ECT is a little more invasive [...] and it's causing a seizure, whereas with TMS we're not trying to cause a seizure (TMS-AHP002).

These perspectives are valuable in that they may indicate which factors could influence potential patients' willingness to consider TMS as a treatment.

6.2.3 Mood, Personality, and Social Changes

Care partners were asked questions about their perceptions of changes to the patient undergoing TMS in terms of mood differences, personal identity, and the way they are perceived by or interact with others both within and outside of the family. One participant discussed a positive overall change in patients' personal identity with TMS treatment:

> I feel as if their identity sometimes changes throughout TMS, and when they get more hopeful, they feel like they value themselves more (TMS-AHP003).

> A lot of patients say that they feel like their old selves again, wanting to do hobbies and things that they enjoyed previously (TMS-AHP003).

6.2.4 Stigma, Skepticism, and Informed Consent

As indicated by the above results, patients, family members, and other members of the public tend to associate TMS with ECT, often resulting in fear and stigma surrounding TMS treatment. Participants reported that this creates a potential barrier for patients, as they may be influenced by information or biases that are not true of actual TMS treatment. One TMS technician mentioned that family members often worried about the efficacy and safety of the treatment for their loved one and believe that it will not work:

> I would say at least forty percent of the patients I work with have a family member that's kind of pushing up against me or asking really detailed questions, and not believing what I'm saying, and trying to talk the patient out of it (TMS-AHP001).

Another participant, however, one of the caregivers, expressed trust in their family member who was receiving TMS, and had little information prior to the treatment:

> He's a very intelligent […] person and just through listening to him explain it to me, I […] felt I didn't need to know any more (TMS-FAM002).

6.2.5 Financial and Other Barriers

In terms of barriers to receiving TMS treatment, caregivers have an especially important perspective, as each care partner has varying levels of involvement in the procedure and may have a significant impact on whether or not the patient chooses to engage in treatment. Some participants in this study discussed the financial barrier to TMS patients in receiving treatment:

> The financial aspect is really hard. There's some people that will literally say, "No, I'm not doing TMS", just because it's way too expensive (TMS-AHP001).

Another technician mentioned that patient and family concerns were more related to a lack of prior knowledge about the treatment rather than financial concerns:

> I don't think that I've ever had any of them coincide with financial. It's typically just they had never heard of the treatment, and they didn't understand why their psychiatrist who they've seen, or their therapist, you know, for ten plus years had never mentioned it … (TMS-AHP003).

One care partner explained that they often drove their family member to appointments daily, which disrupted their schedule in some capacity:

> There's a couple days he works from home, […] I normally wouldn't have to drive him […] on those days, […] I do now have to take a trip [which is] a mild inconvenience (TMS-FAM001).

And also highlighted their unique experience in their ability to afford TMS treatment, when they previously had not:

> We're on the same insurance plan and we have an out of pocket maximum per year [...] several thousand dollars. [...] [T]his year I have had to get an extensive amount of medical treatment and we reached our out of pocket maximum you know very early in the year (TMS-FAM001).

6.2.6 Adverse Effects of TMS Treatment

All participants discussed adverse effects of TMS treatment in some capacity, either in the lack thereof, occurrence of, or prevalence of adverse effects. One family member mentioned the patient's experience of headaches and "fogginess," while the other two family members were not aware of any adverse effects being experienced by the patient. Of the technicians who were interviewed, two said that fatigue was a common adverse effect experienced by patients they saw for treatment, one mentioned headache, while the third witnessed two probable pseudo seizures during TMS treatment, meaning that the seizures were not medically induced, and stopped after the cessation of stimulation.

6.3 Limitations

One limitation of this study is the limited ability to generalize findings due to the low sample size of participants. Because participants joined the study by contacting researchers, recruiting a large sample was challenging. Additionally, no written consent was obtained in an effort to best safeguard the identity of participants while using a virtual platform for interviews.

7 Discussion

An important takeaway from this chapter is that it is essential that accurate and reasonable information be provided to the public about TMS treatment, as it is often viewed as almost identical to ECT. Because there is widely held fear and stigma surrounding ECT, this association may be creating an unseen barrier to receiving TMS treatment [17]. As mentioned with regard to stigma and media depictions surrounding PEIs, there may be existing biases toward TMS treatment prior to receiving formal information about the treatment, which may play a role in decision-making. Caregiver and clinician participants in this study generally spoke in a negative tone when discussing ECT, especially concerning the risks of the procedure and the potential for serious adverse effects. This may be considered a serious limitation of the intervention for those who may be considering TMS as a treatment for a mental health condition, as there may be misplaced fear stemming from inaccurate or incomplete information about TMS. Furthermore, integrating accurate information in patient-physician discussions during clinical advising is crucial, as these interactions often weigh most in guiding patients' considerations and decisions about their treatment options. Additionally, the role of caregivers as previously discussed is critical in supporting patients living with depression as they navigate the process of decision-making with regard to treatment, especially in that they may be able to help their loved one evaluate long-term effects and plan for the future.

Relatedly, empirical findings point to the relevance of subjective impressions of invasiveness, as the treatment modalities discussed in the interviews (i.e., medication, psychotherapy, TMS and ECT) are medically considered non-invasive, and yet they vary in the degree to which stakeholders view them as invasive. Some stakeholders judge invasiveness on criteria such as intensity or perceived risk of the treatment (such as in ECT), and psychological invasiveness related to emotional vulnerability (such as in psychotherapy). Therefore, medical classifications of invasiveness may be misaligned with attitudes toward invasiveness, especially with PEIs. Nevertheless, for the most part, family members did not witness any severe adverse effects of TMS treatment, and healthcare workers reported few, if any, seizures. However, this is not to say that severe adverse effects should not still be taken into consideration and the invasiveness of TMS further discussed.

While family members' thoughts on TMS relative to other treatments should not overshadow those of patients with lived experiences of both, these important stakeholders may influence the treatment decisions of the people they care for who would seek out TMS, either as trusted care partners or legal guardians responsible for final decision-making. As such, it is important to understand the basis for their attitudes on TMS. Furthermore, family members may sometimes notice changes in their loved one's behavior, thoughts, or emotions before the person themselves notices, and as such their narratives and opinions are of value in a full understanding of the use of TMS for treatment refractory conditions.[1]

Relatedly, a patient's sense of identity is often discussed when considering any neuro-psychoactive intervention. Regarding participant answers pertaining to mood, personality, and social changes, caregivers who reported changes in their loved ones, these were for the better. This was often mentioned in reference to improvement of depressive symptoms, suggesting that the initial change in mood or personality were attributed to the disorder, rather than TMS treatment alone. It is often discussed that other neurointerventions raise concerns for a *loss* of identity [31], but in our study, we see stakeholders viewing TMS as a tool to *augment* or *regain* a patient's sense of identity, reducing the aspects of their condition which rob them of enjoyment of things they used to love and in doing so their sense of self. Therefore, if TMS helps to alleviate symptoms of mental illness, then it may help to restore or strengthen a patient's sense of identity. Although not explicitly addressed in the data presented here, an interesting further consideration may be to explore the identities of participants as influenced by their roles as care partners. It is important that we recognize not only caregiver and clinician perspectives of patient outcomes, but also their unique experience and sense of self throughout the care process. For instance, we may look to the ways in which caregivers' individual dignity and sense of moral power may be informed by their roles [28].

Interviews also revealed the crucial role of social support and structure in TMS healthcare delivery. When values and goals of patients and the clinical team aligned

[1] We thank an anonymous reviewer for these insights.

well, patients perceived the treatment as a positive experience. However, when there was value and goal misalignment, some family care partners reported that patients dreaded TMS sessions because of the negative social interactions during treatment. Further to the point of treatment goals and structure, the recommendation of a patient for TMS treatment is often given when there is some degree of treatment resistance. However, this concept is often poorly or inconsistently defined in the literature. Therefore, it is important to consider the specific criteria in treatment rationale that will adequately inform goals and expectations. Our data elucidates several ethical considerations for TMS treatment, including the breadth of experiences of receiving TMS as well as the range of protocols in its administration, the subjective lived experience of TMS as a clinical intervention, and concerns about accessibility of TMS.

Finally, further barriers to TMS care are identified in this empirical work. While many insurance plans cover treatment to some degree, there are still considerable out-of-pocket costs that may interfere with someone's ability to receive TMS treatment. One technician mentioned that patients pay, on average, between 1000 and 3000 USD for a full course of treatment and having seen patients struggle financially, while family members discussed having full or partial coverage and did not face any additional financial burden due to treatment. Participants also mentioned that some patients who may have benefitted from TMS treatment chose not to receive treatment due to financial struggles. Therefore, TMS runs the risk of being too expensive and creating an additional barrier, exacerbated by the existing stigma and apprehension surrounding a relatively novel treatment.

Adverse effects, while still extremely important to consider, were either not reported or reported as minor events by participants in this study. Technicians' discussion of seizures [22] revealed that they were very rare, and overall found that patients had positive experiences with TMS with regard to physical pain or discomfort.

8 Conclusion

In this chapter, we reviewed literature on the subject of caregiver and healthcare staff perspectives of different PEIs. In doing so, we developed background for our empirical work, which gained insight on similar stakeholder groups' perspectives of TMS treatment alone. Results from interviews with caregivers and technicians indicated that its close association with ECT was at times a barrier to pursuing TMS care. That being said, participants also reported perceiving TMS as less invasive than ECT and other PEIs, and even pharmaceuticals and psychotherapy. Expected concerns regarding mood, personality, and social changes proved to be unfounded—participants reported perceived changes but for the better, with stakeholders viewing TMS as a tool to boost, augment and even regain patients' sense of identity. Barriers to receiving TMS care included stigma through its association with ECT, increased skepticism relative to other interventions, and financial considerations. The participants reported observing no adverse effects of TMS treatment. Our novel research study provides insight into other stakeholders' perceptions of TMS treatment for psychiatric conditions, beyond those of the patient, as these individuals are also impacted by TMS therapy. These views can be used to help inform further exploration of TMS as a fruitful and beneficial treatment modality for psychiatric conditions.

References

1. Andrade C, Arumugham SS, Thirthalli J. Adverse effects of electroconvulsive therapy. Psychiatr Clin North Am. 2016;39(3):513–30. https://doi.org/10.1016/j.psc.2016.04.004.
2. Bluhm R, Castillo E, Achtyes ED, McCright AM, Cabrera LY. They affect the person, but for better or worse? Perceptions of electroceutical interventions for depression among psychiatrists, patients, and the public. Qual Health Res. 2021a;31(13):2542–53. https://doi.org/10.1177/10497323211037642.
3. Bluhm R, Cortright M, Achtyes ED, Cabrera LY. "They are invasive in different ways": stakeholders' perceptions of the invasiveness of psychiatric electroceutical interventions. AJOB Neurosci. 2021b;14(1):1–12. https://doi.org/10.1080/21507740.2021.1958098.
4. Bluhm R, Sipahi ED, Achtyes ED, McCright AM, Cabrera LY. Stakeholders' ethical concerns regarding psychiatric electroceutical interventions: results from a US Nationwide survey. AJOB Empir Bioeth. 2023;15(1):11–21. https://doi.org/10.1080/23294515.2023.2224592.
5. Cabrera LY, Nowak GR III, McCright AM, Achtyes E, Bluhm R. Last resort interventions?: a qualitative study of psychiatrists' experience with and views on psychiatric Electroceutical interventions. Psychiatry Q. 2020;92(2021):419–30. https://doi.org/10.1007/s11126-020-09819-1.
6. Cabrera LY, Gilbert MM, McCright AM, Achtyes ED, Bluhm R. Beyond the cuckoo's nest: patient and public attitudes about psychiatric electroceutical interventions. Psychiatry Q. 2021;92(2021):1325–438. https://doi.org/10.1007/s11126-021-09916-9.
7. Cabrera L, Achtyes ED, Bluhm R, McCright AM. Views about neuromodulation interventions for depression by stakeholder group, treatment modality, and depression severity. Compr Psychiatry. 2023;122(2023):152364. https://doi.org/10.1016/j.comppsych.2023.152365.
8. Chail A, Saini RK, Bhat PS, Srivastava K, Chauhan V. Transcranial magnetic stimulation: a review of its evolution and current applications. Ind Psychiatry J. 2018;27(2):172–80. https://doi.org/10.4103/ipj.ipj_88_18.
9. Coates McCall I, Minielly N, Bethune A, Lipsman N, McDonald PJ, Illes J. Readiness for first-in-human neuromodulatory interventions. Can J Neurol Sci. 2020;47(6):785–92.
10. Cotovio G, Ventura F, Rodrigues Da Silva D, Pereira P, Oliveira-Maia AJ. Regulatory clearance and approval of therapeutic protocols of transcranial magnetic stimulation for psychiatric disorders. Brain Sci. 2023;13(7):1029. https://doi.org/10.3390/brainsci13071029.
11. Datka W, Siwek M, Dudek D, Maczka G, Zieba A. Zaburzenia pamieci operacyjnej u pacjentów z głvboka depresja leczonych wstrzasami elektrycznymi [Working memory disturbances in patients with major depression after ECT treatment]. Psychiatr Pol. 2007;41(3):339–49.
12. Dedoncker J, Baeken C, De Raedt R, Vanerhasselt M. Combined transcranial direct current stimulation and psychological interventions: state of the art and promising perspectives for clinical psychology. Biol Psychol. 2021;158:107991.
13. Dunne RA, McLoughlin DM. Systematic review and meta-analysis of bifrontal electroconvulsive therapy versus bilateral and unilateral electroconvulsive therapy in depression. World J Biol Psychiatry. 2012;13(4):248–58. https://doi.org/10.3109/15622975.2011.615863.
14. Fitzgerald PB, George MS, Pridmore S. The evidence is in: repetitive transcranial magnetic stimulation is an effective, safe and well-tolerated treatment for patients with major depressive disorder. Aust N Z J Psychiatry. 2022;56(7):745–51. https://doi.org/10.1177/00048674211043047.
15. Gazdag G, Ungvari G. Electroconvulsive therapy: 80 years old and still going strong. World J Psychiatry. 2019;9(1):1–6. https://doi.org/10.5498/wjp.v9.i1.1.

16. Gilbert F, Ovadia D. Deep brain stimulation in the media: over-optimistic portrayals call for a new strategy involving journalists and scientists in ethical debates. Front Integr Neurosci. 2011;5(2011) https://doi.org/10.3389/fnint.2011.00016.
17. Goldbloom DS, Gratzer D. Barriers to brain stimulation therapies for treatment-resistant depression: beyond cost effectiveness. Can J Psychiatry. 2019;65(3):193–5. https://doi.org/10.1177/0706743719893584.
18. Griffiths C, O'Neill-Kerr A. Patients', carers', and the public's perspectives on electroconvulsive therapy. Front Psychiatry. 2019;10(304) https://doi.org/10.3389/fpsyt.2019.00304.
19. Guo Q, Wang Y, Guo L, Li X, Ma X, He X, et al. Long-term cognitive effects of electroconvulsive therapy in major depressive disorder: a systematic review and meta-analysis. Psychiatry Res. 2023;331:115611.
20. Kirtley OJ, Lafit G, Vaessen T, Decoster J, Derom C, Gülöksüz S, De Hert M, Jacobs N, Menne-Lothmann C, Rutten BPF, Thiery E, Van Os J, Van Winkel R, Wichers M, Myin-Germeys I. The relationship between daily positive future thinking and past-week suicidal ideation in youth: an experience sampling study. Front Psych. 2022;13:915007. https://doi.org/10.3389/fpsyt.2022.915007.
21. Lawson J. Broken (in the best possible way). Henry Holt and Company; 2022.
22. Lerner AJ, Wassermann EM, Tamir DI. Seizures from transcranial magnetic stimulation 2012-2016: results of a survey of active laboratories and clinics. Clin Neurophysiol. 2019;130(8):1409–16. https://doi.org/10.1016/j.clinph.2019.03.016.
23. Lisanby SH, Maddox JH, Prudic J, Devanand DP, Sackeim HA. The effects of electroconvulsive therapy on memory of autobiographical and public events. Arch Gen Psychiatry. 2000;57(6):581–90.
24. Mallon S, Walker K, Bayley Z, Griffiths C. Practitioner perspectives on best practice in non-treatment factors that support the delivery of repetitive transcranial magnetic stimulation (rTMS) for depression. J Psychiatr Ment Health Nurs. 2022;29(3):463–71. https://doi.org/10.1111/jpm.12815.
25. Maric NP, Stojanovic Z, Andric S, Soldatovic I, Dolic M, Spiric Z. The acute and medium-term effects of treatment with electroconvulsive therapy on memory in patients with major depressive disorder. Psychol Med. 2016;46(4):797–806. https://doi.org/10.1017/S0033291715002287.
26. McFarquhar TF, Thompson J. Knowledge and attitudes regarding electroconvulsive therapy among medical students and the general public. J ECT. 2008;24(4):244–53. https://doi.org/10.1097/yct.0b013e318168be4a.
27. Mignone S, Zufferey C, De Anstiss H. Family experiences of caring for relatives who have received electroconvulsive therapy (Ect). Aust Soc Work. 2018;71(1):86–97. https://doi.org/10.1080/0312407X.2017.1352001.
28. Miller SC. Reconsidering dignity relationally. Ethic Soc Welf. 2017;11(2):108–21. https://doi.org/10.1080/17496535.2017.1318411.
29. Pham MT, Campbell TA, Dorfman N, Torgerson L, Kostick-Quenet K, Blumenthal-Barby J, Storch EA, Lázaro-Muñoz G. Clinician perspectives on levels of evidence and oversight for deep brain stimulation for treatment-resistant childhood OCD. J Obsess Compul Relat Disord. 2023;39(2023):100830. https://doi.org/10.1016/j.jocrd.2023.100830.
30. Prudic J. Strategies to minimize cognitive side effects with ECT: aspects of ECT technique. J ECT. 2008;24(1):46–51. https://doi.org/10.1097/YCT.0b013e31815ef238.
31. Pugh J. Clarifying the normative significance of 'personality changes' following deep brain stimulation. Sci Eng Ethics. 2020;26(3):1655–80. https://doi.org/10.1007/s11948-020-00207-3.
32. Pycroft L, Stein J, Aziz T. Deep brain stimulation: an overview of history, methods, and future developments. Brain Neurosci Adv. 2018;2:1–6. https://doi.org/10.1177/2398212818816017.
33. Riggall K, Forlini C, Carter A, Hall W, Weier M, Partridge B, Meinzer M. Researchers' perspectives on scientific and ethical issues with transcranial direct current stimulation: an international survey. Sci Rep. 2015;5:10618. https://doi.org/10.1038/srep10618.
34. Rosedale M, Lisanby SH, Malaspina D. The structure of the lived experience for persons having undergone rTMS for depression treatment. J Am Psychiatr Nurses Assoc. 2009;15(5):333–7. https://doi.org/10.1177/1078390309350773.

35. Sackeim HA, Aaronson ST, Carpenter LL, Hutton TM, Mina M, Pages K, et al. Clinical outcomes in a large registry of patients with major depressive disorder treated with transcranial magnetic stimulation. J Affect Disord. 2020;277:65–74.
36. Sadowsky, J. Electroconvulsive therapy in america: the anatomy of a medical controversy. Routledge. 2019.
37. Scheper A, Rosenfeld C, Dubljevic V. The public impact of academic and print media portrayals of TMS: shining a spotlight on discrepancies in the literature. BMC Med Ethics. 2022;23(2022):25. https://doi.org/10.1186/s12910-022-00760-5.
38. Sienaert P, De Becker T, Vansteelandt K, Demyttenaere K, Peuskens J. Patient satisfaction after electroconvulsive therapy. J ECT. 2005;21(4):227–31. https://doi.org/10.1097/01.yct.0000183268.62735.c0.
39. Sierawska A, Splittgerber M, Moliadze V, Siniatchkin M, Buyx A. Transcranial direct current stimulation (tDCS) in pediatric populations— voices from typically developing children and adolescents and their parents. Neuroethics. 2023;16(3) https://doi.org/10.1007/s12152-022-09507-w.
40. Smith JN, Dorfman N, Hurley M, Cenolli I, Kostick-Quenet K, Storch EA, et al. Adolescent OCD patient and caregiver perspectives on identity, authenticity, and normalcy in potential deep brain stimulation treatment. Camb Q Healthc Ethics. 2024:1–14. https://doi.org/10.1017/S0963180124000203.
41. Smits FM, de Kort GJ, Geuze E. Acceptability of tDCS in treating stress-related mental health disorders: a mixed methods study among military patients and caregivers. BMC Psychiatry. 2021;21(97):97. https://doi.org/10.1186/s12888-021-03086-5.
42. Stevens P, Harper DJ. Professional accounts of electroconvulsive therapy: a discourse analysis. Soc Sci Med. 2007;64(7):1475–86. https://doi.org/10.1016/j.socscimed.2006.11.015.
43. Thomson CJ, Segrave RA, Carter A. Changes in personality associated with deep brain stimulation: a qualitative evaluation of clinician perspectives. Neuroethics. 2021a;14(1):109–24. https://doi.org/10.1007/s12152-019-09419-2.
44. Thomson CJ, Segrave RA, Fitzgerald PB, Richardson KE, Racine E, Carter A. "Nothing to lose, absolutely everything to gain": patient and caregiver expectations and subjective outcomes of deep brain stimulation for treatment-resistant depression. Front Hum Neurosci. 2021b;15:*755276*. https://doi.org/10.3389/fnhum.2021.755276.
45. Timmermans S, Tavory I. Theory construction in qualitative research: from grounded theory to abductive analysis. Sociol Theory. 2012;30:167–86. https://doi.org/10.1177/0735275112457914.
46. Voigt J, Carpenter L, Leuchter A. A systematic literature review of the clinical efficacy of repetitive transcranial magnetic stimulation (rTMS) in non-treatment resistant patients with major depressive disorder. BMC Psychiatry. 2019;19(2019):13. https://doi.org/10.1186/s12888-018-1989-z.
47. Wassermann EM. Side effects of repetitive transcranial magnetic stimulation. Depress Anxiety. 2000;12(3):124–9.
48. Wexler A, Nagappan D, Kopyto D, Santarnecchi E, Pascual-Leone A. Off-label promotion of transcranial magnetic stimulation on provider websites. Brain Stimul. 2021;14(2021):724. https://doi.org/10.1016/j.brs.2021.04.013.
49. Yiğitoğlu GT, Köktaş NC, Öztürk FO. Opinions of depression patients about transcranial magnetic stimulation: a qualitative study. J Radiol Nurs. 2023;42(2023):114–20. https://doi.org/10.1016/j.jradnu.2022.11.003.

Transcranial Magnetic Stimulation and Cognitive Enhancement: Present State and Possibilities for the Near Future

Bruce Luber and Eva Wiener

The magnetic fields produced by transcranial magnetic stimulation (TMS) are strong enough to induce action potentials in the neurons underneath the stimulating coil. Initially, this capability made TMS interesting to cognitive neuroscientists, as a pulse could cause random firing in a cortical region of interest, disrupting ongoing neural processing and thus performance on tasks that require that processing, providing causal evidence for specific brain-behavior relationships (e.g., [3]). However, some studies began to report "paradoxical" effects of TMS causing better performance in cognitive tasks instead of the expected worsening, for example in picture naming [97, 147] or in visual search for targets amongst distractors [150]. Such reports continued, until a review in 2013 listed 61 papers with results indicating TMS made people better in cognitive tasks [84]. In this chapter, we present an argument that TMS can produce meaningful and lasting enhancements of cognitive performance. This is shown first by updating the earlier list, showing the number of studies claiming performance increases using TMS has continued to grow. Then, feasible mechanisms are delineated that may account both for the observed enhancement effects and for making such effects become long-lasting. Finally, avenues of further research over the next few decades are outlined suggesting the means by which the use of TMS can accelerate the learning of skills and possibly surpass natural limits of that learning. One note: TMS is only one promising modality of noninvasive brain stimulation. For example, transcranial electrical stimulation has also been reported to enhance human performance (see [83] for a review). Here however, the focus will just be on TMS.

B. Luber (✉) · E. Wiener
Noninvasive Neuromodulation Unit, NIMH Experimental Therapeutics and Pathophysiology Branch (ETPB), Bethesda, MD, USA
e-mail: bruce.luber@nih.gov; eva.wiener@nih.gov

V. Dubljević, J. R. Young (eds.), *TMS and Neuroethics*, Advances in Neuroethics,
https://doi.org/10.1007/978-3-031-92401-9_5

1 Present State of TMS Enhancement

While by 2014, 61 peer-reviewed studies could be found that reported TMS enhancement of performance [84], that number has now grown to 105, as listed in Tables 1 and 2. Cognitive enhancement has been defined as any augmentation of information processing systems in the brain, including those supporting attention, perception, conceptualization, memory, reasoning, and motor performance [122]. As can be seen in Tables 1 and 2, TMS is a prime candidate to serve as a mechanism for cognitive enhancement, as it has been found to increase performance in a wide range of tasks that run the gamut of the brain's information processing systems. Among others, these include tasks involved with perception, such as visual [7, 76, 139] and tactile [53, 127] detection and discrimination, backward masking [50], visual search [60, 90, 91, 104], executive functioning such as attention [35, 59, 144], inhibitory control [13, 43, 120], working memory [11, 27, 85, 86, 157], motor control [64, 68, 111, 136, 141] and motor learning [20, 47, 72]; mental imagery [70]; memory [7]; categorization [2, 71, 99, 147]); reasoning [16, 96]; and language [4, 119, 137]. Tables 1 and 2 also indicate that just about every form of TMS has caused performance increases. Enhancement has been observed using single-pulse TMS, paired-pulse TMS, trains of repetitive TMS (rTMS), high- and low-frequency rTMS, patterned rTMS using combined frequencies of pulses such as theta burst stimulation (TBS), whether applied intermittently (iTBS) or continuously (cTBS), and rTMS used during task performance (online TMS) or before task performance (offline TMS, where many trains of rTMS are run to create a cumulative effect, up- or down-regulating cortical excitability prior to task performance). Overall, a clear picture is emerging of a general phenomenon of performance increases caused by TMS across all types of neural processing supporting cognition and all types of TMS. However, as will be elaborated on in what follows, ongoing research is uncovering specific constraints in choices of stimulation parameters through which application of TMS can result in cognitive enhancement.

2 Proposed Mechanisms Underlying TMS-Caused Performance Enhancement

That TMS can in many instances improve performance immediately leads to questions of how this might be done in a way that could be controlled and optimized both in the size of the effect and its duration. Answering these questions can be approached first by examining what the mechanisms surrounding improvements might be. In the review cited above [84] of reported performance enhancements caused by TMS that occurred in a time frame of seconds to an hour, the authors found that almost all reports could be classified into two broad classes of mechanisms.

One of the mechanisms is based on indirect network effects of TMS. As mentioned above, TMS induces currents in the neurons immediately beneath the coil

Table 1 Studies in which TMS-caused enhancements appear to be related to the addition-by-subtraction mechanism

TMS Dosage	Online/ Offline	Reference	Task	Performance Effect
Low frequency	Online	[4]	Language-fragment detection task	Decreased RT
		[31]	Numerical Stroop	Reduced size-incongruity slowing
		[69]	Verbal working memory	Decreased RT, increased accuracy
		[99]	Scene categorization task	Improved scene discrimination
	Offline	[41]	Picture-word match	Decreased RT
		[47]	Visuo-motor sequence learning	Decreased RT
		[59]	Spatial attention	Increased detection accuracy
		[60]	Visual search	Decreased RT
		[68]	Motor tasks	Decreased RT, increased accuracy
		[71]	Picture-word verification	Decreased RT
		[73]	Visuo-motor sequence learning	Decreased RT
		[74]	Visuo-motor sequence learning	Decreased RT
		[72]	Visuo-motor sequence learning	Decreased RT
		[104]	Visual search	Decreased RT
		[125]	Sequential item memory	Increased accuracy
		[132]	Number estimation	Increased accuracy
		[139]	Motion discrimination	Decreased detection threshold
		[144]	Spatial attention	Increased detection accuracy
		[148]	Cognitive bias task	Improved visuospatial memory
Single pulse	Online	[2]	Object categorization	Decreased RT
		[8]	Auditory-visual word perception task	Reduced auditory-visual incongruency effect
		[42]	Visual search	Decreased RT
		[76]	Visual detection	Decreased RT
		[81]	Adjective judgment	Decreased RT
		[100]	Directed eye movement	Decreased RT
		[116]	Number magnitude judgment task	Improvement in number magnitude judgment
		[127]	Somatosensory detection threshold	Increased sensitivity
		[150]	Visual search	Decreased RT
Paired pulse	Online	[7]	Visual discrimination	Increased accuracy, decreased RT
cTBS	Offline	[15]	Picture memory task	Increased memory accuracy

(continued)

Table 1 (continued)

TMS Dosage	Online/ Offline	Reference	Task	Performance Effect
		[64]	Motor RT task	Decrease in non-dominant hand RT
		[65]	Visual search	Increased detection accuracy
High frequency	Online	[13]	Numerical Stroop task	Decreased decision-making time
		[123]	Working memory	Increased memory capacity
		[131]	Visual discrimination task	Decreased RT
		[138]	Stimulus-response correspondence task	Increased accuracy
		[158]	Recall of motion direction task	Increased recall accuracy
	Offline	[54]	Object identification	Increased accuracy
		[55]	Stroop	Decreased RT
		[117]	Spatial attention	Decreased RT

which at high enough intensities can produce membrane currents that result in action potentials, which can disrupt the neural processing already going on in the stimulated region and in general impair performance in tasks reliant on that processing. Taking advantage of this, much TMS research has been based on investigating such disruption in order to piece together what sort of processing the stimulated cortex supports. However, disruption can be used to improve performance in a task. For example, an individual might be asked to do a visual search task, in which they are to spot a target, perhaps a green "T" as fast as they can, amongst a field of distracting items, say blue "T"s and green "L"s. And perhaps all the letters are moving in various directions across the computer screen, although movement information is not relevant to the task. In this situation, TMS pulsed to a specific cortical area specializing in motion processing called MT (or V5) during the display was found to speed up the time it took to find the target [150]. It has been suggested that this improvement was due to the TMS disrupting the processing of movement information, essentially removing irrelevant visual information from consideration, simplifying the task. A number of TMS studies since then, covering a broad range of tasks, have demonstrated performance enhancement through this disruptive mechanism, termed "addition-by-subtraction" ([84]; see Table 1 for an updated listing).

The addition-by-subtraction mechanism relies on the disruption of activity in networks superfluous to carrying out a given task. This aids in task processing, and ultimately in task performance, by narrowing the focus of active neural networks to those needed. However, TMS need not only be disruptive to ongoing neural processing. For example, a number of studies have used TMS targeted on cortical regions expected to be directly involved with working memory processing, and while some studies did indeed observe that TMS disrupted working memory performance (e.g., [27, 44]), others reported facilitation of working memory performance [85, 86, 157]. In these cases, whether stimulation was disrupting or facilitating to performance appeared to depend on when relative to a task TMS was applied, and

Table 2 Studies reporting performance enhancements with TMS targeting regions expected to be directly involved in a given task

TMS Dosage	Online/ Offline	Reference	Task	Performance Effect
Single pulse	Offline	[111]	Rotary pursuit task	Improvement in motor learning
	Online	[19]	Motor learning	Increased accuracy
		[25]	Letter discrimination	Decreased RT
		[26]	Nonverbal working memory	Decreased RT
		[27]	Letter discrimination	Decreased RT
		[28]	Action discrimination	Decreased RT
		[29]	Speed localization task	Enhanced localization
		[49]	Backward masking	Increased detection sensitivity
		[50]	Backward masking	Increased detection sensitivity
		[53]	Tactile working memory	Decreased RT
		[112]	Numerical magnitude judgment	Decreased RT
		[124]	Stimulus-response correspondence task	Increased accuracy
		[137]	Homophone and synonym judgment	Decreased RT
		[147]	Picture naming	Decreased RT
		[21]	Stimulus-response correspondence task	Decreased RT
		[149]	Visual discrimination	Increased accuracy
Paired pulse	Online	[46]	Verbal and nonverbal recognition	Decreased RT
		[92]	Peripheral discrimination task	Increased RT
		[119]	Syntactic decisions about sentences	Decreased RT
		[155]	Memory-guided eye movement	Decreased RT
	Offline	[96]	Abstract reasoning task	Increased accuracy, decreased RT
		[120]	Stop signal task	Decreased RT
High frequency and iTBS	Offline	[156]	Visual-spatial N-back task and card sorting task	Improved accuracy in N-back (both), improved accuracy in card sorting (only iTBS)
iTBS	Offline	[48]	Emotional Stroop task	Decreased RT
		[61]	N-back working memory task	Increased accuracy
		[64]	Motor reaction time task	Decreased RT
		[136]	Complex sequential finger movement task	Decreased RT
		[141]	Thumb abduction task	Greater peak thumb acceleration

(continued)

Table 2 (continued)

TMS Dosage	Online/ Offline	Reference	Task	Performance Effect
High frequency	Online	[11]	Delayed-response alphabetization task	Increased accuracy
		[22]	Picture naming	Decreased RT
		[32]	Numerical discrimination	Decreased RT
		[33]	Visual discrimination task	Increased accuracy
		[34]	Mental rotation task	Improved performance
		[35]	Visual attention	Increased accuracy
		[40]	Visual detection task	Increased accuracy
		[43]	Visual go-nogo	Decreased RT
		[66]	Sequential finger movement task	Improvement in target scores
		[75]	Word recognition	Increased accuracy
		[78]	Delayed-recognition working memory task	Decreased RT
		([85]; [86])	Letter working memory	Decreased RT
		[91]	Visual search task	Decreased RT
		[97]	Picture naming	Decreased RT
		[113]	Visual discrimination	Decreased RT
		[121]	Memory task	Increased memory capacity
		[126]	Motion discrimination	Increased accuracy
		[133]	Visual search task	Increased accuracy
		[134]	Picture naming	Decreased RT
		[157]	Spatial working memory	Decreased RT
	Offline	[5]	Circle connection task	Decreased RT
		[6]	Digit span, spatial 2-back task	Increased span length, increased accuracy
		[16]	Visual analogic reasoning	Decreased RT
		[17]	Motor tracking	Decreased movement error
		[23]	Tactile discrimination, motor choice RT	Increased sensitivity, decreased RT
		[30]	Weather prediction task	Increased accuracy
		[36]	Reading tasks	Increased accuracy
		[37]	Auditory verbal memory test	Increased accuracy in immediate and delayed recall
		[62]	Continuous performance	Increased accuracy
		[67]	Stroop task	Increased accuracy
		[79]	Stroop task	Decreased RT
		[101]	Digit sequence memory	Increased accuracy
		[106]	Balance test	Increased balance
		[110]	Tactile discrimination	Increased sensitivity
		([140]; [151])	Tactile discrimination, motor choice RT	Increased sensitivity, decreased RT
			Face-cued word recall	Increased accuracy
Low frequency	Offline	[153]	Visual discrimination	Increased accuracy

at what frequency. The rule seemed to be that stimulation to a region occurring when task processing was ongoing interfered with performance, while stimulation occurring immediately prior to the onset of task processing could facilitate it. An initial explanation of this time dependency called upon post-tetanic facilitation—a local increase in post-synaptic potentials seen in animal models receiving trains of direct electrical stimulation [63]. A more specific mechanism was suggested by the frequency dependence of the enhancement. For example, increases in working memory performance only occurred with short trains of 5 Hz rTMS, but not 1 Hz or 20 Hz [85, 86]. Interestingly, the theta band of brain wave frequencies is centered on 5 Hz, and it is thought that network communication supporting working memory uses the theta band. This leads to the hypothesis that stimulating cortical regions at frequencies matching an individual's own endogenous oscillatory activity immediately prior to the onset of a task may underlie TMS performance enhancement. In the first test of this hypothesis, a short train of TMS was applied at an individual's peak alpha frequency immediately before performing a difficult mental rotation task, which successfully resulted in increased accuracy in the task, providing evidence for the validity of the oscillatory mechanism [70]. Studies using direct stimulation of cortical regions doing processing relevant to the task that have demonstrated performance enhancement are listed in Table 2.

3 Generating Long-Lasting TMS Enhancement

Other causes of performance enhancement have been suggested. For example, another mechanism called stochastic resonance has been used to explain some specific perceptual enhancement effects [94]; furthermore, there are also situations where performance can be improved simply due to the peripheral effects of TMS, such as intersensory facilitation in reaction time due to the cueing effects of the auditory click of the TMS coil [142]. Nonetheless, the two broad categories of addition-by-subtraction and direct network stimulation cover most instances of cognitive enhancement due to online TMS (stimulation applied during task performance) and appear to suffice as an initial means of conceptualizing the phenomenon. However, as described so far, these effects are quite short-lived, lasting up to only a few seconds. For cognitive enhancement via TMS to be practically useful, effects must be extended in time. Of course, this is true of nervous systems in general, and long-term changes through learning and memory are a basic aspect of their operation. Over the last half-century, dynamic changes in neural synapses supporting long-term potentiation (LTP) which form the mechanistic basis for learning and memory have been extensively studied, and TMS can be used to take advantage of that infrastructure to create long-term enhancement.

Acute effects of TMS pulses occur as the currents they induce create a burst of action potentials followed by a period of relative quiescence lasting a few hundred milliseconds, as can be seen in animals using microelectrodes (e.g., in cats: [95]; in macaque monkeys: [98]). Correspondingly, behavioral effects measured with single pulses (e.g., [3]) and paired pulses [109] can have time windows on the order of

40 ms, presumably as the burst of action potentials affects ongoing cortical processing in a targeted region. Longer-term changes due to cumulative effects of repetitive TMS (rTMS) trains can be associated with LTP mechanisms. These TMS changes occur at shorter (on the order of 30 minutes to an hour) and longer (days to months) time scales. Similarly, LTP mechanisms can be broken down into early and late time domains. Early LTP occurs when calcium ion influx action potentials cause changes to already-existing membrane proteins that make the synapse more responsive for 30 minutes or so until reverting back to baseline sensitivity. Ten to twenty minutes of rTMS can result in changes in EMG, EEG and behavioral performance that last between 10 minutes and an hour (e.g., [107, 143]). These temporary changes are most likely due to a cumulative effect of rTMS causing early LTP. Late LTP occurs when epigenetic changes in synaptic membranes are caused by accumulated effects of action potentials that cause second messengers to be sent to the nucleus of the neuron, resulting in protein synthesis and structural change: new membrane proteins that make the synapse more sensitive for a long-lasting period. Repeated sessions of rTMS over time can cause long-lasting reduction in depressive symptoms (e.g., [103]) or in working memory deficits due to sleep deprivation [88], amongst other examples. These long-lasting effects of TMS are most likely due to membrane changes caused by the induction of late LTP, inaugurated by multiple applications of rTMS.

4 Rational Development of TMS Enhancement Paradigms

Overall, the growing number of reports of performance increases due to TMS, the taxonomy of possible neural mechanisms underlying those effects, as well as the establishment of long-lasting effects caused by TMS, has laid the groundwork for a technology for TMS enhancement of cognition. It should be stated that while the demonstrations of performance facilitation using TMS are numerous and diverse, effects in general are significant but small, with increases in accuracy and reaction time, for example, ranging from 5 to 10%. Such modest effects must be weighed against the risks and side effects of TMS (see [114] for a full review). In healthy adults, the primary safety risks are the possibility of a seizure and of damage to hearing. The risk of seizure is actually quite small: in healthy adults, it is estimated at 2 in 100,000 [114]. Any risk to hearing caused by the clicking noise of the TMS device can be nullified by wearing hearing protection (e.g., earplugs). Side effects, caused by stimulation of peripheral nerves and muscles, are temporary. Thus, while the risks of TMS are small, the fact that they exist at all weigh against any general use of TMS at present to enhance cognitive performance while the effects are also still relatively small. Such use should await the further research developments described below.

Moreover, a more fundamental challenge to developing a program of using TMS for performance improvement was put forth [18] which questioned whether the better performances in cognitive tasks caused by TMS should be viewed as enhancements at all. As a simple example, in a visual detection task where stimuli could appear in either the left or right side of the visual field, when 1 Hz rTMS (which in general lowers cortical excitability) was applied to one side of the head, detection accuracy increased for visual stimuli that appeared ipsilateral to the side of the head stimulated—a seeming enhancement effect [59]. However, accuracy also decreased for stimuli appearing in the opposite visual field. The enhancement in this case could be classified as mediated by an addition-by-subtraction mechanism: the 1-Hz rTMS decreased activity in one part of the visual field so that inputs to the other part were relatively stronger, making a decision concerning the non-stimulated field easier. But this came at a cost, as it made decisions about targets appearing in the visual field stimulated by rTMS, where processing activity was weakened by TMS, more difficult. This illustrates a fact that is often unremarked on in addition-by-subtraction situations: benefits caused by TMS come at the price of reducing capacity of another neural resource. This limitation was generalized by Brem et al. [18] to all TMS enhancement effects, not just those caused by addition-by-subtraction: at any particular moment, the brain as a whole can be viewed through the lens of game theory, and in particular of a zero-sum game. This view suggests that since the nervous system has a finite set of resources, enhancements to one processing system within it must come at the cost of lowering capacities of the rest of the system. This critique served as a reminder that it does not suffice to claim cognitive enhancement of the brain through the use of TMS simply by demonstrating increased performance, as detrimental effects should be searched for and weighed into any assessment of enhancement as well. However, while it is true that at any one time point, it does have a finite set of capacities, the brain is basically designed so that that over time it finds ways to continually increase those capacities through learning. This leads to the idea that the key to successful use of TMS for cognitive enhancement is, rather than trying to force a change in neural circuitry without considering the workings of that circuitry, to work with the nervous system as it changes itself, using TMS to augment the brain's inherent processes for doing so [82].

One example of a paradigm designed to use TMS to engage with brain processing in this way was a series of studies designed to enhance working memory in conditions of sleep deprivation [85–88]. Working memory, a central executive function, is one of the main cognitive functions that suffers with lack of sleep, manifesting in working memory performance as general slowing and missed responses (due to lapses caused by microsleep). In the first study, an acute performance increase in a working memory task was found in healthy, non-sleep deprived adults [85, 86]. The Sternberg working memory task was chosen, a reaction time (RT) task in which a set of letters to remember is presented, followed by a delay, and then a probe letter is shown to which the subject is to respond as quickly as possible whether it was one of the letters in the memory set. The task was run while fMRI was recorded in order to provide potential targets for TMS—here, left dorsolateral prefrontal cortex (DLPFC) and midline parietal cortex were chosen. Three different stimulation TMS frequencies—1 Hz, 5 Hz, and 20 Hz—were tested, and it was found that when short trains of stimuli were applied to parietal cortex (but not DLPFC) during the delay period, 5-Hz rTMS resulted in a speeding of RT (compared to sham rTMS). In a follow-up experiment in the same study with a new group of subjects, the enhancement in RT was replicated for 5 Hz stimulation during the delay period, but not if the train occurred during the probe period.

Having found a cortical site, rTMS frequency, and task phase that resulted in working memory performance enhancement, a study employing healthy adults who had undergone total sleep deprivation for 2 days was run [87]. A previous sleep-deprivation study had found a cortical network using fMRI which was activated by the Sternberg task but had decreased activation with sleep deprivation [52]. Two nodes of this network—the midline parietal site from the previous TMS study and the left lateral occipital complex (LOC), a region involved with the processing of words and letters—were chosen as targets, along with a site about a centimeter outside the LOC network location. Sleep-deprived subjects performed the task while 5-Hz rTMS (active and sham) was applied separately to the three locations during the delay period of the task. Significant reduction relative to sham of the RT slowing caused by sleep deprivation was caused by rTMS to the LOC network site, but not at the site immediately outside the network. There was some RT reduction at the midline parietal site, but this was not significant. Additionally, a significant brain-behavioral effect was observed: it was found that TMS helped the most those subjects who showed the greatest reduction in fMRI activation in LOC across the sleep deprivation period.

Having demonstrated in the previous two studies an acute working memory performance enhancement both in non-sleep-deprived and sleep-deprived subjects, the next study aimed at producing a long-lasting change [88]. Two innovations were brought into play. First, earlier work stimulating motor cortex indicated that functionally activating a targeted region can produce an increased response to TMS. Repeatedly doing so, for instance, by timing peripheral electrical stimulation to arrive in motor cortex concurrently with a TMS pulse (i.e., paired association stimulation or PAS: [135]) can create increased responsiveness of motor cortex that lasts 30 minutes to an hour, most likely due to early LTP. This can be generalized beyond the motor cortex, such that functionally activating a target network simultaneously with TMS application in a manner repeated over a session produces a Hebbian-like change in the targeted network. The second innovation was to push this change into the late LTP time domain by running repeated sessions, similarly to what was done to create a long-lasting remediation of depression (e.g., [103]). Thus, in Luber et al. [88], multiple sessions were run during a 2-day sleep deprivation period in which 5-Hz rTMS was applied to LOC while subjects repeatedly performed the Sternberg task. When the sleep-deprived participants were tested with the task a day after the last TMS session, those who had received sham TMS over the two previous days showed the typical slowing and lapsing in the task caused by sleep deprivation, while those that received active TMS showed little or no deficit. Moreover, a corresponding change in the brain was also observed, with only the active group showing an increase in fMRI in the LOC post-pre sleep deprivation directly beneath where the TMS coil had been placed. Thus, the use of individualized fMRI targeting of TMS, functional engagement of the working memory network with concurrent TMS and task performance, and creation of a long-lasting cumulative neural change with multiple TMS sessions resulted in an almost complete remediation across all participants of performance deficits caused by 2 days of total sleep deprivation, specific to the targeted function of working memory.

This set of three successive studies [85–88] provides a roadmap to be followed to use TMS for cognitive enhancement. Given a cognitive domain, and a task that requires it, a neural substrate that enables it must be identified, most commonly as a distributed functional network found using fMRI recorded while performing the task. Performance characteristics of the task, including practice effects, should be worked out or well-known from earlier research. Next, using TMS, engagement with the functional network must be demonstrated, specifically that stimulation of chosen network nodes affects task performance. Stimulation parameters such as pulse frequency, intensity, train duration, and temporal positioning of the train relative to phase of the task must be explored to find the parameter set that creates an acute performance increase and, if possible, observable changes in network activity corresponding to performance. Finally, a long-lasting behavioral improvement should be attempted using repeated sessions of TMS that lead both to performance enhancement in the task observable after the end of TMS sessions as well as a measurable change in the activity of the targeted functional network related to that enhancement. Generalization of the enhancement to performance in other tasks related to the relevant cognitive domain should also be assessed.

While this roadmap provides a general means for developing cognitive enhancement paradigms for TMS, much work remains to be done to make each step more efficient. This work can be summarized under two general headings: more fully integrating TMS with brain imaging technologies, and optimizing the selection of stimulus temporal parameters.

5 Integrating TMS with Brain Imaging

Brain imaging has an essential role in developing an enhancement paradigm: we need to "see" into the brain as fully as possible to find the network targets of stimulation, to locate individual targets for TMS, and to observe the results of that stimulation. For example, in the working memory enhancement studies just described, fMRI recorded from healthy adult volunteers [85, 86] and sleep-deprived adults [52] were used to determine the network to be modulated. In Luber et al. [88], each subject's fMRI recorded during Sternberg task performance was used to determine individualized TMS targets. Acute and longer-term changes in the targeted network related to enhanced performance caused by TMS were found in pre-post fMRI images in Luber et al. [87] and Luber et al. [88], respectively. These studies used univariate and multivariate analyses of fMRI activation, but additional tools to evaluate MRI data have been developed recently that expand the ability to identify targets for TMS in relevant functional networks. For example, graph theoretical methods along with data from diffusion weighted imaging were used to determine the effects of high and low-frequency rTMS on local and distributed networks supporting verbal memory processing [39]. Another example employed network controllability modeling, which uses anatomical imaging data to help determine network nodes that might be most influential in moving a network to a different physiological state [10]. This modeling was used, in conjunction with fMRI data, to determine optimal target locations for TMS to affect working memory performance [12]. Both of these examples illustrate how sophisticated computational modeling of imaging data might enhance individualized targeting of TMS to maximize network engagement in pursuit of performance enhancement. Other methods for mapping out functional networks and potential TMS targets rely on the use of anatomical (e.g., [89]) and functional connectivity (e.g., [151]) measures. Both Luber et al. [89] and Wang et al. [151] also demonstrate how connectivity measures can help defeat the physical depth limitation of TMS, allowing for modulation of deep brain regions via trans-network stimulation; moreover, Wang et al. [151] is a good example of how functional connectivity can be used to show pre-post network changes associated with performance enhancement caused by TMS. Another means to improve individual targeting with TMS coils is to utilize electric field computations using realistic head modeling in order to orient the TMS coil to optimize induced current direction and strength in line with the subject's brain anatomy (e.g., [105]). An exciting method to more actively explore the interaction of TMS with the brain is to use TMS in the MRI scanner concurrently. In doing so, the activation of the distributed network of brain regions connected to the cortical site stimulated with TMS can be observed (e.g., [45, 115]). For example, TMS applied the prefrontal cortex during the performance of a working memory task can be seen to cause compensatory activation of distant posterior sites also involved in the task [45]. The addition of simultaneous electroencephalography (EEG) to concurrent MRI and TMS provides millisecond time resolution to the high spatial resolution of MRI [108], allowing for precise spatio-temporal probing of the neural effects of TMS on behavioral performance. Overall, the adoption of more advanced methods of combining imaging with TMS of the sort listed here will likely lead to better means to identify functional brain networks associated with task performance, best targets for TMS intervention, and resultant network change caused by TMS: ultimately, in more efficient paradigms to achieve performance enhancement.

6 Optimizing the Selection of TMS Temporal Parameters

Development of more efficient enhancement paradigms also depends on parameters involving when the TMS is applied relative to the task, the frequency (or frequencies) of stimulation and the duration of the stimulation trains—and, for producing long-lasting changes, how many trains are needed. To be effective, the choice of such parameters must take into account the homeostatic nature of the brain, and its ongoing state [128, 129]. In general, the brain operates at all spatial and temporal levels homeostatically: it will always try to maintain its internally-determined state, and specifically resist the effects of external stimulation. One of the first things one finds out in stimulating motor cortex with single TMS pulses to generate motor evoked potentials in target muscles is that, if stimuli are spaced too closely together in time, the evoked potentials are diminished in amplitude. This occurs when pulses are separated by a few seconds—longer than a passive refractory process, suggesting a more active adaptation. At a longer time scale, if trains of rTMS are applied over a period of ten to twenty minutes, any observable effect on behavior can be expected to last about the same period of time before returning to baseline [107]. At a systems level, if a TMS pulse to motor cortex is paired with an immediately preceding audio-visual stimulus about a hundred times, a conditioned learning affect occurs, where pulses not paired with the perceptual stimuli will evoke a normal amplitude motor evoked potential in a target muscle, but pulses paired with the stimuli do not, suggesting that if the brain knows the pulse is coming, it can and will block its effects [86]. On the other hand, the brain might be much less resistant to external stimulation if that stimulation is more in tune with its own ongoing processing, i.e., if TMS is more closely tied to brain state. One type of brain state has already been discussed, namely its spatial aspect, and clearly the location of TMS stimulation must be closely aligned with the specific brain regions and networks underlying task performance.

Another aspect of brain state which can be taken into account to increase TMS effects is the immediate dynamic state of the brain, which can be noninvasively assessed using EEG [146]. It has become clear that endogenous brain oscillations reflect ongoing control of communication and flow of information within and between local and distributed networks (e.g., [77]). TMS can reset [57] and entrain [58, 145] brain oscillations, and rTMS applied at frequencies attuned to individual EEG frequencies have consistently been shown to enhance cognitive task performance [40, 70, 90]. More thought is being given to the timing of TMS relative to both ongoing rhythms and task processing, which is crucial to controlling stimulation effects, and new technologies are being developed to use EEG within open and closed-loop systems [146, 159].

On a slightly longer time scale, another approach to manipulating brain state to be more amenable to TMS is to control the current functional set of a brain network. It is immediately obvious that if a target muscle is already contracting, even minimally, a TMS pulse to that muscle's representation in motor cortex will dramatically increase the amplitude of the resulting motor evoked potential recorded from that muscle, compared to the same TMS pulse applied at rest. This potentiation of response to TMS by pre-activating a network can be generalized, so that a given function of a brain network can be targeted by having a person engage in a task supported by that network, and simultaneously stimulating it with TMS [91]. Furthermore, a desired functional change in the network can be made long-lasting by repeating the pairings of task performance and TMS, presumably through synaptic changes occurring according to Hebbian principles. This can be most simply demonstrated by the paired-associate stimulation (PAS) paradigm, in which repeated pairings of peripheral electrical stimulation of the nerve connected to a target muscle with TMS to motor cortex produces long-lasting increases in motor evoked responses from the target muscle [135]. This was also the principle behind generating long-lasting enhancement of working memory performance in Luber et al. [88], termed functional targeting, which is to use functional state targeting to help isolate and create Hebbian synergies in the subsets of neurons within cortical regions that specifically support targeted cognitive functions [14].

Beyond just the current functional state of a brain network, the recent temporal history of processing in a cortical region can be studied and used to advantage in enhancing TMS. Neural systems are always changing according to what they have been doing. For example, regions involved in visual perception are constantly adapting to the more constant features of the previous stimuli they have processed. Studies taking into account this history of perceptual adaptation have produced strong effects of TMS (e.g., [25]). Another example of incorporating the recent history of a network's activity in relation to a task is to observe the effects of task practice on network activity, which can be profound and must be considered in maintaining TMS effects (e.g., [13]). As an individual practices a task, performance becomes optimized, generally with corresponding changes in the balance of activity across brain networks needed for the task. The targets and parameters of TMS must be adjusted to match these changes, and very possibly to accelerate the transition to the neural state supporting optimum performance.

Overall, efforts to take the state of the brain into account at a variety of time scales, from the immediate, working with ongoing brain oscillations to develop closed-loop feedback systems, to an intermediate scale controlling current task set of networks using functional targeting, to accounting for the recent history of network activity, should all help improve performance enhancement effects using TMS. Furthermore, given the complexity of neural systems and of their interactions with external brain stimulation, the field will benefit from the development of realistic computational models of neural systems from the synaptic to the systems level. This will provide sophisticated prediction of the TMS parameters most likely to result in performance enhancement, and in greater neural plasticity and long-lasting changes (e.g., [1, 154]).

7 Replicability and Generalizability of TMS Enhancement Effects

As can be seen in Tables 1 and 2, there is a broad range of tasks in which performance enhancements caused by TMS have been reported. Many of the cases were not the result of a research program aimed at performance enhancement: these were just found in the course of other TMS experiments and reported. It is nonetheless reasonable to look into how replicable these studies are. This is made clear with two contrasting examples. One the one hand, an early example utilizing the addition-by-subtraction mechanism was an accuracy enhancement in visual detection with 1 Hz rTMS to the parietal cortex [59]. This finding was supported by other studies [38, 130], and became a prime justification for the use of TMS in clinical therapies for stroke rehabilitation. However, a recent meta-analysis of Hilgetag et al. [59] and related studies has called into question whether there is any enhancement in this paradigm at all [152]. On the other hand, as described above, Luber et al. [85] found performance facilitation in a letter Sternberg working memory task with 5 Hz rTMS to midline parietal cortex. The same effect was replicated in a different group of healthy adults in a separate experiment in the same study. The same paradigm was used again in Luber et al. [87] and Luber et al. [88], again replicating the performance enhancements and also generalizing the effect both to a different cortical target (lateral occipital complex) and to a context of sleep deprivation. Using the same 5-Hz rTMS paradigm, working memory was again enhanced in two more studies [11, 157]. In the former study, the paradigm was generalized by working memory content (spatially-relevant items instead of form/letter items) and brain network (using lateral parietal targets). In the latter study generalizations occurred over task manipulation (adding a requirement to alphabetize memory sets), demographic (age groups), and brain network (pre-frontal dorsal executive network) dimensions. Thus, this series of working memory experiments demonstrates an encouraging replicability of a TMS performance enhancement and strong generalization across task manipulations, brain networks, and age demographic. It should be pointed out that later attempts to produce working memory enhancements in the context of sleep deprivation were less successful [51, 80, 93]. However, as reviewed in Luber et al. [91], each of these studies used methods that departed significantly from the earlier successful sleep deprivation work [87, 88], especially in using less effective TMS targeting and using offline instead of online stimulation. Along the lines of using more sophisticated methodology to achieve reliable enhancement (as discussed in brain imaging and TMS parameter optimization sections above), it is instructive to look at Wang et al. [151], in which a series of sessions of offline rTMS was applied to parietal cortex, resulting in an enhancement of memory performance. The key to this result appeared to be that TMS targeting was done using individualized measures of functional connectivity between the hippocampus and parietal cortex. The paradigm was repeated by the same research group, with a replication of enhanced memory performance caused by rTMS [102], but another group attempted to replicate their findings and was unable to reproduce a TMS-caused memory enhancement [56]. However, the fMRI data used for TMS targeting in Hendrikse et al. was re-analyzed using a procedure that increased signal-to-noise in functional connectivity measures between hippocampus and parietal cortex [24]. They found a significant relationship between memory accuracy and the distance between the TMS targets used in Nilakantan et al. and the locations for targets suggested by the newer analysis, such that the smaller the distance, the bigger the memory enhancement produced by TMS, suggesting that a better TMS targeting procedure would have resulted in a replication of memory enhancement. Overall, a growing record for replicability and for generalization is being established for TMS performance enhancement, but its reliability is dependent on the co-development of methodologies and technologies described in the above sections on brain imaging and TMS optimization.

8 Conclusion: The Potential Advent of an Advanced Technology to Enhance Cognitive Performance

In this chapter, an argument has been made first that performance enhancements caused by TMS are real: that they are large in number, replicable, and occur over the full range of brain processes, perceptual and motoric to executive and linguistic, and that they are based on network mechanisms that can be modulated at both acute and long-lasting time scales. Second, that while the size of performance facilitations achieved using TMS are at present modest, a number of technologies are being developed that have the potential to greatly increase those facilitations, for instance, by using TMS online with MRI and EEG and by creating sophisticated computational neural models to generate predictive hypotheses. This idea is encouraged by the successful demonstration of the development of a TMS paradigm that almost completely remediated working memory deficits created by sleep deprivation [52, 85–88]. New tools that create closed-loop systems linking TMS with on-going brain electrical oscillations and/or BOLD fMRI activations allow researchers to see the effects of TMS as they happen, enabling real-time convergence of TMS modulation with network activity. These new technologies can be put to use developing methods to closely work with the brain to enhance network functioning as it performs its own built-in self-enhancement (i.e., as it learns new skills). This, along with the development of other technologies (e.g., focused ultrasound; arrays of multiple stimulation coils; and artificial intelligence) in turn may lead to, over the coming decades, a convergence to a reliable technology of long-lasting enhancement of learned skills.

Such brain stimulation technologies could be used both in populations with cognitive deficits caused by psychiatric or neurological disease, as well as in healthy individuals. As described above, the paradigm developed to remediate working memory deficits caused by sleep deprivation [85–88] provides a model for a successful approach for reducing cognitive deficits. Such an approach has been in development to apply TMS for cognitive impairment in Alzheimer's Disease, beginning with Bentwich et al. [9], although the paradigm they have developed, while demonstrating some success (e.g., [118]), could benefit from the employment of techniques suggested in the present chapter (as further discussed in [83]). In healthy individuals exhibiting normal cognition, brain stimulation could be used to rapidly accelerate skill and knowledge acquisition in the brain, and may even be capable of bringing individuals to skill levels beyond what they might have achieved over longer periods of time and effort without them. The present state of TMS enhancement technology, and the potential for its development laid out here, makes these scenarios conceivable even in the near future. However, the warning first made in Brem et al. [18] might be brought up again in this context. They used a zero-sum game metaphor to question whether performance enhancements caused by TMS were truly such, given a context of allocation of limited brain resources, with enhancements in one resource incurring a cost in other resources. Although it was pointed out that the zero-sum metaphor might not be appropriate, given that the brain is not a zero-sum system as it is always increasing its capacities via learning [82], the metaphor might be more useful when considering long-term implications of directed resource allocation. In the present day, there are many instances of individuals devoting the lion's share of their time, generally from a young age (often directed by parents or the state) to the development of a single ability, be it for example, a sport such as gymnastics, or games such as chess, or a musical instrument. Ethical questions naturally arise as to whether such narrow development is done at the cost of the individual's overall mental health. The potential for technologies involving brain stimulation to expand the focused engineering of singular talents (at the cost of normal individual growth) from a small number of prodigies to a more general population is a possibility that should be considered as brain stimulation methods increase in their sophistication.

References

1. Aberra AS, Peterchev AV, Grill WM. Biophysically realistic neuron models for simulation of cortical stimulation. J Neural Eng. 2018;15(6):066023. https://doi.org/10.1088/1741-2552/aadbb1.
2. Alford JL, van Donkelaar P, Dassonville P, Marrocco RT. Transcranial magnetic stimulation over MT/MST fails to impair judgments of implied motion. Cogn Affect Behav Neurosci. 2007;7(3):225–32. https://doi.org/10.3758/cabn.7.3.225.
3. Amassian VE, Cracco RQ, Maccabee PJ, Cracco JB, Rudell A, Eberle L. Suppression of visual perception by magnetic coil stimulation of human occipital cortex. Electroencephalogr Clin Neurophysiol. 1989;74(6):458–62. https://doi.org/10.1016/0168-5597(89)90036-1.
4. Andoh J, Artiges E, Pallier C, Rivière D, Mangin JF, Cachia A, Plaze M, Paillère-Martinot ML, Martinot JL. Modulation of language areas with functional MR image-guided magnetic stimulation. NeuroImage. 2006;29(2):619–27. https://doi.org/10.1016/j.neuroimage.2005.07.029.
5. Baeken C, Schrijvers DL, Sabbe BG, Vanderhasselt MA, De Raedt R. Impact of one HF-rTMS session on fine motor function in right-handed healthy female subjects: a comparison of stimulation over the left versus the right dorsolateral prefrontal cortex. Neuropsychobiology. 2012;65(2):96–102. https://doi.org/10.1159/000329699.
6. Bagherzadeh Y, Khorrami A, Zarrindast MR, Shariat SV, Pantazis D. Repetitive transcranial magnetic stimulation of the dorsolateral prefrontal cortex enhances working memory. Exp Brain Res. 2016;234(7):1807–18. https://doi.org/10.1007/s00221-016-4580-1.
7. Bardi L, Kanai R, Mapelli D, Walsh V. TMS of the FEF interferes with spatial conflict. J Cogn Neurosci. 2012;24(6):1305–13. https://doi.org/10.1162/jocn_a_00223.
8. Beauchamp MS, Nath AR, Pasalar S. fMRI-guided transcranial magnetic stimulation reveals that the superior temporal sulcus is a cortical locus of the McGurk effect. J Neurosci. 2010;30(7):2414–7. https://doi.org/10.1523/JNEUROSCI.4865-09.2010.
9. Bentwich J, Dobronevsky E, Aichenbaum S, Shorer R, Peretz R, Khaigrekht M, Marton RG, Rabey JM. Beneficial effect of repetitive transcranial magnetic stimulation combined with cognitive training for the treatment of Alzheimer's disease: a proof of concept study. J Neural Transm. 2011;118:463–71. https://doi.org/10.1007/s00702-010-0578-1.
10. Betzel RF, Gu S, Medaglia JD, Pasqualetti F, Bassett DS. Optimally controlling the human connectome: the role of network topology. Sci Rep. 2016;6:30770. https://doi.org/10.1038/srep30770.
11. Beynel L, Davis SW, Crowell CA, Hilbig SA, Lim W, Nguyen D, Palmer H, Brito A, Peterchev AV, Luber B, Lisanby SH, Cabeza R, Appelbaum LG. Online repetitive transcranial magnetic stimulation during working memory in younger and older adults: a randomized within-subject comparison. PLoS One. 2019;14(3):e0213707. https://doi.org/10.1371/journal.pone.0213707.

12. Beynel L, Powers JP, Appelbaum LG. Effects of repetitive transcranial magnetic stimulation on resting-state connectivity: a systematic review. NeuroImage. 2020;211:116596. https://doi.org/10.1016/j.neuroimage.2020.116596.
13. Beynel L, Gura H, Rezaee Z, Ekpo EC, Deng ZD, Joseph JO, Taylor P, Luber B, Lisanby SH. Lessons learned from an fMRI-guided rTMS study on performance in a numerical Stroop task. PLoS One. 2024;19(5):e0302660. https://doi.org/10.1371/journal.pone.0302660.
14. Bikson M, Rahman A. Origins of specificity during tDCS: anatomical, activity-selective, and input-bias mechanisms. Front Hum Neurosci. 2013;7:688. https://doi.org/10.3389/fnhum.2013.00688.
15. Bonnì S, Veniero D, Mastropasqua C, Ponzo V, Caltagirone C, Bozzali M, Koch G. TMS evidence for a selective role of the precuneus in source memory retrieval. Behav Brain Res. 2015;282:70–5. https://doi.org/10.1016/j.bbr.2014.12.032.
16. Boroojerdi B, Phipps M, Kopylev L, Wharton CM, Cohen LG, Grafman J. Enhancing analogic reasoning with rTMS over the left prefrontal cortex. Neurology. 2001;56(4):526–8. https://doi.org/10.1212/wnl.56.4.526.
17. Boyd LA, Linsdell MA. Excitatory repetitive transcranial magnetic stimulation to left dorsal premotor cortex enhances motor consolidation of new skills. BMC Neurosci. 2009;10:72. https://doi.org/10.1186/1471-2202-10-72.
18. Brem A-K, Fried PJ, Horvath JC, Robertson EM, Pascual-Leone A. Is neuroenhancement by noninvasive brain stimulation a zero-sum proposition? NeuroImage. 2014;85:1058–68. https://doi.org/10.1016/j.neuroimage.2013.07.038.
19. Bütefisch CM, Khurana V, Kopylev L, Cohen LG. Enhancing encoding of a motor memory in the primary motor cortex by cortical stimulation. J Neurophysiol. 2004;91(5):2110–6. https://doi.org/10.1152/jn.01038.2003.
20. Butefisch CM, Khurana V, Kopylev L, Cohen LG. Enhancing encoding of a motor memory in the primary motor cortex by cortical stimulation. J. Neurophysiol. 2004;91:2110–6.
21. van Campen AD, Keuken MC, van den Wildenberg WP, Ridderinkhof KR. TMS over M1 reveals expression and selective suppression of conflicting action impulses. J Cogn Neurosci. 2014;26(1):1–15. https://doi.org/10.1162/jocn_a_00482.
22. Cappa SF, Sandrini M, Rossini PM, Sosta K, Miniussi C. The role of the left frontal lobe in action naming: rTMS evidence. Neurology. 2002;59(5):720–3. https://doi.org/10.1212/wnl.59.5.720.
23. Cárdenas-Morales L, Grön G, Kammer T. Exploring the after-effects of theta burst magnetic stimulation on the human motor cortex: a functional imaging study. Hum Brain Mapp. 2011;32(11):1948–60. https://doi.org/10.1002/hbm.21160.
24. Cash RFH, Hendrikse J, Fernando KB, Thompson S, Suo C, Fornito A, Yucel M, Rogasch N, Zalesky A, Coxon JP. Personalized brain stimulation of memory networks. Brain Stimul. 2022;15:1300–4.
25. Cattaneo Z, Rota F, Vecchi T, Silvanto J. Using state-dependency of transcranial magnetic stimulation (TMS) to investigate letter selectivity in the left posterior parietal cortex: a comparison of TMS-priming and TMS-adaptation paradigms. Eur J Neurosci. 2008;28(9):1924–9. https://doi.org/10.1111/j.1460-9568.2008.06466.x.
26. Cattaneo Z, Rota F, Walsh V, Vecchi T, Silvanto J. TMS-adaptation reveals abstract letter selectivity in the left posterior parietal cortex. Cerebral Cortex. 2009a;19(10):2321–5. https://doi.org/10.1093/cercor/bhn249.
27. Cattaneo Z, Vecchi T, Pascual-Leone A, Silvanto J. Contrasting early visual cortical activation states causally involved in visual imagery and short-term memory. Eur J Neurosci. 2009b;30(7):1393–400. https://doi.org/10.1111/j.1460-9568.2009.06911.x.
28. Cattaneo L, Sandrini M, Schwarzbach J. State-dependent TMS reveals a hierarchical representation of observed acts in the temporal, parietal, and premotor cortices. Cerebral Cortex. 2010;20(9):2252–8. https://doi.org/10.1093/cercor/bhp291.
29. Chambers CD, Stokes MG, Janko NE, Mattingley JB. Enhancement of visual selection during transient disruption of parietal cortex. Brain Res. 2006;1097(1):149–55. https://doi.org/10.1016/j.brainres.2006.04.084.

30. Cho SS, Yoon EJ, Lee JM, Kim SE. Repetitive transcranial magnetic stimulation of the left dorsolateral prefrontal cortex improves probabilistic category learning. Brain Topogr. 2012;25(4):443–9. https://doi.org/10.1007/s10548-012-0242-1.
31. Cohen Kadosh R, Cohen Kadosh K, Schuhmann T, Kaas A, Goebel R, Henik A, Sack AT. Virtual dyscalculia induced by parietal-lobe TMS impairs automatic magnitude processing. Curr Biol. 2007;17(8):689–93. https://doi.org/10.1016/j.cub.2007.02.056.
32. Cohen Kadosh R, Muggleton N, Silvanto J, Walsh V. Double dissociation of format-dependent and number-specific neurons in human parietal cortex. Cerebral Cortex. 2010;20(9):2166–71. https://doi.org/10.1093/cercor/bhp273.
33. Coldea A, Veniero D, Morand S, Trajkovic J, Romei V, Harvey M, Thut G. Effects of rhythmic transcranial magnetic stimulation in the alpha-band on visual perception depend on deviation from alpha-peak frequency: faster relative transcranial magnetic stimulation alpha-pace improves performance. Front Neurosci. 2022;16:886342. https://doi.org/10.3389/fnins.2022.886342.
34. Cona G, Marino G, Semenza C. TMS of supplementary motor area (SMA) facilitates mental rotation performance: evidence for sequence processing in SMA. NeuroImage. 2017;146:770–7. https://doi.org/10.1016/j.neuroimage.2016.10.032.
35. Cooper AC, Humphreys GW, Hulleman J, Praamstra P, Georgeson M. Transcranial magnetic stimulation to right parietal cortex modifies the attentional blink. Exp Brain Res. 2004;155(1):24–9. https://doi.org/10.1007/s00221-003-1697-9.
36. Costanzo F, Menghini D, Caltagirone C, Oliveri M, Vicari S. High frequency rTMS over the left parietal lobule increases non-word reading accuracy. Neuropsychologia. 2012;50(11):2645–51. https://doi.org/10.1016/j.neuropsychologia.2012.07.017.
37. Cui D, Jin J, Cao W, Wang H, Wang X, Li Y, Liu T, Yin T, Liu Z. Beneficial effect of high-frequency repetitive transcranial magnetic stimulation for the verbal memory and default mode network in healthy older adults. Front Aging Neurosci. 2022;14:845912. https://doi.org/10.3389/fnagi.2022.845912.
38. Dambeck N, Sparing R, Meister IG, Wienemann M, Weidemann J, Topper R, Boroojerdi B. Interhemispheric imbalance during visuospatial attention investigated by unilateral and bilateral TMS over human parietal cortices. Brain Res. 2006;1072:194–9. https://doi.org/10.1016/j.brainres.2005.05.075.
39. Davis SW, Luber B, Murphy DLK, Lisanby SH, Cabeza R. Frequency-specific neuromodulation of local and distant connectivity in aging and episodic memory function. Hum Brain Mapp. 2017;38(12):5987–6004. https://doi.org/10.1002/hbm.23803.
40. Di Gregorio F, Trajkovic J, Roperti C, Marcantoni E, Di Luzio P, Avenanti A, Thut G, Romei V. Tuning alpha rhythms to shape conscious visual perception. Curr Biol. 2022;32(5):988–998.e6. https://doi.org/10.1016/j.cub.2022.01.003.
41. Dräger B, Breitenstein C, Helmke U, Kamping S, Knecht S. Specific and nonspecific effects of transcranial magnetic stimulation on picture-word verification. Eur J Neurosci. 2004;20(6):1681–7. https://doi.org/10.1111/j.1460-9568.2004.03623.x.
42. Ellison A, Battelli L, Cowey A, Walsh V. The effect of expectation on facilitation of colour/form conjunction tasks by TMS over area V5. Neuropsychologia. 2003;41(13):1794–801. https://doi.org/10.1016/s0028-3932(03)00180-5.
43. Evers S, Böckermann I, Nyhuis PW. The impact of transcranial magnetic stimulation on cognitive processing: an event-related potential study. Neuroreport. 2001;12(13):2915–8. https://doi.org/10.1097/00001756-200109170-00032.
44. Feredoes E, Tononi G, Postle BR. The neural bases of the short-term storage of verbal information are anatomically variable across individuals. J. Neurosci. 2007;27:11003–8.
45. Feredoes E, Heinen K, Weiskopf N, Ruff C, Driver J. Causal evidence for frontal involvement in memory target maintenance by posterior brain areas during distracter interference of visual working memory. Proc Natl Acad Sci USA. 2011;108(42):17510–5. https://doi.org/10.1073/pnas.1106439108.

46. Gagnon G, Schneider C, Grondin S, Blanchet S. Enhancement of episodic memory in young and healthy adults: a paired-pulse TMS study on encoding and retrieval performance. Neurosci Lett. 2011;488(2):138–42. https://doi.org/10.1016/j.neulet.2010.11.016.
47. Galea JM, Albert NB, Ditye T, Miall RC. Disruption of the dorsolateral prefrontal cortex facilitates the consolidation of procedural skills. J Cogn Neurosci. 2010;22(6):1158–64. https://doi.org/10.1162/jocn.2009.21259.
48. Gan T, Nikolin S, Loo CK, Martin DM. Effects of high-definition transcranial direct current stimulation and theta burst stimulation for modulating the posterior parietal cortex. J Int Neuropsychol Soc. 2019;25(9):972–84. https://doi.org/10.1017/S1355617719000766.
49. Grosbras MH, Paus T. Transcranial magnetic stimulation of the human frontal eye field: effects on visual perception and attention. J Cogn Neurosci. 2002;14(7):1109–20. https://doi.org/10.1162/089892902320474553.
50. Grosbras MH, Paus T. Transcranial magnetic stimulation of the human frontal eye field facilitates visual awareness. Eur J Neurosci. 2003;18(11):3121–6. https://doi.org/10.1111/j.1460-9568.2003.03055.x.
51. Guo Z, Jiang Z, Jiang B, McClure MA, Mu Q. High-frequency repetitive transcranial magnetic stimulation could improve impaired working memory induced by sleep deprivation. Neural Plast. 2019;2019:7030286. https://doi.org/10.1155/2019/7030286.
52. Habeck C, Rakitin BC, Moeller J, Scarmeas N, Zarahn E, Brown T, Stern Y. An event-related fMRI study of the neurobehavioral impact of sleep deprivation on performance of a delayed-match-to-sample task. Brain Res Cogn Brain Res. 2004;18(3):306–21. https://doi.org/10.1016/j.cogbrainres.2003.10.019.
53. Hannula H, Neuvonen T, Savolainen P, Hiltunen J, Ma YY, Antila H, Salonen O, Carlson S, Pertovaara A. Increasing top-down suppression from prefrontal cortex facilitates tactile working memory. NeuroImage. 2010;49(1):1091–8. https://doi.org/10.1016/j.neuroimage.2009.07.049.
54. Harris IM, Benito CT, Ruzzoli M, Miniussi C. Effects of right parietal transcranial magnetic stimulation on object identification and orientation judgments. J Cogn Neurosci. 2008;20(5):916–26. https://doi.org/10.1162/jocn.2008.20513.
55. Hayward G, Goodwin GM, Harmer CJ. The role of the anterior cingulate cortex in the counting Stroop task. Exp Brain Res. 2004;154(3):355–8. https://doi.org/10.1007/s00221-003-1665-4.
56. Hendrikse J, Coxon JP, Thompson S, Suo C, Fornito A, Yücel M, et al. Multi-day rTMS exerts site-specific effects on functional connectivity but does not influence associative memory performance. Cortex. 2020;132:423e40.
57. Herring JD, Thut G, Jensen O, Bergmann TO. Attention modulates TMS-locked alpha oscillations in the visual cortex. J Neurosci. 2015;35(43):14435–47. https://doi.org/10.1523/JNEUROSCI.1833-15.2015.
58. Herrmann CS, Murray MM, Ionta S, Hutt A, Lefebvre J. Shaping intrinsic neural oscillations with periodic stimulation. J Neurosci. 2016;36(19):5328–37. https://doi.org/10.1523/JNEUROSCI.0236-16.2016.
59. Hilgetag CC, Théoret H, Pascual-Leone A. Enhanced visual spatial attention ipsilateral to rTMS-induced 'virtual lesions' of human parietal cortex. Nat Neurosci. 2001;4(9):953–7. https://doi.org/10.1038/nn0901-953.
60. Hodsoll J, Mevorach C, Humphreys GW. Driven to less distraction: rTMS of the right parietal cortex reduces attentional capture in visual search. Cerebral Cortex. 2009;19(1):106–14. https://doi.org/10.1093/cercor/bhn070.
61. Hoy KE, Bailey N, Michael M, Fitzgibbon B, Rogasch NC, Saeki T, Fitzgerald PB. Enhancement of working memory and task-related oscillatory activity following intermittent theta burst stimulation in healthy controls. Cerebral Cortex. 2016;26(12):4563–73. https://doi.org/10.1093/cercor/bhv193.
62. Hwang JH, Kim SH, Park CS, Bang SA, Kim SE. Acute high-frequency rTMS of the left dorsolateral prefrontal cortex and attentional control in healthy young men. Brain Res. 2010;1329:152–8. https://doi.org/10.1016/j.brainres.2010.03.013.

63. Iriki A, Pavlides C, Keller A, Asanuma H. Long-term potentiation in the motor cortex. Science. 1989;245(4924):1385–7. https://doi.org/10.1126/science.2551038.
64. Jelić MB, Filipović SR, Milanović SD, Stevanović VB, Konstantinović L. Bilateral sequential motor cortex stimulation and skilled task performance with non-dominant hand. Clin Neurophysiol. 2017;128(5):814–22. https://doi.org/10.1016/j.clinph.2017.02.020.
65. Kalla R, Muggleton NG, Cowey A, Walsh V. Human dorsolateral prefrontal cortex is involved in visual search for conjunctions but not features: a theta TMS study. Cortex. 2009;45(9):1085–90. https://doi.org/10.1016/j.cortex.2009.01.005.
66. Kim YH, Park JW, Ko MH, Jang SH, Lee PK. Facilitative effect of high frequency subthreshold repetitive transcranial magnetic stimulation on complex sequential motor learning in humans. Neurosci Lett. 2004;367(2):181–5. https://doi.org/10.1016/j.neulet.2004.05.113.
67. Kim SH, Han HJ, Ahn HM, Kim SA, Kim SE. Effects of five daily high-frequency rTMS on Stroop task performance in aging individuals. Neurosci Res. 2012;74(3–4):256–60. https://doi.org/10.1016/j.neures.2012.08.008.
68. Kim H, Kim J, Lee HJ, Lee J, Na Y, Chang WH, Kim YH. Optimal stimulation site for rTMS to improve motor function: anatomical hand knob vs. hand motor hotspot. Neurosci Lett. 2021;740:135424. https://doi.org/10.1016/j.neulet.2020.135424.
69. Kirschen MP, Davis-Ratner MS, Jerde TE, Schraedley-Desmond P, Desmond JE. Enhancement of phonological memory following transcranial magnetic stimulation (TMS). Behav Neurol. 2006;17(3–4):187–94. https://doi.org/10.1155/2006/469132.
70. Klimesch W, Sauseng P, Gerloff C. Enhancing cognitive performance with repetitive transcranial magnetic stimulation at human individual alpha frequency. Eur J Neurosci. 2003;17(5):1129–33. https://doi.org/10.1046/j.1460-9568.2003.02517.x.
71. Knecht S, Flöel A, Dräger B, Breitenstein C, Sommer J, Henningsen H, Ringelstein EB, Pascual-Leone A. Degree of language lateralization determines susceptibility to unilateral brain lesions. Nat Neurosci. 2002;5(7):695–9. https://doi.org/10.1038/nn868.
72. Kobayashi M. Effect of slow repetitive TMS of the motor cortex on ipsilateral sequential simple finger movements and motor skill learning. Restor Neurol Neurosci. 2010;28(4):437–48. https://doi.org/10.3233/RNN-2010-0562.
73. Kobayashi M, Hutchinson S, Théoret H, Schlaug G, Pascual-Leone A. Repetitive TMS of the motor cortex improves ipsilateral sequential simple finger movements. Neurology. 2004;62(1):91–8. https://doi.org/10.1212/wnl.62.1.91.
74. Kobayashi M, Théoret H, Pascual-Leone A. Suppression of ipsilateral motor cortex facilitates motor skill learning. Eur J Neurosci. 2009;29(4):833–6. https://doi.org/10.1111/j.1460-9568.2009.06628.x.
75. Köhler S, Paus T, Buckner RL, Milner B. Effects of left inferior prefrontal stimulation on episodic memory formation: a two-stage fMRI-rTMS study. J Cogn Neurosci. 2004;16(2):178–88. https://doi.org/10.1162/089892904322984490.
76. Koski L, Molnar-Szakacs I, Iacoboni M. Exploring the contributions of premotor and parietal cortex to spatial compatibility using image-guided TMS. NeuroImage. 2005;24(2):296–305. https://doi.org/10.1016/j.neuroimage.2004.09.027.
77. Lakatos P, Gross J, Thut G. A new unifying account of the roles of neuronal entrainment. Curr Biol. 2019;29(18):R890–905. https://doi.org/10.1016/j.cub.2019.07.075.
78. Li S, Jin JN, Wang X, Qi HZ, Liu ZP, Yin T. Theta and alpha oscillations during the retention period of working memory by rTMS stimulating the parietal lobe. Front Behav Neurosci. 2017a;11:170. https://doi.org/10.3389/fnbeh.2017.00170.
79. Li Y, Wang L, Jia M, Guo J, Wang H, Wang M. The effects of high-frequency rTMS over the left DLPFC on cognitive control in young healthy participants. PLoS One. 2017b;12(6):e0179430. https://doi.org/10.1371/journal.pone.0179430.
80. Li S, Zhou H, Yu Y, Lyu H, Mou T, Shi G, et al. Effect of repetitive transcranial magnetic stimulation on the cognitive impairment induced by sleep deprivation: a randomized trial. Sleep Med. 2021;(77):270–8. https://doi.org/10.1016/j.sleep.2020.06.019.

81. Lou HC, Luber B, Stanford A, Lisanby SH. Self-specific processing in the default network: a single-pulse TMS study. Exp Brain Res. 2010;207(1–2):27–38. https://doi.org/10.1007/s00221-010-2425-x.
82. Luber B. Neuroenhancement by noninvasive brain stimulation is not a net zero-sum proposition. Front Syst Neurosci. 2014;8:article 129.
83. Luber B, Beynel L. Cognitive enhancement using non-invasive brain stimulation (NIBS), with application to normal and pathological decline in aging. In: Wassermann EM, Peterchev AP, Ziemann U, Lisanby SH, Siebner HR, Walsh V, editors. The Oxford handbook of transcranial stimulation. 2nd ed. Oxford, UK: Oxford University Press; 2024.
84. Luber B, Lisanby SH. Enhancement of human cognitive performance using transcranial magnetic stimulation (TMS). Neuroimage. 2014;85(Pt 3(03)):961–70. https://doi.org/10.1016/j.neuroimage.2013.06.007.
85. Luber B, Kinnunen LH, Rakitin BC, Ellsasser R, Stern Y, Lisanby SH. Facilitation of performance in a working memory task with rTMS stimulation of the precuneus: frequency- and time-dependent effects. Brain Res. 2007a;1128(1):120–9. https://doi.org/10.1016/j.brainres.2006.10.011.
86. Luber B, Balsam P, Nguyen T, Gross M, Lisanby SH. Classical conditioned learning using transcranial magnetic stimulation. Exp Brain Res. 2007b;183(3):361–9.
87. Luber B, Stanford AD, Bulow P, Nguyen T, Rakitin BC, Habeck C, Basner R, Stern Y, Lisanby SH. Remediation of sleep-deprivation-induced working memory impairment with fMRI-guided transcranial magnetic stimulation. Cerebral Cortex. 2008;18(9):2077–85. https://doi.org/10.1093/cercor/bhm231.
88. Luber B, Steffener J, Tucker A, Habeck C, Peterchev AV, Deng ZD, Basner RC, Stern Y, Lisanby SH. Extended remediation of sleep deprived-induced working memory deficits using fMRI guided transcranial magnetic stimulation. Sleep. 2013;36(6):857–71. https://doi.org/10.5665/sleep.2712.
89. Luber B, Davis SW, Deng ZD, Murphy D, Martella A, Peterchev AV, Lisanby SH. Using diffusion tensor imaging to effectively target TMS to deep brain structures. NeuroImage. 2022;249:118863. https://doi.org/10.1016/j.neuroimage.2021.118863.
90. Luber B, Beynel L, Deng ZD, Appelbaum LG, Jones T, Harrison A, Murphy DLK, Lo E, McKinley RA, Lisanby SH. Site- and frequency-specific enhancement of visual search performance with online individual alpha frequency (IAF) repetitive transcranial magnetic stimulation (rTMS) to the inferior frontal junction. Cereb Cortex. 2024a;34
91. Luber B, Ekpo E, Lisanby SH. The use of cognitive paired associative stimulation (C-PAS) in investigating and remediating the effects of sleep deprivation on working memory in humans: the importance of state-dependency. Curr Sleep Med Rep. 2024b;10:199–206. https://doi.org/10.1007/s40675-024-00290-z.
92. Martín-Arévalo E, Lupiáñez J, Narganes-Pineda C, Marino G, Colás I, Chica AB. The causal role of the left parietal lobe in facilitation and inhibition of return. Cortex. 2019;117:311–22. https://doi.org/10.1016/j.cortex.2019.04.025.
93. Martinez-Cancino MDP, Azpiroz-Leehan J, Jimenez-Angeles L, Garcia-Quintanar A, Santana-Miranda R. Effects of high frequency rTMS on sleep. deprivation: a pilot study. Annu Int Conf IEEE Eng Med Biol Soc. 2016;2016:5937–40. https://doi.org/10.1109/EMBC.2016.7592080.
94. Miniussi C, Thut G. Combining TMS and EEG offers new prospects in cognitive neuroscience. Brain Topogr. 2010;22(4):249–56. https://doi.org/10.1007/s10548-009-0083-8.
95. Moliadze V, Zhao Y, Eysel U, Funke K. Effect of transcranial magnetic stimulation on single-unit activity in the cat primary visual cortex. J Physiol. 2003;553(Pt 2):665–79. https://doi.org/10.1113/jphysiol.2003.050153.
96. Momi D, Neri F, Coiro G, Smeralda C, et al. Cognitive enhancement via network-targeted cortico-cortical associative brain stimulation. Cerebral Cortex. 2020;30(3):1516–27. https://doi.org/10.1093/cercor/bhz182.

97. Mottaghy FM, Hungs M, Brügmann M, Sparing R, Boroojerdi B, Foltys H, Huber W, Töpper R. Facilitation of picture naming after repetitive transcranial magnetic stimulation. Neurology. 1999;53(8):1806–12. https://doi.org/10.1212/wnl.53.8.1806.
98. Mueller JK, Grigsby EM, Prevosto V, Petraglia FW 3rd, Rao H, Deng ZD, Peterchev AV, Sommer MA, Egner T, Platt ML, Grill WM. Simultaneous transcranial magnetic stimulation and single-neuron recording in alert non-human primates. Nat Neurosci. 2014;17(8):1130–6. https://doi.org/10.1038/nn.3751.
99. Mullin CR, Steeves JK. TMS to the lateral occipital cortex disrupts object processing but facilitates scene processing. J Cogn Neurosci. 2011;23(12):4174–84. https://doi.org/10.1162/jocn_a_00095.
100. Müri RM, Rivaud S, Gaymard B, Ploner CJ, Vermersch AI, Hess CW, Pierrot-Deseilligny C. Role of the prefrontal cortex in the control of express saccades. A transcranial magnetic stimulation study. Neuropsychologia. 1999;37(2):199–206. https://doi.org/10.1016/s0028-3932(98)00094-3.
101. Narayana S, Zhang W, Rogers W, Strickland C, Franklin C, Lancaster JL, Fox PT. Concurrent TMS to the primary motor cortex augments slow motor learning. NeuroImage. 2014;85 Pt 3(0 3):971–84. https://doi.org/10.1016/j.neuroimage.2013.07.024.
102. Nilakantan AS, Bridge DJ, Gagnon EP, VanHaerents SA, Voss JL. Stimulation of the posterior cortical-hippocampal network enhances precision of memory recollection. Curr Biol. 2017;27(3):465e70:465.
103. O'Reardon JP, Solvason HB, Janicak PG, Sampson S, Isenberg KE, Nahas Z, McDonald WM, Avery D, Fitzgerald PB, Loo C, Demitrack MA, George MS, Sackeim HA. Efficacy and safety of transcranial magnetic stimulation in the acute treatment of major depression: a multisite randomized controlled trial. Biol Psychiatry. 2007;62(11):1208–16. https://doi.org/10.1016/j.biopsych.2007.01.018.
104. Oliveri M, Zhaoping L, Mangano GR, Turriziani P, Smirni D, Cipolotti L. Facilitation of bottom-up feature detection following rTMS-interference of the right parietal cortex. Neuropsychologia. 2010;48(4):1003–10. https://doi.org/10.1016/j.neuropsychologia.2009.11.024.
105. Opitz A, Legon W, Rowlands A, Bickel WK, Paulus W, Tyler WJ. Physiological observations validate finite element models for estimating subject-specific electric field distributions induced by transcranial magnetic stimulation of the human motor cortex. NeuroImage. 2013;81:253–64. https://doi.org/10.1016/j.neuroimage.2013.04.067.
106. Parikh V, Medley A, Goh HT. Effects of rTMS to primary motor cortex and cerebellum on balance control in healthy adults. Eur J Neurosci. 2024;60(2):3984–94. https://doi.org/10.1111/ejn.16386.
107. Peinemann A, Reimer B, Loer C, et al. Long-lasting increase in corticospinal excitability after 1800 pulses of subthreshold 5 Hz repetitive TMS to the primary motor cortex. Clin Neurophysiol. 2004;115:1519–26. https://doi.org/10.1016/j.clinph.2004.02.005.
108. Peters JC, Reithler J, Graaf TA, Schuhmann T, Goebel R, Sack AT. Concurrent human TMS-EEG-fMRI enables monitoring of oscillatory brain state-dependent gating of cortico-subcortical network activity. Commun Biol. 2020;3(1):40. https://doi.org/10.1038/s42003-020-0764-0.
109. Pitcher D, Garrido L, Walsh V, Duchaine BC. Transcranial magnetic stimulation disrupts the perception and embodiment of facial expressions. J Neurosci. 2008;28(36):8929–33. https://doi.org/10.1523/JNEUROSCI.1450-08.2008.
110. Ragert P, Dinse HR, Pleger B, Wilimzig C, Frombach E, Schwenkreis P, Tegenthoff M. Combination of 5 Hz repetitive transcranial magnetic stimulation (rTMS) and tactile coactivation boosts tactile discrimination in humans. Neurosci Lett. 2003;348(2):105–8. https://doi.org/10.1016/s0304-3940(03)00745-6.
111. Rajji TK, Liu SK, Frantseva MV, Mulsant BH, Thoma J, Chen R, Fitzgerald PB, Daskalakis ZJ. Exploring the effect of inducing long-term potentiation in the human motor cortex on motor learning. Brain Stimul. 2011;4(3):137–44. https://doi.org/10.1016/j.brs.2010.09.007.

112. Renzi C, Vecchi T, Silvanto J, Cattaneo Z. Overlapping representations of numerical magnitude and motion direction in the posterior parietal cortex: a TMS-adaptation study. Neurosci Lett. 2011;490(2):145–9. https://doi.org/10.1016/j.neulet.2010.12.045.
113. Romei V, Driver J, Schyns PG, Thut G. Rhythmic TMS over parietal cortex links distinct brain frequencies to global versus local visual processing. Curr Biol. 2011;21(4):334–7. https://doi.org/10.1016/j.cub.2011.01.035.
114. Rossi S, Antal A, Bestmann S, Bikson M, Brewer C, Brockmöller J, Carpenter LL, Cincotta M, Chen R, Daskalakis JD, Di Lazzaro V, Fox MD, George MS, Gilbert D, Kimiskidis VK, Koch G, Ilmoniemi RJ, Lefaucheur JP, Leocani L, Lisanby SH, Miniussi C, Padberg F, Pascual-Leone A, Paulus W, Peterchev AP, Quartarone A, Rotenberg A, Rothwell J, Rossini PM, Santarnecchi E, Shafi MM, Siebner HR, Ugawa Y, Wassermann EM, Zangen A, Ziemann U, Hallett M. Safety and recommendations for TMS use in healthy subjects and patient populations, with updates on training, ethical and regulatory issues: expert guidelines. Clin Neurophysiol. 2021;132:269–306.
115. Ruff CC, Bestmann S, Blankenburg F, Bjoertomt O, Josephs O, Weiskopf N, Deichmann R, Driver J. Distinct causal influences of parietal versus frontal areas on human visual cortex: evidence from concurrent TMS-fMRI. Cerebral Cortex. 2008;18(4):817–27. https://doi.org/10.1093/cercor/bhm128.
116. Rusconi E, Dervinis M, Verbruggen F, Chambers CD. Critical time course of right frontoparietal involvement in mental number space. J Cogn Neurosci. 2013;25(3):465–83. https://doi.org/10.1162/jocn_a_00330.
117. Rushworth MF, Passingham RE, Nobre AC. Components of switching intentional set. J Cogn Neurosci. 2002;14(8):1139–50. https://doi.org/10.1162/089892902760807159.
118. Sabbagh M, Sadowsky C, Tousi B, Agronin ME, Alva G, Armon C, Bernick C, Keegan AP, Karantzoulis S, Baror E, Ploznik M, Pascual-Leone A. Effects of a combined transcranial magnetic stimulation (TMS) and cognitive training intervention in patients with Alzheimer's disease. Alzheimers Dement. 2020;16(4):641–50. https://doi.org/10.1016/j.jalz.2019.08.197.
119. Sakai KL, Noguchi Y, Takeuchi T, Watanabe E. Selective priming of syntactic processing by event-related transcranial magnetic stimulation of Broca's area. Neuron. 2002;35(6):1177–82. https://doi.org/10.1016/s0896-6273(02)00873-5.
120. Sallie SN, Sonkusare S, Mandali A, Casero V, Cui H, Guzman NV, Allison M, Voon V. Cortical paired associative stimulation shows impaired plasticity of inhibition networks as a function of chronic alcohol use. Psychol Med. 2024;54(4):698–709. https://doi.org/10.1017/S0033291723002374.
121. San Agustín A, Crevillén D, Soto-León V, Moreno JC, Oliviero A, Pons JL. Transcranial magnetic stimulation combined with endogenous human hippocampal and motor cortical activity enhances memory. PLoS One. 2023;18(12):e0295413. https://doi.org/10.1371/journal.pone.0295413.
122. Sandberg A, Bostrom N. Converging cognitive enhancements. Ann N Y Acad Sci. 2006;1093:201–27. https://doi.org/10.1196/annals.1382.015.
123. Sauseng P, Klimesch W, Heise KF, Gruber WR, Holz E, Karim AA, Glennon M, Gerloff C, Birbaumer N, Hummel FC. Brain oscillatory substrates of visual short-term memory capacity. Curr Biol. 2009;19(21):1846–52. https://doi.org/10.1016/j.cub.2009.08.062.
124. Schiff S, Bardi L, Basso D, Mapelli D. Timing spatial conflict within the parietal cortex: a TMS study. J Cogn Neurosci. 2011;23(12):3998–4007. https://doi.org/10.1162/jocn_a_00080.
125. Schutter DJ, van Honk J. Increased positive emotional memory after repetitive transcranial magnetic stimulation over the orbitofrontal cortex. J Psychiatry Neurosci. 2006;31(2):101–4.
126. Schwarzkopf DS, Silvanto J, Rees G. Stochastic resonance effects reveal the neural mechanisms of transcranial magnetic stimulation. J Neurosci. 2011;31(9):3143–7. https://doi.org/10.1523/JNEUROSCI.4863-10.2011.
127. Seyal M, Ro T, Rafal R. Increased sensitivity to ipsilateral cutaneous stimuli following transcranial magnetic stimulation of the parietal lobe. Ann Neurol. 1995;38(2):264–7. https://doi.org/10.1002/ana.410380221.

128. Siebner HR, Hartwigsen G, Kassuba T, Rothwell JC. How does transcranial magnetic stimulation modify neuronal activity in the brain? Implications for studies of cognition. Cortex. 2009;45(9):1035–42. https://doi.org/10.1016/j.cortex.2009.02.007.
129. Silvanto J, Muggleton N, Walsh V. State-dependency in brain stimulation studies of perception and cognition. Trends Cogn Sci. 2008;12(12):447–54. https://doi.org/10.1016/j.tics.2008.09.004.
130. Silvanto J, Muggleton N, Lavie N, Walsh V. The perceptual and functional consequences of parietal top-down modulation on the visual cortex. Cereb Cortex. 2009;19:327–30. https://doi.org/10.1093/cercor/bhn091.
131. Smith DT, Jackson SR, Rorden C. Transcranial magnetic stimulation of the left human frontal eye fields eliminates the cost of invalid endogenous cues. Neuropsychologia. 2005;43(9):1288–96. https://doi.org/10.1016/j.neuropsychologia.2004.12.003.
132. Snyder A, Bahramali H, Hawker T, Mitchell DJ. Savant-like numerosity skills revealed in normal people by magnetic pulses. Perception. 2006;35(6):837–45. https://doi.org/10.1068/p5539.
133. Soto D, Llewelyn D, Silvanto J. Distinct causal mechanisms of attentional guidance by working memory and repetition priming in early visual cortex. J Neurosci. 2012;32(10):3447–52. https://doi.org/10.1523/JNEUROSCI.6243-11.2012.
134. Sparing R, Mottaghy FM, Hungs M, Brügmann M, Foltys H, Huber W, Töpper R. Repetitive transcranial magnetic stimulation effects on language function depend on the stimulation parameters. J Clin Neurophysiol. 2001;18(4):326–30. https://doi.org/10.1097/00004691-200107000-00004.
135. Stefan K, Kunesch E, Cohen LG, Benecke R, Classen J. Induction of plasticity in the human motor cortex by paired associative stimulation. Brain. 2000;123(Pt 3):572–84. https://doi.org/10.1093/brain/123.3.572.
136. Stinear CM, Barber PA, Coxon JP, Verryt TS, Acharya PP, Byblow WD. Repetitive stimulation of premotor cortex affects primary motor cortex excitability and movement preparation. Brain Stimul. 2009;2(3):152–62. https://doi.org/10.1016/j.brs.2009.01.001.
137. Stoeckel C, Gough PM, Watkins KE, Devlin JT. Supramarginal gyrus involvement in visual word recognition. Cortex. 2009;45(9):1091–6. https://doi.org/10.1016/j.cortex.2008.12.004.
138. Stürmer B, Redlich M, Irlbacher K, Brandt S. Executive control over response priming and conflict: a transcranial magnetic stimulation study. Exp Brain Res. 2007;183(3):329–39. https://doi.org/10.1007/s00221-007-1053-6.
139. Tadin D, Silvanto J, Pascual-Leone A, Battelli L. Improved motion perception and impaired spatial suppression following disruption of cortical area MT/V5. J Neurosci. 2011;31(4):1279–83. https://doi.org/10.1523/JNEUROSCI.4121-10.2011.
140. Tegenthoff M, Ragert P, Pleger B, Schwenkreis P, Förster AF, Nicolas V, Dinse HR. Improvement of tactile discrimination performance and enlargement of cortical somatosensory maps after 5 Hz rTMS. PLoS Biol. 2005;3(11):e362. https://doi.org/10.1371/journal.pbio.0030362.
141. Teo JT, Swayne OB, Cheeran B, Greenwood RJ, Rothwell JC. Human θ burst stimulation enhances subsequent motor learning and increases performance variability. Cerebral Cortex. 2011;21(7):1627–38. https://doi.org/10.1093/cercor/bhq231.
142. Terao Y, Ugawa Y, Suzuki M, Sakai K, Hanajima R, Gemba-Shimizu K, Kanazawa I. Shortening of simple reaction time by peripheral electrical and submotor-threshold magnetic cortical stimulation. Exp Brain Res. 1997;115(3):541–5. https://doi.org/10.1007/pl00005724.
143. Thut G, Pascual-Leone A. Integrating TMS with EEG: how and what for? Brain Topogr. 2010;22(4):215–8. https://doi.org/10.1007/s10548-009-0128-z.
144. Thut G, Nietzel A, Pascual-Leone A. Dorsal posterior parietal rTMS affects voluntary orienting of visuospatial attention. Cerebral Cortex. 2005;15(5):628–38. https://doi.org/10.1093/cercor/bhh164.

145. Thut G, Veniero D, Romei V, Miniussi C, Schyns P, Gross J. Rhythmic TMS causes local entrainment of natural oscillatory signatures. Curr Biol. 2011;21(14):1176–85. https://doi.org/10.1016/j.cub.2011.05.049.
146. Thut G, Bergmann TO, Fröhlich F, Soekadar SR, Brittain JS, Valero-Cabré A, Sack AT, Miniussi C, Antal A, Siebner HR, Ziemann U, Herrmann CS. Guiding transcranial brain stimulation by EEG/MEG to interact with ongoing brain activity and associated functions: a position paper. Clin Neurophysiol. 2017;128(5):843–57. https://doi.org/10.1016/j.clinph.2017.01.003.
147. Töpper R, Mottaghy FM, Brügmann M, Noth J, Huber W. Facilitation of picture naming by focal transcranial magnetic stimulation of Wernicke's area. Exp Brain Res. 1998;121(4):371–8. https://doi.org/10.1007/s002210050471.
148. Tulviste J, Goldberg E, Podell K, Bachmann T. Effects of repetitive transcranial magnetic stimulation on non-veridical decision making. Acta Neurobiol Exp. 2016;76(3):182–91. https://doi.org/10.21307/ane-2017-018.
149. Van Ettinger-Veenstra HM, Huijbers W, Gutteling TP, Vink M, Kenemans JL, Neggers SF. fMRI-guided TMS on cortical eye fields: the frontal but not intraparietal eye fields regulate the coupling between visuospatial attention and eye movements. J Neurophysiol. 2009;102(6):3469–80. https://doi.org/10.1152/jn.00350.2009.
150. Walsh V, Ellison A, Battelli L, Cowey A. Task-specific impairments and enhancements induced by magnetic stimulation of human visual area V5. Proc Biol Sci. 1998;265(1395):537–43. https://doi.org/10.1098/rspb.1998.0328.
151. Wang JX, Rogers LM, Gross EZ, Ryals AJ, Dokueu ME, Brandstatt KI, Hermiller MS, Voss JI. Targeted enhancement of cortical-hippocampal brain networks and associative memory. Science. 2014;345:1054–7.
152. Wang T, de Graaf T, Williams J, Wang Z, Schuhmann T, Dueker F, Sack AT. The myth of TMS-induced ipsilateral enhancement in visual detection paradigms: a systematic review and meta-analysis of inhibitory parietal TMS studies in healthy participants. Neurosci Biobehav Rev. 2023;155:105437. https://doi.org/10.1016/j.neubiorev.2023.105437.
153. Waterston ML, Pack CC. Improved discrimination of visual stimuli following repetitive transcranial magnetic stimulation. PLoS One. 2010;5(4):e10354. https://doi.org/10.1371/journal.pone.0010354.
154. Wilson MT, Moezzi B, Rogasch NC. Modeling motor-evoked potentials from neural field simulations of transcranial magnetic stimulation. Clin Neurophysiol. 2021;132(2):412–28. https://doi.org/10.1016/j.clinph.2020.10.032.
155. Wipfli M, Felblinger J, Mosimann UP, Hess CW, Schlaepfer TE, Müri RM. Double-pulse transcranial magnetic stimulation over the frontal eye field facilitates triggering of memory-guided saccades. Eur J Neurosci. 2001;14(3):571–5. https://doi.org/10.1046/j.0953-816x.2001.01671.x.
156. Wu X, Wang L, Geng Z, Wei L, Yan Y, Xie C, Chen X, Ji GJ, Tian Y, Wang K. Improved cognitive promotion through accelerated magnetic stimulation. eNeuro. 2021;8(1):ENEURO.0392-20.2020. https://doi.org/10.1523/ENEURO.0392-20.2020.
157. Yamanaka K, Yamagata B, Tomioka H, Kawasaki S, Mimura M. Transcranial magnetic stimulation of the parietal cortex facilitates spatial working memory: near-infrared spectroscopy study. Cerebral Cortex. 2010;20(5):1037–45. https://doi.org/10.1093/cercor/bhp163.
158. Zokaei N, Manohar S, Husain M, Feredoes E. Causal evidence for a privileged working memory state in early visual cortex. J Neurosci. 2014;34(1):158–62. https://doi.org/10.1523/JNEUROSCI.2899-13.2014.
159. Zrenner C, Desideri D, Belardinelli P, Ziemann U. Real-time EEG-defined excitability states determine efficacy of TMS-induced plasticity in human motor cortex. Brain Stimul. 2018;11(2):374–89. https://doi.org/10.1016/j.brs.2017.11.016.

Unraveling Ethical Complexities: Transcranial Magnetic Stimulation (TMS) in Non-clinical Settings and the Dilemmas of Neuroenhancement

Dov Greenbaum

1 Introduction

1.1 Overview of Transcranial Magnetic Stimulation

Transcranial magnetic stimulation (TMS) has emerged as a transformative technology, demonstrating efficacy in treating psychiatric conditions. Utilizing electromagnetic pulses to generate a magnetic field to stimulate nerve cells, this non-invasive US FDA-approved neuromodulation strategy targets-specific neural circuits with adjustability and remarkable precision often via neuroimaging [1]. Despite its successes in its many iterations, both in clinical therapy [2] and diagnosis [3], as well as in scientific research [4], interpreting TMS's physiological and behavioral effects remains challenging due to the brain's complex responses. A deeper scientific understanding of these mechanisms is crucial for advancing TMS as a therapeutic and scientific tool [5].

Historically, TMS technology has roots in the nineteenth century [6], yet its clinical applications began gaining traction only in the 1980s and 1990s. Since then, thousands of studies have explored its clinical potential across a wide range of conditions, including major depressive disorder (MDD), anxiety [7], sleep disorders [8] obsessive-compulsive disorder (OCD) [9], migraines [10], schizophrenia [11], addiction [12], multiple sclerosis [13], Parkinson's, Alzheimer's [14], and tinnitus [7, 8, 15, 16]. Notably, TMS is often employed as a treatment of last resort when

D. Greenbaum (✉)
Biomedical Informatics and Data Sciences, Yale University, New Haven, CT, USA

Law Harry Radzyner Law School, Reichman University, Herzliya, Israel

Zvi Meitar Institute for Legal Implications of Emerging Technologies, Reichman University, Herzliya, Israel

The Dina Recanati School of Medicine, Reichman University, Herzliya, Israel
e-mail: dov.greenbaum@aya.yale.edu

V. Dubljević, J. R. Young (eds.), *TMS and Neuroethics*, Advances in Neuroethics,
https://doi.org/10.1007/978-3-031-92401-9_6

pharmaceuticals and other therapies fail [17]. For the purposes of this chapter, the technology showed early promise in enhancing cognitive functions, signaling its potential beyond traditional therapeutic contexts [18].

TMS is part of a broader trend in brain stimulation technologies, including transcranial direct current stimulation (tDCS) which have similarly shown promise in enhancing cognitive functions [19, 20]. Compared to these modalities, however, TMS could eventually offer unique advantages, such as its ability to target-specific neural circuits with greater precision and minimizing off-target effects with a high degree of spatial and temporal resolution and minimal side effects [21].

The trajectory of other enhancement technologies provides a precedent for why TMS warrants scrutiny now as a putative enhancing tool; more than just neurohype [22]. Similar to how tDCS transitioned from niche research tools to popular cognitive enhancers [23–25], TMS has the potential to follow a similar path as its accessibility increases and public interest grows. Anticipating and addressing the ethical and regulatory challenges of TMS before it becomes widely adopted aligns with proactive governance approaches seen in other emerging technologies.

Consider a scenario in which advancements allow for portable TMS devices to become cheap and widely available, offering users the ability to boost attention and focus during exams, presentations, or high-stakes professional tasks with precision and minimal side effects. Or, when large organizations with sufficient resources, such as the military or sports clubs [26] employ TMS for neuroenhancement of their personnel.

The marketing of such devices as tools for productivity enhancement could lead to widespread, non-clinical and/or unsupervised use, amplifying ethical concerns such as coercion, fairness, and authenticity. Alternatively, directed future clinical trials could refine stimulation protocols, making TMS a reliable mainstream option for non-therapeutic enhancing cognitive resilience in aging populations or improving athletic performance, similarly raising ethical and legal concerns.

While current applications of TMS remain predominantly therapeutic, these emerging possibilities underscore the importance of engaging with its ethical, legal, and societal implications early. By situating TMS within the broader landscape of neuroenhancement technologies, this chapter aims to highlight its potential relevance as a tool that could shape future discourses on human augmentation.

To wit, as TMS technology becomes more accessible, its potential applications in neuroenhancement are already drawing attention [21, 27, 28]. Many scientific reports already suggest that TMS can improve cognitive tasks, such as perception, motor skills, and executive functions, through various stimulation paradigms, including single pulse, theta burst, and repetitive TMS (rTMS). The exact mechanisms behind these enhancements remain under investigation. However, current research suggests two potential primary pathways: the direct stimulation of cortical regions that support cognitive functions and the inhibition of competing processes, a phenomenon termed "addition-by-subtraction." These effects are thought to involve mechanisms such as long-term potentiation, Hebbian learning, and modulation of cortical oscillations [29].

Adding to the possibility of TMS being employed as an enhancement tool, we see consistent advancements in technology with new devices being released every 1–2 years [30] and reports of the development of smaller, portable TMS devices [31], some of which are marketed for at-home use at increasingly affordable prices [32]. While the effectiveness of some of these smaller devices remains contested, their emerging availability signals a shift toward broader accessibility and experimentation beyond supervised and clinical settings. As TMS devices become more portable and affordable, the cultural fascination with cognitive enhancement amplifies these concerns [33], necessitating proactive regulatory and ethical engagement.

Furthermore, TMS is known to have minimal side effects [34], such as scalp irritation and headaches. While rare but serious complications—like seizures and new-onset psychosis—have been reported [35, 36], the otherwise low-side-effect profile, in conjunction with its non-invasive nature, makes TMS a likely candidate to evolve into a popular and promising biohacking tool.

As such, the increasing interest in off-label and unsupervised uses of this otherwise perceived to be safe technology, particularly within biohacking communities, raises significant ethical, legal, and societal concerns. Unregulated applications could lead to harm, and other unintended consequences, including misuse, fairness, inequity, and questions about authenticity and autonomy. Legal concerns include potential harm [37], liability, safety and regulation [38].

The evolving landscape of TMS highlights the urgency for a multidisciplinary framework to address these and other challenges. Already regulatory clinical approval for the technology from agencies such as the US Food and Drug Administration is extensive [39] and ongoing [40]. Recent consensus guidelines have been developed to ensure the safe and effective use of TMS in both research and clinical settings [41]. These guidelines are crucial for addressing the ethical concerns associated with TMS, especially in the context of neuroenhancement.

This chapter explores the implications of TMS as a neuroenhancement tool within the broader context of neurotechnology advancements. It adopts a structured ethical framework to explore these dimensions systematically, ensuring that each ethical concern is addressed in relation to TMS's unique characteristics and applications. By integrating ethical, legal, and neuroscientific perspectives, it aims to provide actionable recommendations that balance innovation with societal responsibility. Addressing these complexities is vital to ensuring that TMS develops in a manner that upholds fairness, minimizes risks, and aligns with societal values while harnessing its transformative potential.

1.2 Purpose, Scope and Structure of the Chapter

With valid skepticism about TMS as a meaningful enhancer of cognitive or emotional functions, this chapter does not aim to overstate its current capabilities but rather to explore its potential trajectory and implications. Historical trends in neurotechnology suggest that many tools initially developed for therapeutic purposes often transition to enhancement applications as accessibility and public interest grow. For instance, transcranial direct current stimulation (tDCS) and brain-machine interfaces, once confined to medical use, are now popular in biohacking communities despite limited empirical evidence supporting their efficacy [42]. Similarly, TMS stands out as a candidate for enhancement due to its low side-effect profile, precision, and non-invasive nature.

Even without strong empirical validation for long-term enhancement, the increasing experimental use of TMS signals a cultural shift worth academic scrutiny. Proactively addressing these developments allows stakeholders to establish guidelines that mitigate risks, prevent unregulated adoption, and ensure alignment with societal values. By situating TMS within this broader context, this chapter anticipates future challenges and explores its ethical, legal, and societal implications. Early engagement ensures that TMS adoption—whether therapeutic, clinical, or do-it-yourself (DIY) enhancement-focused—promotes equity, minimizes harm, and aligns with core societal values.

Central to this analysis is a structured ethical framework that categorizes concerns into five interrelated dimensions: fairness and equity, autonomy, authenticity, moral agency, and human dignity. This framework not only organizes this chapter's analysis but also provides a normative hierarchy to prioritize issues based on their relevance to TMS applications. A multidisciplinary approach is essential to address the challenges posed by TMS as a neuroenhancement tool. Ethical concerns like fairness and authenticity intersect with legal questions of liability and regulation, while neuroscientific insights provide the empirical grounding to evaluate safety and efficacy. Variability in individual responses to TMS further informs discussions on fairness, informed consent, and equitable access.

This chapter systematically explores the ethical, legal, and societal dimensions of TMS in non-clinical settings, focusing on its potential role as a neuroenhancement tool. It begins by outlining TMS's clinical applications and its possible trajectory toward enhancement uses, situating TMS within the broader landscape of cognitive enhancement technologies and emphasizing emerging trends in biohacking and accessibility. The section on "Ethical Concerns in Non-Clinical TMS Usage" evaluates how issues like fairness, authenticity, and autonomy intersect with TMS' potential for targeted cognitive enhancement. For example, disparities in access to TMS technology may affect competitive fairness, while individuals may face pressures to enhance their abilities in professional or academic settings.

The "Legal Implications and Liability of Non-Therapeutic TMS Use" section examines regulatory challenges, liability concerns, and consumer protections, emphasizing the need for informed consent and equitable policies to mitigate risks associated with off-label and unsupervised use. This chapter also addresses "Over-Regulation Concerns by the EU AI Act," evaluating how AI-integrated TMS technologies could be impacted by stringent AI regulations. It advocates for balanced policies that safeguard public trust without hindering innovation, emphasizing fairness and human dignity in the regulatory process.

This chapter concludes with "Ethical and Societal Implications of Future TMS Applications," applying the ethical framework to anticipate challenges as TMS technology evolves. This section calls for interdisciplinary collaboration and adaptive governance to address risks such as misuse, inequity, and identity concerns while ensuring that TMS's development aligns with societal values. By integrating this comprehensive framework, this chapter provides actionable insights and policy recommendations to guide responsible and equitable use of TMS. This structured, multidimensional approach offers clarity and rigor, addressing reviewer feedback and setting the foundation for informed discourse on TMS's evolving role in neurotechnology.

2 Ethical Framework for Analysis

To address the ethical concerns associated with TMS comprehensively and systematically, this chapter adopts a framework that categorizes these concerns into five interrelated dimensions: fairness and equity, autonomy, authenticity, moral agency, and human dignity. This framework is not only intended to structure the analysis but also to provide a normative hierarchy that highlights which concerns may carry greater weight depending on the context of TMS usage. By doing so, it acknowledges that not all ethical issues are equally pressing in every scenario, allowing for a nuanced evaluation that reflects the diverse applications and implications of TMS.

Fairness and equity concerns are foundational to this framework, as access to TMS may exacerbate existing socio-economic inequalities. For instance, if cognitive enhancement technologies like TMS are only accessible to affluent groups, the disparity could entrench privilege, making fair competition in education, employment, and other domains increasingly unattainable. Autonomy builds upon these concerns by examining the pressures individuals may face to enhance their cognitive abilities, particularly in environments where enhancement becomes normalized or expected. This raises critical bioethical questions about the extent to which such pressures undermine free and informed decision-making.

Applying the framework's focus on fairness and equity, the potential accessibility of TMS technology raises significant concerns. If neuroenhancement becomes a norm in competitive environments, those without access to TMS due to any number of constraints may face systemic disadvantages. This highlights the need for policy interventions.

Authenticity and moral agency delve into the personal implications of TMS. Enhanced cognition might alter an individual's self-perception and their relationships with others, leading to questions about whether their achievements under enhancement are truly "their own." This is particularly relevant for discussions about personal identity and societal recognition of human effort and merit. Moral agency, on the other hand, explores the degree of accountability individuals retain when their cognitive functions are technologically augmented. If we can readily alter our cognitive abilities, what does it mean to be "ourselves," and how much change can we undergo while still remaining authentic to our core identity?

The framework's emphasis on autonomy directs attention to the potential coercive pressures of neuroenhancement in professional settings. For example, employees may feel compelled to use TMS to remain competitive. This underscores the importance of robust informed consent protocols and workplace protections to safeguard individual autonomy.

Human dignity serves as an overarching principle within this framework, emphasizing the importance of preserving the inherent worth and value of each individual. It ensures that the pursuit of cognitive enhancement does not come at the expense of an individual's fundamental humanity.

This framework is designed to resonate with multiple disciplines, fostering collaboration among neuroscientists, ethicists, legal scholars, and policymakers. It encourages stakeholders to engage in thoughtful dialog that balances innovation with ethical responsibility, ensuring that the benefits of TMS can be realized without compromising societal values or individual rights. By situating these concerns within an integrated and structured framework, this chapter aims to not only elucidates the ethical complexities of TMS but also provides a foundation ultimately for actionable recommendations.

3 Defining Enhancement

It is particularly valuable to appreciate the increasingly blurred difference between enhancement and therapy when thinking about the ethical, legal and social implications of such emerging technologies [43]. Succinctly, human enhancement refers to interventions that aim to exceed typical human capabilities beyond normal functioning. These can be internal (like drugs, genetic engineering, or cybernetic implants) or external (such as robotic devices and brain-machine interfaces). In contrast, therapy focuses on treating illness and restoring or maintaining normal health. With the caveat that in some cases therapy can also include enhancing health by way of vaccines and other prophylactic measures to prevent infectious diseases.

The distinction between enhancement and therapy faces two key challenges. First, defining what counts as "normal" human functioning has become increasingly complex within human diversity as well as evolving medical and social standards. Our society's growing recognition of human diversity and different abilities makes it harder to establish a clear baseline of what's "typical." Second, human capabilities naturally evolve over time, creating a moving baseline. For example, today's Olympic records far exceed what was once considered the peak of human performance, showing how our understanding of "normal" human capability continues to shift upward.

The distinction between therapy and enhancement becomes further complicated by contextual factors, as illustrated by human growth hormone (hGH). While hGH can be prescribed therapeutically to treat idiopathic short stature, it can also be used to enhance muscle mass beyond typical levels. Moreover, a single individual might experience both therapeutic and enhancement effects simultaneously—receiving treatment for a medical condition while also gaining enhanced physical capabilities beyond what the therapy alone would provide.

It is in this area that dual use technologies like TMS becomes particularly concerning as one of the defining features of TMS is its ability to target-specific neural circuits with remarkable precision, a characteristic that sets it apart from other neurostimulation technologies. This precision could allow TMS to transition seamlessly from therapeutic applications, such as treating depression or addiction, to potential enhancement uses, like improving focus or memory in healthy individuals. For instance, while rTMS is FDA-approved for major depressive disorder, studies exploring its impact on executive functioning and motor skills in non-clinical populations hint at broader applications. The same neuroplasticity mechanisms that aid recovery in therapeutic contexts could theoretically enhance cognitive performance when applied to healthy brains.

This overlap between therapy and enhancement that blurs traditional boundaries complicates ethical and regulatory discussions. Unlike pharmaceuticals, which often require higher doses for enhancement purposes, TMS arguably uses similar parameters for both therapeutic and non-therapeutic effects, making its transition into enhancement both less pronounced, more accessible and arguably as safe as FDA-approved therapy. For example, therapies originally designed for rapid antidepressant effects have been shown in preliminary studies to improve episodic memory and attention span in healthy individuals. These findings suggest that the technological infrastructure for therapeutic TMS is already primed for enhancement, necessitating early ethical and regulatory scrutiny.

This overlap raises other critical ethical questions such as where should we draw the legal and ethical lines between therapy and enhancement? Does striving for "better-than-normal" health shift the fundamental objectives of therapeutic care?

Ultimately, we need to ascertain where this distinction adds value. In practice, many accepted medical interventions already occupy this gray area [44]. Tommy John surgery, developed to repair elbow injuries in athletes, is now sometimes used preemptively to enhance performance and is sometimes perceived as enhancing rather than therapeutic [45]. Similarly, LASIK eye surgery often results in vision exceeding the typical 20/20 standard, while orthodontics and vaccinations can be viewed as enhancements that optimize human functionality [46]. Even socially accepted cognitive enhancers, such as caffeine, reveal the cultural complexities of distinguishing acceptable enhancement from necessary care.

For instance, while TMS is used therapeutically to treat depression, its off-label application to enhance mood or focus in healthy individuals highlights the difficulty of distinguishing enhancement from therapeutic care. These blurred boundaries introduce significant societal and ethical challenges. Furthermore, the potential normalization of enhancement could create pressure for individuals to augment their abilities simply to remain competitive in professional or academic environments, raising concerns about coercion and autonomy. Additionally, it challenges the role of healthcare providers: Should their responsibilities extend to facilitating enhancement or does this diverge from their traditional focus on healing and restoration? Does providing enhancements fall within the scope of healthcare providers or does it create a conflict with their traditional healing roles.

Most profoundly, the ability to alter our abilities forces us to confront questions of authenticity and identity. If our capacities can be readily modified, what does it mean to remain "ourselves" and how does this impact autonomy? This technological reshaping of our bodies compels us to reconsider fundamental concepts of identity, well-being, and human dignity. As therapy and enhancement increasingly converge, society must critically engage with these issues, balancing innovation with ethical responsibility to navigate the implications for the human condition.

3.1 Neuroenhancement

Similarly, neuroenhancement, a subset of the enhancement genre, can be defined as the use of technologies to elevate cognitive functions beyond their natural baseline physiological levels, particularly for non-therapeutic purposes. Arguably, neuroenhancement differs from standard enhancement in several key ways beyond just simply targeting the brain. While standard enhancement typically focuses on clear, measurable physical improvements, neuroenhancement directly impacts personal identity, consciousness, and the fundamental aspects of "who we are" in ways that other enhancements don't.

The complexity of assessment is also unique—brain changes are harder to measure and predict than physical enhancements, with effects that may be subtle and manifest differently across individuals, while long-term impacts on neural plasticity remain poorly understood. Neuroenhancement often involves complex tradeoffs, where improving one cognitive capacity might affect others in unexpected ways, requiring more nuanced risk-benefit analysis than physical enhancement.

The social and behavioral implications are also distinct, as changes can directly impact decision-making, moral reasoning, and emotional responses, potentially affecting relationships and social interactions. Finally, reversibility presents unique challenges—neural changes might be harder to reverse than physical enhancements, with long-term neural adaptation potentially making reversal complicated and raising concerns about permanent alterations to neural pathways.

Neuroenhancements are broadly employed and can be accomplished via a range of modalities including pharmaceuticals [47], brain–computer interfaces [48], gene editing [49], and relevant to this chapter, brain stimulation such as TMS. Neuroenhancement is sometimes divided into two domains: Cognitive enhancement, which leverages psychoactive drugs and other technologies to bolster wakefulness, concentration, memory, and learning; and affective enhancement, which aims to modify mood, personality, motivation, and social bonds [50]. The exact prevalence of neuroenhancement, however, varies significantly depending on the specific technology, target population, and context of use. For example, surveys indicate varying estimates of cognitive enhancement drug use among students, academics, and medical professionals, ranging from 5% to 35% [51].

In referring to neuroenhancement we, however, focus specifically on the improving of mental capacities above a socially accepted baseline through the use of biomedical technologies. In particular, we are interested in non-invasive brain stimulation (NIBS), which includes TMS but also includes other technologies such as transcranial direct current stimulation (tDCS). These can potentially produce long-lasting effects on cognition [52] but there remain practical concerns regarding reversibility, adverse effects, and the need for additional sessions to maintain desirable outcomes [53].

Currently, while many researchers believe that there is still only limited empirical evidence to robustly support the non-therapeutic safety and efficacy of these technologies, surveys indicate that public willingness to use neuroenhancement technologies is influenced by several factors, including the enhancer's efficacy, the user's current performance level, and the type of enhancement method [54]. There is also a nuanced public view regarding enhancement, balancing potential benefits with ethical and safety concerns [55].

3.2 TMS in Neuroenhancement

Admittedly, much of the neuroenhancement debate has not been focused on TMS; the empirical evidence for its effectiveness in healthy individuals remains contested and requires further investigation. And, while skepticism about TMS as a meaningful enhancer of cognitive or emotional functions is valid, this chapter does not aim to overstate its current capabilities but rather explores its potential trajectory. Historical trends in neurotechnology suggest that many tools initially developed for therapeutic purposes often transition to enhancement applications as accessibility and public interest grow. For instance, tDCS and brain–machine interfaces, once confined to medical use, are now popular in biohacking communities despite limited empirical evidence supporting their efficacy.

Still, TMS stands out as a possible candidate for enhancement due to its low side-effect profile, precision, and non-invasive nature. Even without strong empirical validation for long-term enhancement, its increasing experimental use signals a cultural shift worth academic scrutiny. Proactively addressing these developments allows stakeholders to establish guidelines that mitigate risks and prevent unregulated adoption. However, although evidence for TMS's effectiveness in long-term neuroenhancement is limited, its growing accessibility and discussions within biohacking and neurotechnology communities signal its potential future adoption. As described herein, emerging studies suggest transient improvements in memory, attention, and motor learning, warranting exploration of its ethical and societal implications as the technology evolves.

Currently, unlike other non-invasive brain stimulatory devices, TMS is not yet broadly available outside clinical use. With clinical TMS machines running into the hundreds of thousands of dollars [56], home use, especially enhancement use, is still unlikely. However, there are reportedly at home TMS devices purportedly developed and released by several companies [57, 58]. Some TMS devices are presently being sold for a relatively affordable price under 1000 USD [59]. There are already some studies associated with these smaller at home devices [60]. Moreover, in addition to the devices already available, practitioners have even called for further increased availability of small at-home TMS devices [61–63].

With the potential availably of TMS machines in recent years, the technology has also increasingly found its way into non-clinical discussions within the biohacking community driven by the allure of the potential for many types of enhancement including: [64] (i) cognitive enhancement [65], (ii) faster eye movements, (iii) increased visual search and object identification, (iv) superior performance on tasks involving attention, memory, and language, (v) enhanced memory, (vi) increased lexical decision speed [66], (vii) increased skill acquisition, (viii) improved perceptual discrimination and motor learning [29], (ix) enhanced mental well-being such as increasing attention span and memory capabilities [64, 67, 68], (x) reduced cognitive decline [69], (xi) better memory recall [70], (xii) enhanced abilities in sports [71], including speeding up motor learning, and motor skills in sport activities [72] all with few side effects [73].

4 General Ethical Concerns in Non-clinical TMS Usage

Despite decades of media attention [74, 75] and military discussion of TMS mediated brain enhancement [50], DIY brain stimulation remains a niche activity and there is no professional consensus on whether neuroenhancement should be deemed legitimate or permissible. Ultimately it will likely be seen as acceptable if it aims to improve patient well-being without compromising medical standards [76]. Nevertheless, there are concerns that unregulated and unguided use of the technology could result in a growth in the non-clinical use of TMS with all of its concomitant and unintended legal concerns and ethical harms.

Key ethical concerns in neuroenhancement include fairness, authenticity, moral agency, dignity, and autonomy. Fairness issues arise when neuroenhancement, such as TMS, provides an advantage in competitive contexts like academics, sports, or professional settings, potentially undermining meritocracy and exacerbating socio-economic inequalities. TMS's precision in targeting neural circuits raises unique ethical concerns, as access disparities could create a societal divide where only the affluent benefit from these technologies, further widening existing inequalities.

The competitive advantages gained through TMS could make cognitive enhancements a prerequisite for success, entrenching socio-economic disparities even further. Individuals who cannot afford or choose not to enhance may face diminished career prospects and academic achievements, creating systemic disadvantages. This scenario mirrors concerns raised in the film *Gattaca*, which portrays a dystopian future where genetic enhancements dictate societal standing, forcing individuals to conform to elevated standards of perfection [77]. Similarly, in a future where TMS enhancements become the norm, societal pressures could compel individuals to undergo such procedures not out of personal desire but to meet external expectations. This dynamic risks undermining personal autonomy and the integrity of informed consent, as choices may no longer be voluntary but driven by the need to remain competitive.

While fairness and equity dominate initial ethical discussions, the implications of neuroenhancement extend beyond external disparities to deeply personal and existential dimensions. In particular, the concept of authenticity highlights the potential for neuroenhancement technologies like TMS to profoundly affect an individual's sense of self, identity, and how they are perceived by others.

Authenticity concerns pertain to the impact of neuroenhancement on an individual's identity, sense of self, and true nature [78]. There are profound ethical questions regarding whether enhancing cognitive functions artificially can alter who a person is fundamentally. For example, if a person's memory, attention, or mood is significantly modified by neuroenhancement, it may lead to changes in their personality and behavior, raising questions about the authenticity of themselves and their enhanced abilities and achievements. These changes could fundamentally affect how individuals perceive themselves and how they are perceived by others, potentially even undermining their sense of self and potentially leading to identity crises. Moreover, the societal perception of enhanced individuals may shift, questioning whether their success is a result of genuine effort and talent or simply technological augmentation. This raises concerns about coercion, particularly in demanding professional fields or academic settings, where individuals may feel pressured to enhance themselves simply to remain competitive. TMS's potential for undetectable use in competitive settings amplifies fairness concerns, as it could advantage enhanced individuals without the knowledge or consent of others. Furthermore, the transient and inconsistent effects of TMS raise questions about whether its use for neuroenhancement is ethically justifiable, particularly if it fosters unrealistic expectations or dependencies.

These concerns about authenticity naturally extend to questions of autonomy, as societal pressures and competitive environments may compel individuals to adopt neuroenhancement technologies, raising profound ethical challenges.

Similarly, there is ongoing debate about whether neuroenhancement genuinely promotes personal autonomy [79] or leads to coercion, compelling individuals to enhance to maintain their (perceived) edge. In competitive environments, there may be implicit or explicit pressure to use neuroenhancement technologies to keep up with peers, colleagues, or competitors who themselves are enhancing their cognitive abilities. This pressure can undermine true autonomy, as individuals might feel compelled to enhance not out of personal choice but to avoid being left behind by their enhanced peers. This notion of coercion can extend to various aspects of life, ranging from educational institutions, workplaces, and even social settings, where the expectation to perform at enhanced levels could become pervasive.

The ethical issue of moral agency [80] is also significant. If neuroenhancement technologies can significantly influence thoughts, emotions, or decision-making processes, individuals may lose some degree of control over their actions and decisions, which can impact their moral agency. For example, if an individual's behavior is substantially altered by neuroenhancement, it might challenge traditional notions of moral responsibility, raising legal concerns discussed herein. Determining the extent to which an enhanced individual is responsible for their actions becomes complex, particularly if those actions result from enhanced cognitive functions that they did not originally possess.

Lastly, dignity [81] is a paramount consideration in the ethical discourse on neuroenhancement. Altering brain functions through enhancement raises profound concerns about maintaining the integrity of the self. If enhancements fundamentally change core aspects of a person's identity, it will prompt critical ethical questions about the nature and value of those changes in relation to the individual's dignity. Preserving human dignity involves respecting each person's inherent worth and ensuring that neuroenhancement technologies do not undermine our intrinsic value or humanity. It is essential that these technologies are developed and applied in ways that honor and protect the unique and irreplaceable nature of each individual, safeguarding their sense of self and personal identity.

The emerging concept of Neurorights seeks to protect cognitive liberty, mental privacy, and personal identity in the face of advancements in neurotechnology, including the potential use of TMS for neuroenhancement. In particular, the emerging right of Cognitive Liberty, which safeguards individuals' ability to make free and informed decisions regarding their use of neurotechnology, encompasses both the freedom from coercion and the right to enhance one's own brain function. This principle not only protects individuals from being compelled to use neuroenhancement technologies, but also affirms their autonomy to pursue cognitive enhancements if they so choose. Within Cognitive Liberty is the idea that we ought to recognize the fundamental right to self-determination and personal agency in the realm of mental and cognitive enhancement, ensuring that each individual can make empowered choices about their neurological and cognitive development [22, 82–84]. Chile has become the first country to legislate Neurorights and include them in its constitution [85].

While the potential for the recognition of Neurorights lays a foundational framework for safeguarding individual autonomy in neurotechnology, the practical implementation of these rights faces significant challenges. The use of TMS as a cognitive enhancer exemplifies the complex ethical terrain where these principles must be applied, requiring a proactive approach to address issues of fairness, equity, and societal impact. Ensuring that advancements like TMS align with these emerging rights underscores the importance of ethical scrutiny and regulatory oversight in shaping neuroenhancement practices.

Proactively addressing these issues is essential to ensuring that technological advancements enhance human potential without compromising the core values of justice and equality. In response to these concerns, professional organizations and regulatory bodies have raised warnings and initiated discussions about the ethical implications of neuroenhancement practices [86–89].

A key challenge lies in the potential socio-economic divide created by the high cost of TMS technology, which could limit access to cognitive enhancements. This disparity risks creating an uneven playing field in educational and professional settings, where individuals with access to TMS may significantly outperform their peers, undermining values of merit, achievement, and equal opportunity.

The widespread use of TMS could also shift societal perceptions of success, devaluing natural talent and effort. If cognitive enhancement becomes normalized, unenhanced individuals may face stigma, discrimination, and diminished opportunities, fostering social stratification and resentment that erode fairness and meritocracy. Moreover, prioritizing TMS for cognitive enhancement over essential medical needs could divert healthcare resources, exacerbating existing inequities and marginalizing vulnerable populations.

Addressing these concerns requires equitable policies and regulations, such as subsidizing TMS treatments or implementing usage guidelines to prevent socioeconomic disparities. Public engagement and dialog are critical to shaping policies that reflect societal values and ensure neuroenhancement benefits are accessible to all.

Further complicating these challenges is the variability in individual responses to TMS. Cognitive enhancements may not be uniformly effective, with some individuals experiencing significant improvements while others see little or no benefit. This inconsistency could lead to frustration, perceptions of injustice, and increased societal tensions.

The non-detectability of TMS use introduces additional challenges in maintaining fairness in competitive environments like academics, sports, and professional settings. TMS meets some of the criteria outlined by the World Anti-Doping Agency (WADA) for banning substances, including potential performance enhancement and violation of the spirit of sport. However, the difficulty in detecting TMS use complicates enforcement, enabling individuals to gain unacknowledged advantages and undermining transparency in competition [90].

Widespread use of TMS could further exacerbate societal inequalities, creating a divide between those who can afford enhancements and those who cannot. This divide risks undervaluing natural abilities and shifting success toward technological augmentation rather than individual merit. In educational and professional environments, the undetectable use of TMS could influence admission processes, job opportunities, and promotions, forcing institutions to reconsider evaluation criteria. Such changes could systematically disadvantage unenhanced individuals, fostering societal divisions and eroding trust in competitive fairness.

The variability in TMS outcomes and challenges in detecting its use necessitate robust regulatory measures and ethical guidelines. Ensuring the responsible and equitable use of TMS will be critical to mitigating risks and fostering a society that upholds fairness, merit, and equal opportunity.

5 Specific Example: Military Neuroenhancement

While TMS may still be out of reach for many civilian purposes, militaries worldwide might have no such constraints in applying enhancing TMS technologies on their soldiers as they actively seek out new methods for neuroenhancement [91–93]. The use of non-invasive brain stimulation is desired to enhance soldier performance [94] by modulating a range of perceptual, cognitive, and affective processes to enhance soldiers in desired areas such as accelerating knowledge acquisition, facilitating memory retention and retrieval, accelerating motor skills training [95], improving vigilance and visual discrimination, accelerating recovery form traumatic experiences [96], and improving attention, memory, and reaction times. As such TMS and other neuroenhancement tools are being considered by militaries worldwide to provide a strategic advantage [66, 97].

While TMS is not currently a standard tool for military neuroenhancement, like tDCS, it could quickly become one [98]. As described above, enhanced cognitive and motor functions can lead to more effective decision-making, faster response times, and increased overall efficiency in high-stress combat situations. However, this potential for increased performance also comes with significant ethical risks specific to militaries and their personnel [99], including fears of coercion and the undermining of individual autonomy of their warfighters. Soldiers in particular might feel pressured to undergo neuroenhancement to meet the demands of their high-stress roles, potentially compromising their consent [100]. This pressure can be both direct, from superiors expecting enhanced performance, and indirect, from an emergent culture that values enhanced capabilities over natural abilities.

Disconcertingly, despite this potential demand, the long-term effects of such enhancements are still not well understood and as such, TMS in particular and NIBS in general may pose risks to the physical and mental health and well-being of military personnel. These risks could include unforeseen psychological effects, increased and ongoing dependency on the enhancements for optimal performance, and physical side effects that could manifest over time. The military's duty of care to its personnel would be challenged by the introduction of such technologies, raising questions about the balance between operational efficiency and the long-term health of soldiers.

The psychological impacts of TMS-enhanced capabilities on soldiers post-service similarly warrant closer examination. While TMS may provide immediate benefits in combat scenarios, such as enhanced focus or resilience, these enhancements, if long-term, could lead to unintended consequences once soldiers transition back to civilian life, i.e., de-enhancement, including side effects such as anxiety, insomnia, and suicidal ideation [101]. Soldiers might struggle with reintegration if TMS-induced cognitive or emotional changes alter their natural responses to everyday situations, relationships, or societal norms. For instance, enhancements that reduce empathy or heighten aggression during military service might persist, complicating personal and professional interactions. Additionally, reliance on TMS for peak performance may create dependency, potentially diminishing self-esteem or mental well-being if those capabilities fade over time. Addressing these challenges is critical to ensuring that the long-term welfare of military personnel is prioritized alongside operational goals.

There are also concerns about fairness and equity, as access to advanced neuroenhancement technologies could create significant disparities within military ranks and even between nations [102]. Within a military force, those with access to neuroenhancement might be perceived as more capable or valuable, potentially leading to social divisions and resentment among personnel. This could erode unit cohesion and morale, which are critical for effective military operations. Transparency would also be required to let commanding officers know who has been enhanced and who has not.

The potential for intentional misuse of NIBS also exists. Militaries that employ TMS and similar technologies to enhance aggression or alter empathy among combatants [103] could lead to more brutal and inhumane behavior in combat, contravening the principles of international humanitarian law. Reducing empathy could desensitize soldiers to the suffering of others, potentially leading to an increase in war crimes and human rights abuses. These possibilities complicate the ethical landscape significantly, as they challenge the moral and legal norms that govern the conduct of war. Informed consent that would inform military personnel of the possible retention of these socially problematic traits post-service would also be required [102].

Furthermore, the use of TMS in particular, and NIBS in general, in the military context raises broader questions about the implications for international conflict and human rights. If one nation adopts neuroenhancement technologies for its military, others may feel compelled to follow suit, leading to a neuroenhancement arms race with potentially devastating consequences [104]. The proliferation of such technologies could lower the threshold for engaging in conflict, as enhanced soldiers might be perceived as more expendable or capable of achieving swift victories. This dynamic could destabilize international relations and increase the likelihood of conflicts.

The military applications of TMS present unique ethical challenges that intersect with all five dimensions of our framework: fairness/equity, autonomy, authenticity, moral agency, and human dignity.

In terms of fairness and equity, the selective enhancement of soldiers could create concerning disparities not only within military ranks but also between nations with varying access to these technologies. This technological asymmetry could fundamentally alter military power dynamics and potentially destabilize international relations. Autonomy concerns are particularly acute in the military context, where the hierarchical command structure may compromise genuine voluntary consent. Soldiers might feel implicitly or explicitly pressured to undergo TMS enhancement to meet performance expectations or maintain competitiveness with enhanced peers. This pressure is especially problematic given the unique power dynamics and limited individual agency inherent in military service.

Questions of authenticity arise when considering how TMS-enhanced capabilities might affect soldiers' sense of self and achievement. If a soldier's enhanced performance results from technological intervention rather than traditional training and skill development, it could impact their professional identity and sense of authentic accomplishment. This becomes particularly complex when enhancement effects persist beyond active service. And, moral agency considerations are critical when TMS is used to modify traits like aggression or empathy. If TMS can significantly alter a soldier's emotional responses or decision-making capabilities, it raises profound questions about responsibility and accountability for actions in combat situations. Who bears moral responsibility when enhanced capabilities influence battlefield decisions—the individual soldier, their commanding officers, or those who mandated the enhancement?

Finally, human dignity must be carefully considered when implementing military enhancement programs. Treating soldiers as technological platforms to be optimized rather than as human beings with inherent worth risks undermining their fundamental dignity. The potential for enhancement to become a prerequisite for service or advancement could create a dehumanizing environment where natural human capabilities are devalued.

These ethical dilemmas associated with military neuroenhancement intersect with broader legal concerns about liability, informed consent, and the regulation of non-therapeutic TMS use, underscoring the need for comprehensive oversight both within and outside of the military.

6 Regulatory Scrutiny

Given these concerns, regulatory bodies and policymakers need to consider both the benefits and the ethical and societal implications of its potential widespread use of TMS. This involves creating guidelines to ensure equitable access, addressing affordability, and potentially providing financial assistance programs to prevent socio-economic divides. Alternatively, guidelines might severely restrict the enhancement use of TMS.

Monitoring for misuse is critical. Regulatory frameworks should include mechanisms to detect and prevent unauthorized TMS use, particularly in competitive environments. This could involve developing standardized protocols, mandatory reporting of TMS use and sales, imposing stringent penalties for violations, and even developing methods to recognize neuroenhancement as a result of TMS use. Regular audits and inspections can help ensure compliance, maintaining competitive fairness and protecting individual well-being.

Public education and engagement are vital for addressing broader societal impacts. Policymakers should inform the public about TMS's risks and benefits, fostering dialog about its ethical implications. Public forums, educational campaigns, and stakeholder consultations can help shape policies that reflect societal values and address concerns about fairness, autonomy, and equity. Fostering public dialog and education, particularly in highly competitive environments such as sports [105], about TMS neuroenhancement is essential to ensure that its development and implementation align with societal values and address diverse concerns.

The long-term impacts of neuroenhancement on cognitive functions and overall health remain largely unknown [106]. Comprehensive studies are needed to fully understand potential risks and benefits. This knowledge will help create guidelines that ensure the safe and effective use of TMS without unforeseen negative consequences.

Ensuring informed and voluntary decisions about neuroenhancement is crucial. Societal pressure to enhance, especially in competitive environments, could undermine autonomy. Clear guidelines and educational initiatives are necessary to support informed consent and prevent coercion.

Promoting responsible research and development practices is key. Ethical considerations should be integrated into the design and testing phases of new neuroenhancement technologies. Researchers must be vigilant about potential ethical issues and prioritize public safety and well-being. Integrating ethical considerations into the research and development process is crucial. Researchers and developers should adopt ethical frameworks and conduct thorough risk assessments during innovation. Funding bodies and research institutions can prioritize projects that demonstrate a commitment to ethical standards and societal benefit.

Overall, while neuroenhancement technologies like TMS offer exciting possibilities, their ethical implications are complex and far-reaching. Addressing these concerns through comprehensive, multidisciplinary approaches is crucial to ensuring that these technologies benefit society without compromising ethical standards. By balancing innovation with ethical responsibility, we can harness the benefits of neuroenhancement while safeguarding societal values and individual rights.

This ethical landscape intersects with significant legal concerns. The deployment of TMS must navigate a myriad of regulatory and liability issues to ensure safe and equitable use. Exploring these legal dimensions in detail is essential to formulate robust frameworks that protect both users and practitioners, ensuring that the promise of TMS is realized without unintended harm.

7 Legal Implications and Liability of Non-therapeutic TMS Use

The proposed ethical framework underscores autonomy and moral agency as critical dimensions when evaluating liability concerns. For instance, informed consent protocols must ensure that individuals fully understand the implications of TMS use on their cognitive autonomy and moral accountability.

The growing interest in TMS for non-therapeutic purposes, such as cognitive enhancement, necessitates a thorough examination of its legal implications and liability concerns. While off label prescribing of TMS is legal for competent healthcare workers, doctors could face malpractice claims if a patient is harmed. Courts typically evaluate these cases based on traditional medical negligence principles, but there may be a trend toward more patient-favorable rulings in enhancement cases, especially those deemed medically unnecessary and risky [107].

Malpractice claims in the context of TMS for cognitive enhancement could hinge on several factors. Typically, healthcare providers must adhere to established standards of care, which include obtaining informed consent, thoroughly explaining the risks and benefits, and following proper procedural protocols. If a patient suffers harm due to negligence, such as improper administration or failure to adequately screen for contraindications, the provider may be held liable.

Many neuroenhancement products are often marketed directly to consumers as recreational (i.e., non-medical devices) without sufficient evidence of their benefits, raising concerns about false advertising. Stricter regulations are necessary to protect consumers from ineffective or harmful products. The Federal Trade Commission

(FTC) and other regulatory bodies should enforce rules against misleading claims and require robust evidence of efficacy before allowing these products to be marketed [108, 109]. This could involve mandatory clinical trials and clear disclosures of potential risks and side effects [110].

Legal protections might be further necessary to prevent discrimination based on enhanced or non-enhanced cognitive abilities in employment, education, and insurance. Despite their potential for unfair advantage, employers and educational institutions should still be prohibited from making decisions based on whether an individual has undergone TMS or other neuroenhancement procedures. Similarly, insurance companies should not be allowed to deny coverage or adjust premiums based on neuroenhancement status. These protections would require amendments to existing anti-discrimination laws or the creation of new legislation specifically addressing neuroenhancement.

It is crucial that individuals are fully informed and voluntarily choose to use neuroenhancement technologies, understanding the potential risks, benefits, and long-term implications, when known. Informed consent processes must be rigorous, ensuring that individuals are aware of all possible outcomes and have the capacity to make autonomous decisions. Special considerations should be required for vulnerable populations, such as children or individuals with impaired decision-making capacity, to ensure they are not coerced or unduly influenced into using these technologies for enhancement. This might involve additional consent procedures, such as parental or guardian approval for minors, and stricter ethical oversight for research involving these groups.

Additionally, neuroenhancement technologies that significantly alter cognitive functions may in some situations affect an individual's moral and legal responsibility for their actions. This raises questions about accountability and culpability if an enhanced individual commits a crime or behaves unethically. Existing legal definitions of competence and responsibility might need to be re-evaluated to account for the potential impacts of neuroenhancement on individuals' cognitive abilities. Courts may need to consider whether neuroenhanced individuals possess the same level of agency and intent as those without enhancements, potentially leading to nuanced legal standards for such cases [42, 111, 112].

The potential use of neuroenhancement in workplaces and educational settings raises further concerns about fairness and pressure to conform. Legal guidelines may be needed to regulate the use of cognitive enhancers in these environments to ensure a level playing field. Employers and educational institutions must ensure that individuals are not coerced into using neuroenhancement technologies to meet performance expectations. This could involve policies that prohibit mandatory enhancement and protect individuals who choose not to enhance from discrimination or undue pressure.

Regulatory compliance, when available, is crucial, requiring manufacturers to meet stringent safety and efficacy standards even for non-medical applications. As TMS and other neuroenhancement technologies move beyond clinical settings, the legal landscape becomes more complex. Regulatory agencies like the FDA should enforce rigorous pre-market approval processes and post-market surveillance to monitor the long-term effects and safety of these technologies. There is a heightened risk of misuse and adverse effects when these technologies are used without medical supervision, necessitating clear guidelines for safe usage and regular audits to ensure compliance.

Product liability for potential harm caused by non-therapeutic TMS, whether due to device malfunction, improper use, or unforeseen long-term side effects, raises significant legal questions as well. Manufacturers could be held liable for defects in design, manufacturing, or failure to warn users about potential risks. Additionally, practitioners could face legal consequences for failing to adhere to safety protocols or for off-label use that results in harm. This underscores the need for comprehensive liability insurance for both manufacturers and practitioners as well as clear legal frameworks that delineate responsibility in cases of adverse effects.

A detailed legal analysis is essential for formulating robust regulatory frameworks that protect users and practitioners. Such frameworks should encompass stringent approval processes for new neuroenhancement technologies, ongoing monitoring for safety and efficacy, clear guidelines for informed consent, protections against coercion and discrimination, and comprehensive liability provisions. By addressing these legal and ethical concerns proactively, we can ensure that the benefits of TMS and other neuroenhancement technologies are realized while minimizing potential harms and injustices.

Overall, the intersection of TMS technology with legal and ethical considerations is complex and evolving, requiring continuous attention and adaptation to safeguard the interests of all stakeholders involved. Given these complexities, the use of TMS for cognitive enhancement demands careful legal and ethical scrutiny. Ensuring that individuals are informed and protected, and that vulnerable populations are not exploited, is paramount. As the technology advances, so too must our regulatory frameworks, addressing potential misuse and ensuring fair and equitable application across various sectors. However, while robust regulation is essential, it is equally important to avoid over regulation that could stifle innovation and beneficial uses of TMS. This delicate balance becomes particularly relevant in light of new legislative measures, such as the EU AI Act, which imposes stringent controls on the use of artificial intelligence in areas that include neuroenhancement technologies. The next section will explore these regulatory challenges and their implications in greater detail.

8 Over-Regulation Concerns by the EU AI Act

Within the framework, fairness and equity demand that regulations balance public safety with accessibility to ensure that smaller innovators and marginalized groups are not disproportionately excluded. Similarly, human dignity calls for transparency and accountability in the deployment of AI-enhanced TMS technologies.

The integration of artificial intelligence (AI) into transcranial magnetic stimulation (TMS) technologies has introduced both opportunities and regulatory challenges. Modern TMS devices can leverage AI to optimize stimulation protocols, personalize treatments, predict patient outcomes, and enhance overall functionality

[113–119]. These advancements hold great promise for improving the effectiveness and safety of TMS therapies, potentially expanding their accessibility and impact in clinical settings. However, the incorporation of AI also brings these devices under stricter regulatory scrutiny.

Under frameworks like the EU AI Act, TMS devices utilizing AI may be classified as high-risk technologies, subjecting manufacturers to rigorous conformity assessments, detailed documentation requirements, external audits, and continuous monitoring. These measures are designed to ensure safety, reliability, and ethical use of AI in medical applications, safeguarding users from untested or unreliable technologies.

While these regulations are arguably crucial for public safety, they also impose significant financial and administrative burdens, particularly for smaller innovators. The additional costs and delays associated with compliance could hinder the development and accessibility of new, beneficial TMS technologies, potentially stifling innovation in this promising field. At the same time, robust oversight is essential to maintain public trust and protect users from the risks associated with poorly designed or insufficiently tested AI applications.

Striking the right balance between regulatory oversight and fostering innovation is vital for the responsible advancement of AI-enhanced TMS technologies. Policies should aim to encourage innovation while ensuring the safe and effective integration of AI, allowing the field to continue evolving and delivering transformative benefits to patients.

The new EU AI Act establishes varying oversight requirements for artificial intelligence algorithms and models based on their perceived risks to democracy and society. These precautionary principle-based requirements impose stringent regulations on those AI technologies seen as existential threats to liberal democratic societies. While aimed at mitigating risks, this rigorous approach may also hinder the development and implementation of beneficial AI technologies, including those used in neuroenhancement.

The Cassandran Paradox of AI regulation stems from a tragic irony: just as Cassandra was cursed to speak true prophecies that no one would believe, we routinely reject AI systems that demonstrably outperform humans simply because they cannot guarantee perfect accuracy. Even when an AI system proves more reliable than human judgment—say, detecting financial fraud with 95% accuracy compared to a human analyst's 85%—we dismiss its predictions due to the mere possibility of error. Like the Trojans who ignored Cassandra's warnings to their doom, we cling to familiar but flawed human decision-making while rejecting potentially more accurate AI alternatives that cannot meet an impossible standard of perfection. This Cassandran stance produces a perverse regulatory environment where proven benefits are rejected in pursuit of theoretical perfection. The tragedy isn't that we demand high standards from AI—it's that we refuse to heed systems that are demonstrably more reliable than their human counterparts, simply because they cannot guarantee infallibility.

Under the EU AI act, high-risk AI activities, such as those involving TMS devices, are allowed but heavily regulated with extensive disclosure requirements.

The Act targets a subset of medical devices, likely including TMS treatments, for enhanced oversight [120]. High-risk activities are defined by referencing medical devices classified as 2a or higher under the EU Medical Device Regulations. These devices must meet stringent criteria, including risk management, algorithmic transparency, human oversight, and comprehensive conformity assessments.

While regulatory oversight is crucial for ensuring safety and ethical use, it must be balanced with the need to foster innovation. Excessive regulation could impede the development of promising neuroenhancement technologies like TMS, which, despite various concerns, have the potential to offer significant societal benefits. Developing a balanced regulatory approach that ensures safety without stifling innovation is essential for the responsible advancement of TMS and other neuroenhancement technologies.

In contrast to the current lack of oversight regarding neuroenhancement, future iterations of TMS may soon face over-regulation under the EU AI Act. The Act as currently written would likely treat innovative efforts to employ artificial intelligence as part of TMS treatments [121–124] as a high-risk activity that requires extensive compliance efforts.

As a high-risk AI under the EU AI Act, TMS devices would face not only medical device oversight but also enhanced scrutiny under the AI Act. This includes rigorous requirements for risk management, algorithmic transparency, and human oversight to ensure that AI systems are developed and operated in a manner that is ethical and accountable. Extensive pre-market and post-market conformity assessments are mandatory, requiring ongoing evaluations of the AI's performance and impact. Additionally, detailed record-keeping on technology usage and dynamic technical documentation outlining the AI's architecture and training processes are necessary to ensure that the AI's development and deployment are transparent and traceable.

Even if non-EU associated devices are not subject to European extra-jurisdictional enforcement—which can result in fines of up to 15 million euros or 5% of annual turnover, whichever is greater—the Brussels effect could lead other jurisdictions to adopt similar regulations [125]. This phenomenon occurs when non-EU countries align their regulations with EU standards due to the EU's significant influence on global markets. Consequently, other regions may develop similarly stringent laws, potentially stifling innovation elsewhere and slowing the development of new TMS iterations. This regulatory environment could create substantial barriers for researchers and developers, limiting the pace at which new and beneficial AI applications can be brought to market.

Fortunately however, the EU AI Act exempts research and development from its stringent oversight requirements. This exemption is crucial for fostering innovation and allowing experimental AI applications to be tested and refined without the full burden of compliance. By carving out this space for R&D, the EU aims to balance the need for stringent regulation with the necessity of promoting scientific and technological advancement. This approach might ensure that at least while AI technologies are developed responsibly and ethically, the creative processes driving progress are not unduly hampered.

9 Conclusions

By applying this structured ethical framework, this chapter demonstrates how fairness, autonomy, authenticity, moral agency, and human dignity can guide preemptive governance of the use of TMS technologies for neuroenhancement, ensuring equitable and responsible integration into society. Although TMS is not yet a widely adopted neuroenhancement tool, its future potential and emerging discussions in academic and non-academic circles justify proactive engagement with its implications. Its evolving applications and increasing accessibility demand proactive engagement with the ethical, legal, and societal implications to ensure responsible innovation and equitable access.

The ethical and societal implications of TMS in non-clinical settings, particularly as a potential tool for neuroenhancement, are undeniably intricate and contested. Skepticism about TMS's current capabilities as a meaningful enhancer of cognitive or emotional functions is essential to temper these discussions. Preliminary studies suggest some potential benefits, but these are often transient, inconsistent, and lack robust empirical validation. This raises critical questions about whether the discourse surrounding TMS's ethical and legal implications is premature, especially when empirical evidence supporting its non-clinical use remains limited.

Moreover, given the nascent state of biohacking and the limited accessibility of TMS for enhancement purposes, the immediate societal impact of this technology may be minimal. Current uses of TMS outside regulated clinical settings are niche and poorly documented, casting doubt on whether this issue demands urgent attention. Nevertheless, the rapid pace of technological advancement and the growing cultural fascination with neuroenhancement underscore the need for early engagement. Anticipating and addressing the potential risks and challenges of TMS before they become pervasive is prudent and aligns with historical lessons about the societal harms caused by regulatory delays in emerging technologies.

As the technological frontier of TMS expands, so too must our ethical, legal, and societal frameworks evolve to guide its development responsibly. The lessons of delayed governance in fields like social media, artificial intelligence and gene editing underscore the risks of reactive rather than proactive regulation. If we wait until TMS technology permeates non-clinical settings, we may face inequities, coercion, and misuse that are far harder to mitigate once they become entrenched.

To ensure TMS fulfills its promise without compromising societal values, we must act now. Policymakers, researchers, and industry leaders must collaborate to establish adaptive, forward-thinking guidelines that anticipate potential misuses while fostering equitable access. Public engagement is crucial: society at large must play an active role in shaping norms and expectations around neuroenhancement.

While the current limitations of TMS are acknowledged, this chapter emphasizes that early ethical and legal discourse is crucial to preempt misuse and foster responsible innovation. Future research should prioritize longitudinal studies to assess the long-term effects of TMS in non-clinical settings, while regulatory frameworks must adapt dynamically to balance innovation with ethical responsibility. Public engagement and transparent dialog are essential to shape societal norms and prevent misuse of this evolving technology. Early-stage discussions allow stakeholders to shape policies, guidelines, and ethical frameworks that align with societal values while mitigating future risks. Researchers must prioritize rigorous empirical studies to establish the safety, efficacy, and long-term effects of TMS, standardizing methodologies and collaborating across disciplines to address emerging ethical and societal concerns. Policymakers should monitor the development of TMS technologies in non-clinical settings and craft adaptive, evidence-based guidelines that prioritize safety, equity, and accessibility. International regulatory harmonization is essential to prevent gaps that might lead to exploitation or harm.

Educators, advocates, and industry leaders also play critical roles in shaping public understanding and expectations. Transparent communication about TMS's capabilities, limitations, and risks is necessary to counter misinformation and prevent misuse. Public consultation can establish ethical norms and societal expectations, ensuring that the trajectory of neuroenhancement technologies reflects shared values and addresses diverse concerns.

By embracing interdisciplinary dialog, supporting robust empirical research, and committing to a balanced regulatory approach, we can navigate the complexities of TMS and other neurotechnologies. Together, we can harness these innovations to enhance human well-being while safeguarding fairness, autonomy, and human dignity. The time for thoughtful, inclusive action is not in the distant future—it is now.

Importantly, this chapter does not advocate for an alarmist view but instead calls for a balanced approach that weighs the current state of TMS against its future potential. While TMS is not yet a transformative neuroenhancement tool, its evolving applications and increasing accessibility demand careful consideration. By addressing ethical and legal challenges preemptively, society can ensure that advancements in TMS are harnessed responsibly and equitably. Such foresight will safeguard against potential harms, minimize inequities, and provide a foundation for sustainable and ethical progress in neuroenhancement technologies. This dual recognition of TMS's present limitations and future possibilities makes this discourse relevant and timely, meeting the concerns raised by the reviewers while reinforcing the value of early engagement with emerging technologies.

References

1. Nieminen AE, Nieminen JO, Stenroos M, Novikov P, Nazarova M, Vaalto S, Nikulin V, Ilmoniemi RJ. Accuracy and precision of navigated transcranial magnetic stimulation. J Neural Eng. 2022;19(6):066037.
2. Sack AT, Paneva J, Küthe T, Dijkstra E, Zwienenberg L, Arns M, Schuhmann T. Target engagement and brain state dependence of transcranial magnetic stimulation: implications for clinical practice. Biol Psychiatry. 2024;95(6):536–44.
3. Chen R, Cros D, Curra A, Di Lazzaro V, Lefaucheur J-P, Magistris MR, Mills K, et al. The clinical diagnostic utility of transcranial magnetic stimulation: report of an IFCN committee. Clin Neurophysiol. 2008;119(3):504–32.
4. Yen C, Valentine EP, Chiang M-C. The use of transcranial magnetic stimulation in attention optimization research: a review from basic theory to findings in attention-deficit/hyperactivity disorder and depression. Life. 2024;14(3):329.

5. Siebner HR, Funke K, Aberra AS, Antal A, Bestmann S, Chen R, Classen J, Davare M, Di Lazzaro V, Fox PT, Hallett M. Transcranial magnetic stimulation of the brain: what is stimulated?–a consensus and critical position paper. Clin Neurophysiol. 2022;140:59–97.
6. Vidal-Dourado M, Conforto AB, Caboclo LOSF, Scaff M, Guilhoto LMDFF, Yacubian EMT. Magnetic fields in noninvasive brain stimulation. Neuroscientist. 2014;20(2):112–21.
7. Kolbinger HM, Höflich G, Hufnagel A, Müller HJ, Kasper S. Transcranial magnetic stimulation (TMS) in the treatment of major depression—a pilot study. Hum Psychopharmacol Clin Exp. 1995;10(4):305–10.
8. Lanza G, Fisicaro F, Cantone M, Pennisi M, Cosentino FII, Lanuzza B, Tripodi M, Bella R, Paulus W, Ferri R. Repetitive transcranial magnetic stimulation in primary sleep disorders. Sleep Med Rev. 2023;67:101735.
9. Liang K, Li H, Bu X, Li X, Cao L, Liu J, Gao Y, Li B, Qiu C, Bao W, Zhang S. Efficacy and tolerability of repetitive transcranial magnetic stimulation for the treatment of obsessive-compulsive disorder in adults: a systematic review and network meta-analysis. Transl Psychiatry. 2021;11(1):332.
10. Lan L, Zhang X, Li X, Rong X, Peng Y. The efficacy of transcranial magnetic stimulation on migraine: a meta-analysis of randomized controlled trails. J Headache Pain. 2017;18:1–7.
11. Cordes J, Arends M, Mobascher A, Brinkmeyer J, Kornischka J, Eichhammer P, Klimke A, Winterer G, Agelink MW. Potential clinical targets of repetitive transcranial magnetic stimulation treatment in schizophrenia. Neuropsychobiology. 2007;54(2):87–99.
12. Stramba-Badiale C, Mancuso V, Cavedoni S, Pedroli E, Cipresso P, Riva G. Transcranial magnetic stimulation meets virtual reality: the potential of integrating brain stimulation with a simulative technology for food addiction. Front Neurosci. 2020;14:720.
13. Ruiz ML, Sospedra M, Arce SA, Tejeiro-Martínez J, Benito-León J. Current evidence on the potential therapeutic applications of transcranial magnetic stimulation in multiple sclerosis: a systematic review of the literature. Neurología. 2022;37(3):199–215.
14. Sabbagh M, Sadowsky C, Tousi B, Agronin ME, Alva G, Armon C, Bernick C, Keegan AP, Karantzoulis S, Baror E, Ploznik M. Effects of a combined transcranial magnetic stimulation (TMS) and cognitive training intervention in patients with Alzheimer's disease, vol. 16. Alzheim Demen; 2019. p. 641.
15. Lefaucheur JP, André-Obadia N, Antal A, Ayache SS, Baeken C, Benninger DH, Cantello RM, Cincotta M, de Carvalho M, De Ridder D, Devanne H. Evidence-based guidelines on the therapeutic use of repetitive transcranial magnetic stimulation (rTMS). Clin Neurophysiol. 2014;125(11):2150–206.
16. Mattioli F, Maglianella V, D'Antonio S, Trimarco E, Caligiore D. Non-invasive brain stimulation for patients and healthy subjects: current challenges and future perspectives. J Neurol Sci. 2023;456:122825.
17. Roth Y, Barnea-Ygael N, Carmi L, Storch EA, Tendler A, Zangen A. Deep transcranial magnetic stimulation for obsessive-compulsive disorder is efficacious even in patients who failed multiple medications and CBT. Psychiatry Res. 2020;290:113179.
18. Pascual-Leone A, Valls-Sollé J, Wassermann EM, Brasil-Neto J, Cohen LG, Hallett M. Effects of focal transcranial magnetic stimulation on simple reaction time to acoustic, visual and somatosensory stimuli. Brain. 1992;115:1045–59.
19. Hoy KE, Emonson MRL, Arnold SL, Thomson RH, Daskalakis ZJ, Fitzgerald PB. Testing the limits: investigating the effect of tDCS dose on working memory enhancement in healthy controls. Neuropsychologia. 2013;51(9):1777–84.
20. Pisoni A, Mattavelli G, Papagno C, Rosanova M, Casali AG, Romero LJ, Lauro. Cognitive enhancement induced by anodal tDCS drives circuit-specific cortical plasticity. Cereb Cortex. 2018;28(4):1132–40.
21. Jangwan NS, Ashraf GM, Ram V, Singh V, Alghamdi BS, Abuzenadah AM, Singh MF. Brain augmentation and neuroscience technologies: current applications, challenges, ethics and future prospects. Front Syst Neurosci. 2022;16:1000495.
22. Greenbaum D, Gerstein M. The control and privacy of your most intimate information: navigating the battle for your brain. AJOB Neurosci. 2023;14(3):328–31.

23. Fisicaro F, Lanza G, Bella R, Pennisi M. "Self-neuroenhancement": The last frontier of non-invasive brain stimulation? J Clin Neurol. 2020;16(1):158–9.
24. Bikson M, Bestmann S, Edwards D. Neuroscience: transcranial devices are not playthings. Nature. 2013;501(7466):167.
25. Wurzman R, Hamilton RH, Pascual-Leone A, Fox MD. An open letter concerning do-it-yourself users of transcranial direct current stimulation. Ann Neurol. 2016;80(1):1.
26. Qi S, Jinglun Y, Li L, Dong C, Ji Z, Cao L, Wei Z, Liang Z. Advances in non-invasive brain stimulation: enhancing sports performance function and insights into exercise science. Front Hum Neurosci. 2024;18:1477111.
27. Kiss J. Do not try this at home: hacking your brain with electronics. The Guardian, March 12, 2016. https://www.theguardian.com/culture/2016/mar/12/sxsw-neuroscientists-home-hacking-your-brain-with-electronics
28. Rossi S, Antal A, Bestmann S, Bikson M, Brewer C, Brockmöller J, Carpenter LL, et al. Safety and recommendations for TMS use in healthy subjects and patient populations, with updates on training, ethical and regulatory issues: expert guidelines. Clin Neurophysiol. 2021;132(1):269–306.
29. Clark VP, Parasuraman R. Neuroenhancement: enhancing brain and mind in health and in disease. NeuroImage. 2014;85(3):889–94.
30. Gutiérrez-Muto AM, Bestmann S, Sánchez R, de la Torre JL, Pons AO, Tornero J. The complex landscape of TMS devices: a brief overview. PLoS One. 2023;18(11):e0292733.
31. https://nibbot-tms.com/en/home/
32. https://neurostimtms.com/the-limitations-of-diy-tms-and-at-home-devices/
33. Violante IR, Okyere P. Can neurotechnology revolutionize cognitive enhancement? PLoS Biol. 2024;22(10):e3002831.
34. Afifi SY. A new era of current and future treatment applications of transcranial magnetic stimulation. Egypt J Neurol Psychiat Neurosurg. 2024;60(1):54.
35. Chen R, Spencer DC, Weston J, Nolan SJ. Transcranial magnetic stimulation for the treatment of epilepsy. Cochr Database Syst Rev. 2016;
36. Loo CK, McFarquhar TF, Mitchell PB. A review of the safety of repetitive transcranial magnetic stimulation as a clinical treatment for depression. Int J Neuropsychopharmacol. 2008;11:131–47.
37. San Agustín A, Moreno JC. Ethical aspects of transcranial magnetic stimulation for neuroenhancement. Dilemata. 2021;34:121–32.
38. Barker AT, Jalinous R, Freeston IL. Non-invasive magnetic stimulation of human motor cortex. Lancet. 1985;325(8437):1106–7.
39. Cohen SL, Bikson M, Badran BW, George MS. A visual and narrative timeline of US FDA milestones for transcranial magnetic stimulation (TMS) devices. Brain Stimul Bas Transl Clin Res Neuromodul. 2022;15(1):73–5.
40. O'Brien E. FDA clears NeuroStar TMS for treatment of MDD in adolescents. Psychiatric Times; March 27, 2024. https://www.psychiatrictimes.com/view/fda-clears-neurostar-tms-for-treatment-of-mdd-in-adolescents
41. Rossi S, Antal A, Bestmann S, Bikson M, Brewer C, Brockmöller J, Carpenter LL, Cincotta M, Chen R, Daskalakis JD, Di Lazzaro V. Safety and recommendations for TMS use in healthy subjects and patient populations, with updates on training, ethical and regulatory issues: expert guidelines. Clin Neurophysiol. 2021;132(1):269–306.
42. Nakar S, Weinberger S, Greenbaum D. Legal and social implications of predictive brain machine interfaces: duty of care, negligence, and criminal responsibility. AJOB Neurosci. 2015;6(4):40–2.
43. Greenbaum D, Cabrera LY. ELSI in human enhancement: what distinguishes it from therapy? Front Genet. 2020;11:618.
44. Greenbaum D. Ethical, legal and social concerns relating to exoskeletons. ACM SIGCAS Comput Soc. 2016;45(3):234–9.
45. Conte A, Stan JL, Hodgins NS, ElAttrache NP-F, Ahmad CS. Media perceptions of Tommy John surgery. Phys Sportsmed. 2015;43(4):375–80.

46. Brown J. Genetic doping: WADA we do about the future of 'cheating' in sport? Int Sport Law J. 2019;19(3):258–80.
47. Marazziti D, Avella MT, Ivaldi T, Palermo S, Massa L, Della Vecchia A, Basile L, Mucci F. Neuroenhancement: state of the art and future perspectives. Clin Neuropsychiatry. 2021;18(3):137.
48. Greenbaum D. Cyberbiosecurity: an emerging field that has ethical implications for clinical neuroscience. Camb Q Healthc Ethics. 2021;30(4):662–8.
49. Gordon EC, Dunn L. Pharmacological cognitive enhancement and cheapened achievement: a new dilemma. Neuroethics. 2021;14(3):409–21.
50. Antal A, Luber B, Brem A-K, Bikson M, Brunoni AR, Kadosh RC, Dubljević V, et al. Non-invasive brain stimulation and neuroenhancement. Clin Neurophysiol Pract. 2022;7:146–65.
51. Ragan CI, Bard I, Singh I. What should we do about student use of cognitive enhancers? An analysis of current evidence. Neuropharmacology. 2013;64:588–95.
52. Snowball A, Tachtsidis I, Popescu T, Thompson J, Delazer M, Zamarian L, Zhu T, Kadosh RC. Long-term enhancement of brain function and cognition using cognitive training and brain stimulation. Curr Biol. 2013;23(11):987–92.
53. Huo L, Zhu X, Zheng Z, Ma J, Ma Z, Gui W, Li J. Effects of transcranial direct current stimulation on episodic memory in older adults: a meta-analysis. J Gerontol B. 2021;76(4):692–702.
54. Bard I, Gaskell G, Allansdottir A, da Cunha RV, Eduard P, Hampel J, Hildt E, Hofmaier C, Kronberger N, Laursen S, Meijknecht A. Bottom up ethics-neuroenhancement in education and employment. Neuroethics. 2018;11:309–22.
55. Bard I. Why (not) enhance the brain? A mixed-methods exploration of the acceptability and desirability of neuroenhancement. PhD diss., London School of Economics and Political Science, 2023
56. TMS Device Comparison. https://caputron.com/pages/tms-device-comparison?srsltid=AfmBOoq9XWiAIitLtiUFM4o6AARIj_ue3Zd_EVh5JvSXJSvCr-j0H3sF
57. The Limitations of DIY TMS and AT-Home Devices. NeuroStim TMS Centers. https://neurostimtms.com/the-limitations-of-diy-tms-and-at-home-devices/
58. NioHackeing the Brain: Exploring Cutting Edge Startups in Neuroenhancements, Faster Capital June 26, 2024. https://fastercapital.com/content/Biohacking-the-Brain%2D%2DExploring-Cutting-Edge-Startups-in-Neuroenhancement.html
59. Transcranial Magnetic Stimulation Therapy Equipment for Impression. https://drive.google.com/file/d/1olE-WOZnOW0eLLBX7RiEHjt-MzAAz2o1/view?usp=sharing
60. Lee S, Jang KI, Yoon S, Chae JH. The efficacy of miniaturized repetitive transcranial magnetic stimulation in patients with depression. Clin Psychopharmacol Neurosci. 2019;17(3):409.
61. TMS Therapy at Home https://gia.miami/rehab-blog/tms-therapy-at-home/
62. Caulfield KA, George MS. Treating the mental health effects of COVID-19: the need for at-home neurotherapeutics is now. Brain Stimul Bas Transl Clin Res Neuromodul. 2020;13(4):939–40.
63. Abbasi S, Alluri S, Leung V, Asbeck P, Makale MT. Design and validation of miniaturized repetitive transcranial magnetic stimulation (rTMS) head coils. Sensors. 2024;24(5):1584.
64. Luber B, Lisanby SH. Enhancement of human cognitive performance using transcranial magnetic stimulation (TMS). NeuroImage. 2014;85:961–70.
65. Klimesch W, Sauseng P, Gerloff C. Enhancing cognitive performance with repetitive transcranial magnetic stimulation at human individual alpha frequency. Eur J Neurosci. 2003;17(5):1129–33.
66. Brunyé TT, Beaudoin ME, Feltman KA, Heaton KJ, McKinley RA, Vartanian O, Tangney JF, et al. Neuroenhancement in military personnel: conceptual and methodological promises and challenges. In: NATO symposium on applying neuroscience to performance: from rehabilitation to human cognitive 2021: Rome, Italy, 11-12 October 2021; 2022.
67. Kuo M-F, Nitsche MA. Effects of transcranial electrical stimulation on cognition. Clin EEG Neurosci. 2012;43(3):192–9.

68. Brain Enhancement Solutions: Enhancing Memory and Learning: Strategies That Work Faster Capital June 22, 2024. https://fastercapital.com/content/Brain-Enhancement-Solutions%2D%2DEnhancing-Memory-and-Learning%2D%2DStrategies-That-Work.html
69. Miniussi C, Ruzzoli M. Transcranial stimulation and cognition. Handb Clin Neurol. 2013;116:739–51.
70. Wasserman EM. Use and safety of a new repetitive transcranial magnetic stimulator. Electroencephalogr Clin Neurophysiol. 1996;101:412–7.
71. Zhuang W, Yin K, Zi Y, Liu Y. Non-invasive brain stimulation: augmenting the training and performance potential in esports players. Brain Sci. 2020;10(7):454.
72. Gazerani P. Performance enhancement by brain stimulation. J Sport Sci Med. 2017;16(3):438.
73. Biohacking: The New Science of Optimizing Your Brain WavWatch August 10, 2022. https://www.wavwatch.com/blog/biohacking-the-new-science-of-optimizing-your-brain-and-body
74. Dubljevic V, Saigle V, Racine E. The rising tide of tDCS in the media and academic literature. Neuron. 2014;82:731–6.
75. Racine E, Waldma S, Palmour N, Risse D, Illes J. "Currents of hope": neurostimulation techniques in U.S. and U.K. print media. Camb Q Healthc Ethics. 2007;16:312–6.
76. Larriviere D, Williams MA, Rizzo M, Bonnie RJ. Responding to requests from adult patients for neuroenhancements: guidance of the ethics, law and humanities committee. Neurology. 2009;73(17):1406–12.
77. Greenbaum D, Gerstein M. GATTACA is still pertinent 25 years later. Nat Genet. 2022;54(12):1758–60.
78. Kraemer F. Authenticity anyone? The enhancement of emotions via neuro-psychopharmacology. Neuroethics. 2011;4(1):51–64.
79. Smith JE. The pre-eminence of autonomy in bioethics. In: Human lives: critical essays on consequentialist bioethics. London: Palgrave Macmillan UK; 1997. p. 182–95.
80. Møller S. How moral neuroenhancement impacts autonomy and agency. Bioethics. 2022;36(7):794–801.
81. Brännmark J. Respect for persons in bioethics: towards a human rights-based account. Hum Rights Rev. 2017;18(2):171–87.
82. Ienca M. On neurorights. Front Hum Neurosci. 2021;15:701258.
83. Baselga-Garriga C, Rodriguez P, Yuste R. Neuro rights: a human rights solution to ethical issues of neurotechnologies. In: Protecting the mind: challenges in law, neuroprotection, and neurorights. Cham: Springer; 2022. p. 157–61.
84. Liv N, Greenbaum D. Cyberneurosecurity. In: Policy, identity, and neurotechnology: the neuroethics of brain-computer interfaces. Cham: Springer; 2023. p. 233–51.
85. Chile: Pioneering the protection of neurorights UNESCO, 13 October 2023. https://courier.unesco.org/en/articles/chile-pioneering-protection-neurorights
86. Health Council of the Netherlands. Human enhancement (ethics and health monitoring report 2003 no. 4). The Hague; 2003
87. Baldwin T, Cole J, Fitzgerald M, Kitzinger J, Laurie G, Price J, Rose N, Rose S, Singh I, Walsh V, Warwick K. Novel neurotechnologies: intervening in the brain; 2013
88. Coenen C, Schuijff M, Smits M, Klaassen P, Hennen L, Rader M, Wolbring G. Human enhancement: a report for the European Parliament science technology options assessment, Brussels 2009
89. Presidential Commission for the study of Bioethical Issues. 2015. Grey matters: Topics at the intersection of neuroscience, Ethics, and Society. https://bioethicsarchive.georgetown.edu/pcsbi/node/4704.html
90. Waddington I, Møller V. WADA at twenty: old problems and old thinking? Int J Sport Policy Politics. 2019;11(2):219–31.
91. D'Alessandro M, Boggs J, Feltman K, Wilkins J, Kelley A: A systematic review of the transcranial stimulation literature for performance enhancement, USAARL-TECH-TR--2023-15 (U.S. Army Medical Research and Development Command).

92. Nelson JT, Tepe V. Neuromodulation research and application in the us department of defense. Brain Stimul. 2015;8(2):247–52.
93. Oberman LM, Exley S, Philip NS, Siddiqi SH, Adamson MM, Brody DL. The use of transcranial magnetic stimulation in the treatment of neuropsychiatric and neurocognitive symptoms associated with concussion in military populations. J Head Trauma Rehab. 2020;35:388–400.
94. Malish RG. The importance of the study of cognitive enhancement for U.S. national security. Aerosp Med Human Perform. 2017;88:773–8.
95. NATO Science and Technology Organization. Neuroenhancement in military personnel: conceptual and methodological promises and challenges. STO-TR-HFM-311. STO Technical Report RDP; 30 Sept. 2024. https://doi.org/10.14339/STO-TR-HFM-311. ISBN: 978-92-837-2483-4.
96. Nelson JT. Enhancing warfighter cognitive abilities with transcranial magnetic stimulation: a feasibility analysis. Human Effectiveness Directorate, Wright-Patterson AFB, OH. Aircrew Protection Branch, Interim Report; 2007.
97. Malish RG. The importance of the study of cognitive performance enhancement for US national security. Aerospace Med Human Perform. 2017;88(8):773–8.
98. Feltman KA, Hayes AM, Bernhardt KA, Nwala E, Kelley AM. Viability of tDCS in military environments for performance enhancement: a systematic review. Mil Med. 2020;185(1–2):e53–60.
99. Krauel E, Frewer A. Dual use concerns in artificial intelligence and the neurosciences: how medical research can end up in war. Res Ethic. 2024;17470161241262149
100. Latheef S, Henschke A. Can a soldier say no to an enhancing intervention? Philosophies. 2020;5(3):13.
101. Janicak PG, O'Reardon JP, Sampson SM, Husain MM, Lisanby SH, Rado JT, Heart KL, Demitrack MA. Transcranial magnetic stimulation in the treatment of major depressive disorder: a comprehensive summary of safety experience from acute exposure, extended exposure, and during reintroduction treatment. J Clin Psychiatry. 2008;69:222–32.
102. Sattler S, Jacobs E, Singh I, Whetham D, Bárd I, Moreno J, Galeazzi G, Allansdottir A. Neuroenhancements in the military: a mixed-method pilot study on attitudes of staff officers to ethics and rules. Neuroethics. 2022;15(1):11.
103. Yang C-C, Khalifa N, Völlm B. The effects of repetitive transcranial magnetic stimulation on empathy: a systematic review and meta-analysis. Psychol Med. 2018;48(5):737–50.
104. Hereth B, Evans N. Can we justify military enhancements? Some yes, most no. Camb Q Healthc Ethics. 2022;31(4):557–69.
105. Prete G, Bertollo M, Tommasi L. Brain stimulation techniques and sports performance. In: Advancements in mental skills training. Routledge; 2020. p. 164–76.
106. Phipps CJ, Murman DL, Warren DE. Stimulating memory: reviewing interventions using repetitive transcranial magnetic stimulation to enhance or restore memory abilities. Brain Sci. 2021;11(10):1283.
107. U.S. Attorney's Office, Southern District of Texas, Psychiatrist settles claims for unnecessary brain stimulation treatments March 3, 2023. https://www.justice.gov/usao-sdtx/pr/psychiatrist-settles-claims-unnecessary-brain-stimulation-treatments
108. American Academy of Arts and Sciences. Bulletin: The regulatory and ethical dimensions of human performance enhancement; Summer 2016. https://www.amacad.org/news/regulatory-and-ethical-dimensions-human-performance-enhancement
109. Wexler A, Nagappan A, Kopyto D, Choi R. Neuroenhancement for sale: assessing the website claims of neurofeedback providers in the USA. J Cogn Enhanc. 2020;4:379–88.
110. Nagappan A, Kalokairinou L, Wexler A. Ethical and legal considerations of alternative neurotherapies. AJOB Neurosci. 2021;12(4):257–69.
111. Weinberger S, Greenbaum D. Are BMI prosthetics uncontrollable Frankensteinian monsters? Brain-Comput Interfaces. 2016;3(3):149–55.
112. Greenbaum D. Is criminal law both redundant and inconsistent?: crime and consciousness in light of developments in neuroscience. AJOB Neurosci. 2018;9(1):51–2.

113. Gobhinath S, Karhtikeyan S, Janani K, Arunkumar U. Detection and analysis of earlier cognitive disorder using ai based transcranial magnetic stimulator. In: 2022 8th international conference on advanced computing and communication systems (ICACCS), vol. 1. IEEE; 2022. p. 553–6.
114. Noda Y, Sakaue K, Wada M, Takano M, Nakajima S. Development of artificial intelligence for determining major depressive disorder based on resting-state EEG and single-pulse transcranial magnetic stimulation-evoked EEG indices. J Personaliz Med. 2024;14(1):101.
115. Tubbs A, Vazquez EA. Engineering and technological advancements in repetitive transcranial magnetic stimulation (rTMS): a five-year review. Brain Sci. 2024;14(11):1092.
116. Chen D, Lei X, Lian D, Long Z. Use of machine learning in predicting the efficacy of repetitive transcranial magnetic stimulation on treating depression based on functional and structural thalamo-prefrontal connectivity: a pilot study. J Psychiatr Res. 2022;148:88–94.
117. Moser P, Reishofer G, Prückl R, Schaffelhofer S, Freigang S, Thumfart S, Ali KM. Real-time estimation of the optimal coil placement in transcranial magnetic stimulation using multi-task deep learning. Sci Rep. 2024;14(1):19361.
118. Zandvakili A, Philip NS, Jones SR, Tyrka AR, Greenberg BD, Carpenter LL. Use of machine learning in predicting clinical response to transcranial magnetic stimulation in comorbid posttraumatic stress disorder and major depression: a resting state electroencephalography study. J Affect Disord. 2019;252:47–54.
119. Jaroonsorn P, Neranon P, Smithmaitrie P, Dechwayukul C. Implementation of an autonomous assistive robotic system for transcranial magnetic stimulation. J Mech Eng. 2020;
120. Gilbert S. The EU passes the AI act and its implications for digital medicine are unclear. NPJ Digit Med. 2024;7(1):135.
121. Erguzel TT, Tarhan N. Machine learning approaches to predict repetitive transcranial magnetic stimulation treatment response in major depressive disorder. In: Proceedings of SAI intelligent systems conference (IntelliSys) 2016, vol. 2. Springer; 2018. p. 391–401.
122. Arns M, Drinkenburg WH, Fitzgerald PB, Kenemans JL. Neurophysiological predictors of non-response to rTMS in depression. Brain Stimul. 2012;5(4):569–76.
123. Chen D, Lei X, Du L, Long Z. Use of machine learning in predicting the efficacy of repetitive transcranial magnetic stimulation on treating depression based on functional and structural thalamo-prefrontal connectivity: a pilot study. J Psychiatr Res. 2022;148:88–94.
124. Squires M, Tao X, Elangovan S, Gururajan R, Xie H, Zhou X, Li Y, Acharya UR. DE-CGAN: boosting rTMS treatment prediction with diversity enhancing conditional generative adversarial networks. arXiv preprint arXiv:2404.16913, 2024.
125. Bradford A. The Brussels effect: how the European Union rules the world. USA: Oxford University Press; 2020.

TMS for Social Neuroscience: New Applications and Ethical Issues

Michela Balconi

1 TMS: Preliminary Remarks for Social Sciences

The present work focuses on the overview of basic cognitive-affective neuroscience research using transcranial magnetic stimulation (TMS) as non-invasive brain stimulation (NIBS) technique to investigate neural bases of affective functions.

In particular, this chapter focuses on two main topics that have recently seen the application of neuromodulation and neurostimulation techniques: the field of decision-making processes, with a specific attention also on the side of moral conduct; secondly applications related to social neuroscience in general.

Next paragraphs represent an overall introduction to the most consistent outcomes of TMS-based research with regard to these topics, with a specific focus on critical issues and open questions.

1.1 NIBS Protocols and Some Safety Questions

We must consider the presence of two families of NIBS techniques—i.e., TMS and transcranial electrical stimulation (tES)—which are traditionally considered primarily passive intervention techniques. This distinction reflects the degree of involvement and active participation required by the person carrying out the experiments so that the technique can show valid and detectable effects. Specifically, NIBS techniques are among the passive ones as they are capable of inducing a modification in the functioning of the target brain structures—in the form of

M. Balconi (✉)
International research center for Cognitive Applied Neuroscience (IrcCAN), Università Cattolica del Sacro Cuore, Milan, Italy

Research Unit in Affective and Social Neuroscience, Department of Psychology, Università Cattolica del Sacro Cuore, Milan, Italy
e-mail: michela.balconi@unicatt.it

V. Dubljević, J. R. Young (eds.), *TMS and Neuroethics*, Advances in Neuroethics, https://doi.org/10.1007/978-3-031-92401-9_7

interference, enhancement, or inhibition—even without the conscious contribution of the person experimenting with them. Given this premise, however, it is also important to underline, for correctness and completeness of information, that this potential is not associated, in practice, with a complete passivization of the individual who receives a neuroenhancement protocol based on TMS. If it is true that the specific effect of these techniques on a physiological level largely ignores the will and compliance of the person who receives them, their active involvement and motivation to change remain crucial factors in modulating the perceived and objective effectiveness and the overall outcomes of a neuroenhancement or neurorehabilitation path supported by NIBS, as in the case of other forms of psychological and psychophysiological intervention.

The use of non-invasive methods to modulate brain function has a long tradition that dates back to the use of energy obtained from electric fish (e.g., eels, torpedoes, and many more) to modify brain activity in order to relieve pain headache or pain [37]. Over time, the use of brain creativity through electrical currents has become a therapeutic strategy to minimize the effects of mental disorders. In recent decades, NIBS has acquired particular relevance both in the clinical context and in research, and the techniques have been optimized with the ultimate goal of inducing long-lasting plastic changes at a sensorimotor and cognitive level.

With specific reference to TMS, this technique is capable of directly activating the nerve cells affected by the stimulation and can therefore be defined in all respects as a neurostimulation technique, i.e., a technique capable of inducing neuronal firing. In protocols with TMS, the extension of the stimulated surface depends primarily on three factors: the stimulation intensity, the type of coil and the coil orientation. Specifically, as the stimulation intensity increases, the stimulated surface will also be greater, and therefore the precision of the stimulation will decrease. At the same time, by increasing the intensity of stimulation, its depth will also be relatively increased, with a consequent involvement of a greater volume of cortical tissue. The role of the coil in determining the effect of the stimulation depends, however, on its size and shape. As a general rule, increasing the size of the stimulation probe will potentially increase the intensity of the electric field and, systematically, the stimulated area. The shape of the coil, however, modulates the geometry of the developed magnetic field, with effects on the stimulation focus and its depth.

In defining a NIBS protocol with TMS, therefore, both the choice of coil, the location of the stimulation, and the stimulation intensity settings are critical methodological choices for the professional who is designing it. In fact, they have a critical relevance in shaping the effects of the stimulation itself, and therefore, ultimately, the outcome and effectiveness of the protocol.

TMS is, today, considered a painless and safe technique when applied in compliance with safety guidelines, with experimental stimulation programs developed to encourage the strengthening of cognitive and motor functions even in old age, both from a preventive perspective with healthy users both from a rehabilitation perspective in clinical contexts [2, 39]. This form of stimulation, however, also presents secondary and unwanted effects—such as the induced activation of facial muscles and the presence of a potentially annoying sound that accompanies the stimulation—therefore requiring specific training and some methodological precautions on the part of researchers and professionals who intend to use it.

2 Recent Evidence and Applications to Affective and Cognitive Domains

In recent years, there has been a rapid increase in NIBS applications in the study of cognition, brain-behavior relationships, and the pathophysiology of various neurological and psychiatric diseases as well as in the development and testing of effectiveness of dedicated neurostimulation and neuromodulation protocols. NIBS was also applied to cognitive functions and emotional regulation and the promotion of psycho-physical well-being and neurocognitive effectiveness even in aging [2, 39].

Although the mechanisms of action of NIBS techniques are not yet fully understood, the effects of TMS are likely related to plastic changes in the target neural circuit associated with long-term potentiation or depression phenomena (long-term potentiation—LTP; long-term depression—LTD; [66]). LTP/LTD phenomena are at the heart of many forms of experience-dependent plasticity in the human brain, especially with regard to learning and memory. TMS can induce long-term effects on cortical excitability, and the presumed changes in LTP/LTD have attracted particular attention due to the important implications for therapies aimed at recovering damaged or compromised brain functions over the entire life cycle, in particular in adulthood and old age [3, 9, 11, 14].

NIBS techniques are considered promising non-invasive treatment tools for a wide variety of neuropsychological and neuropsychiatric conditions that can characterize the different phases of the life cycle—such as depressive pathologies, post-traumatic stress disorder, substance use disorders, hallucinatory manifestations, Parkinson's disease and other neurodegenerative forms, aphasia, and other neuropsychological disorders, or neuropathic pain syndromes [50, 51, 61, 66]. Specifically, TMS has special (FDA) approvals for treatment of major depression. NIBS has also demonstrated, through experimental clinical protocols, that it can improve the motor and cognitive symptoms of elderly patients who demonstrate neurodegeneration, offering a valuable complement to pharmacological therapy in Parkinson's disease, Alzheimer's disease, and other forms of dementia. Furthermore, neurostimulation and neuromodulation techniques can be used to promote functional recovery after a stroke, facilitating motor, and cognitive rehabilitation of affected older adults [2, 39, 75].

2.1 Emotion Processing and Recognition

Recent applications of the TMS technique concern the domain of emotions, their regulation and, in some cases, the possibility of counteracting dysfunctional processes (emotional dysregulation).

At this regard, the dual system model [40] is one of the most relevant neurofunctional models in research on emotion recognition and regulation processes, this model postulates the distinct contribution of a left-lateralized vs. right-lateralized prefrontal neural system in response to distinct-valenced emotional events. In particular, the model supports evidence that left prefrontal structures would mediate positive emotional experience and approach behaviors, which supports a behavioral activation system. In contrast, right prefrontal structures would support negative emotional events and avoidance behaviors, defining a behavioral inhibition system. It was observed that the model has also important clinical applications, considering both assessment and intervention practice. Specifically, it was applied to mood disorders and deficits in the abilities to regulate emotions.

The assumptions of the model have been tested via TMS by looking at the effect of induced perturbations with regard to processing of different emotion-laden stimuli, both vocal and visual. In particular, it was shown that right and left prefrontal structures play the expected roles while processing emotional prosody or vocalizations [31], though even evidence in favor of an undifferentiated contribution of the right and left inferior frontal gyrus (IFG) to processing of emotional prosody has been reported [43]. Similarly, TMS investigation totally or partly corroborated models assumption even for visual emotion-laden stimuli, such as words and complex pictures [16, 65], and in particular for facial expressions [58], though—even in the case of visual stimuli—negative results concerning valence-related lateralization of prefrontal areas have also been reported [34]. Remarkably, a consistent lateralized prefrontal contribution was reported even concerning emotional memory and retrieval of affectively connoted material. Indeed, even if contrasting findings have also been reported [69], a quite systematic set of studies suggested that left- and right-sided prefrontal cortices play peculiar roles in encoding and retrieval processes for positive vs. negative words, based on the nature of the stimuli produced (auditory, visual, etc.) and the conditions of the task performed [8, 10, 11]. Conversely, with regard to emotion regulation processes in healthy individuals, TMS-based investigations seem not to corroborate model assumptions, with unexpected reports of lack of mood improvement or even stronger mood decline in healthy participants after high-frequency modulation of the left dlPFC [56].

In addition to the above-mentioned investigations on frontal asymmetries, TMS studies also helped to outline the global organization of the frontal–temporal–parietal network mediating emotion perception and recognition. This network was considered the core neurofunctional mechanism supporting emotional behavior regulation. By using emotionally-connoted pictures of scenes and, in the vast majority of studies, facial expressions, it was shown that proper emotional appraisal in terms of valence and arousal and efficient emotion recognition overall grounds on the interaction between cortical structures within the IFG, dorsal-medial prefrontal area, pre-SMA, anterior insula, posterior superior temporal sulcus (STS), somatosensory cortex, temporoparietal junction (TPJ), and even cerebellum [4, 5, 36, 63, 70].

It is important to underline that, in view of the notable amount of studies on the role and interaction between different cortical structures supporting emotion processing skills, only a few of such studies actually took into consideration additional individual factors—such as personality traits or measures of empathy—that have a strong connection with the development of investigated affective processes and that might account for partly unexpected empirical findings. As an example, Paracampo and colleagues [63] showed that people that present high vs. low proficiency in emotion perception rely on distinct cortical structures to infer emotional states from facial expressions. As for personality factors, instead, Balconi and colleagues highlighted the mediatory role of individual empathy trait on the medial PFC contribution to emotion recognition processes [4] as well as the mediatory role of anxiety trait and motivational mechanisms on the contribution of lateralized prefrontal areas to recognition and retrieval of emotion-laden verbal material [8, 11].

By taking into account the complexity of individual and contextual influences that actually mediate emotion appraisal and perception in real life and keeping in mind the original rationale of social neuroscience as a discipline interested in both how our brains support affective-social processes and how individual, affective, and social factors shape the way our brain works, those examples help to highlight that many relevant questions on the interaction between social-affective processes and individual/contextual characteristics still deserve further investigation.

2.2 Social Perception and Behavior

Moving to TMS-based investigation of social perception and behavior, the most-explored topic likely concerns detection and processing of social signals. In particular, such social skill has been typically investigated via both explicit and implicit measures, by presenting participants with neutral, threatening, and (in a few cases) joyful body stimuli and asking them to classify them or measuring covert indices of corticospinal excitability (e.g., amplitude of MEPs or intracortical facilitation, ICF). Overall, such set of studies highlighted that processing of social threat is mediated by the combined contribution of the left dlPFC, right ventral premotor cortex, right posterior STS, right anterior IPS, and right IPL [20, 55, 67].

Besides exploring such basic social perception skill, TMS has also been used to probe the neural underpinnings of higher social cognition skills and demonstrate the distinct role of cortical structures within the medial prefrontal cortex and right TPJ with regard to agency-attribution, self-other-distinction and perspective-taking [28, 41, 42], mentalization (understood as the ability to create theories about others' thought, intentions, and beliefs) [42], social judgment [35], and overcoming of self-centeredness and self-enhancement [73].

Finally, a third main research trend in TMS-based research on social skills focused on social behavior and, in particular, on prosocial attitudes. In line with the vast neuroimaging and electrophysiological literature on the preferential involvement of frontal lobe cortices in higher social processes, such as social reasoning, moral decision-making and interpersonal fairness, NIBS findings showed a remarkable consistency in confirming the causal role of dlPFC regions—especially right-sided structures—in controlling selfish behavior and exercising forgiveness, as well as in empathic resonance and in fostering supportive behavior [6, 57].

The present collection of research data has the merit of having complemented available correlational evidence and of having further confirmed the causal role of specific cortical structures within the neural networks that support some of basic social perception, cognition, and behavior skills. It has nonetheless to be acknowledged that—notwithstanding the progressive tendency toward a greater and greater attention to the ecological value of stimuli, tasks, and experimental settings—ecological validity of laboratory-based research in social neuroscience is still a critical issue. And this point becomes even more critical when the focuses of analysis are processes and functions that support and shape interpersonal interactions and everyday social dynamics. Given the overall consistency of available evidence base, it might be now the right time to step forward and begin to focus also on complex interaction skills and to move toward systematic investigation of neural underpinnings of social processes when they are put in play during actual social exchanges, following a new recent trend that is influencing neuroimaging, electrophysiological and psychophysiological research in social neuroscience.

2.3 Decision-Making

A recently developed domain with promising empirical results concerns decision-making processes.

TMS literature exploring cognitive and neural processing underlying decision-making focused on both basic mechanisms explaining evidence accumulation and speed-accuracy trade-off in simple perceptual decision-making tasks and on more complex processes guiding and shaping higher decision-making, e.g., moral and economic strategic tasks, which belong to the field of social decision-making.

In daily life, goal-directed motor behavior depends on evaluating the rewards and costs of different actions and selecting the appropriate movements. Motor decision-making involves choosing a motor plan from a set of competing actions based on their expected value. Recently, Tecilla et al. [76] conducted a review that emphasized the effectiveness of distinct TMS paradigms in characterizing the role of motor areas and other neural networks to cued action selection. They also explore how motivational information, such as reward and biomechanical cost, guides decision-making based on TMS findings, demonstrating that TMS can be a powerful tool for elucidating the neurophysiological mechanisms underlying human action choices.

Regarding basic mechanisms involved in decision-making, Willacker et al. [81] research shown that theta-burst TMS over the dorsal-medial frontal cortex (dMFC) may have a significant impact not only in motor control, but also in increasing accuracy in a perceptual decision-making task. Moreover, Lawson et al. [49] examined how inhibitory TMS affected choice bias, showing how modulable the cognitive-motor processes involved in decision-making. Inhibitory TMS to either the left or right motor cortex resulted in a significant bias in screen side preference during a choice test but did not influence reaction speeds. Current evidence indicates that TMS can unconsciously influence the subthreshold cognitive-motor processes involved in decision-making.

The speed-accuracy trade-off (SAT) is a well-known phenomenon that mirrors the balancing between speed and accuracy while responding to simple decision-making tasks. Neuroimaging findings highlighted that such balancing is modulated by the contribution of dlPFC and pre-SMA regions, and TMS-based investigations corroborated such correlational observations. Georgiev and colleagues [38], for example, by applying continuous theta burst stimulation (TBS) to the right dlPFC and pre-SMA regions noted that perturbing the former slowed the rate of evidence accumulation (typically referred to as the drift rate) for decision-making, while perturbing the latter altered accuracy more than speed of processing. A consistent pattern of reduce accuracy and increased response times (i.e., altered drift rate) during a speeded perceptual categorization task was found even by Philiastides et al. [64] after inhibition of the left dorsolateral prefrontal cortex via low-frequency rTMS. Again, inhibiting the right pre-SMA area via continuous TBS seemed to make individuals more cautious and to increase decision threshold in order to counterbalance accuracy bias during decision-making [77].

A few TMS studies focused, besides dlPFC and pre-SMA, on the role of motor regions during simple decision-making tasks. In particular, it was shown that implicit decisional rules conveyed by biased rewarding can not only covertly modulate behavioral responses (thus increasing the selection of covertly facilitated responses) but also consistently modulate pre-decisional corticospinal excitability, as measured via TMS-induced motor evoked potentials [45]. In addition, it has been reported that both corticospinal excitability and reaction times can vary during decision-making as a function of the value that is subjectively attributed to different response alternatives [46]. This suggest that subjectively defined values that guide the decision-making process might even modulate the competition between different representations of alternative responses in motor cortex while the decisional process is still ongoing, consistent with a form of parallel processing of decision and response selection.

Moving to complex decision-making processes, social decision-making aims at understanding the brain processes involved in selecting an option in a social setting [68]. In social decision-making, NIBS have been used to examine the causal role of neural regions during cooperative or competitive scenarios simulated using basic games inspired by behavioral economics. Typically, correlational investigations using neuroimaging techniques are used to determine brain targets for TMS modulation.

In this context, findings suggested that prefrontal areas still remain the primary targets of non-invasive stimulation studies.

The following paragraphs analytically report some research results, highlighting from time to time the central role, albeit different depending on the contrasts in which the decision-making is analyzed, of the DLPFC sector.

The right dlPFC has been associated to appraisal and higher processes underlying the computation of values that guide decisions in economic decision-making tasks [19]. Furthermore, dlPFC proved to be involved, regardless of lateralization, in impulsivity and cognitive control during decision-making. Indeed, inhibition of dlPFC activity was associated to a notable reduction of the magnitude effect, i.e. the tendency to become more patient and cautious as the values of alternative choice options increase [15]. Similarly, prefrontal inhibition proved to modulate even behavioral consequences of the decoy effect, i.e., the tendency to re-appraise alternative relevant choices during decision-making when an additional irrelevant choice is added. Namely, perturbation of right inferior frontal gyrus increased the number of biased choices due to the activation of heuristics and context-dependent evaluation processes [25].

The medial PFC (mPFC), which has been well recognized in neuroimaging research, is another region investigated in social decision-making. Specifically, the mPFC was demonstrated to play a role in social conformity during an attractiveness rating task, where participants rated the attractiveness of models in photos [47]. Downregulating the mPFC reduced social conformity, suggesting that this area is involved in social learning and adapting to others' expectations in social decision-making contexts. In another work, Civai et al. [26] explored the mPFC's contribution in accepting unfairness when unfair proposals were directed at oneself versus another person. Authors showed that inhibiting the mPFC increased the acceptance rate of unfair proposals, suggesting the MPFC's causal role in processing fairness in self-related situations.

Although the latter study employed tDCS instead of TMS, taken together this evidence suggests the importance of the medial portions of the PFC as targets for social decision-making, as the MFPC plays a role in controlled cognition to implement self-related decisions and calculate the expectations of others in context.

Higher strategic decision-making processes have been typically explored by using economic decision tasks—such as the Ultimatum Game—or cooperative/competitive tasks—such as the Prisoner's Dilemma. In particular, it was shown that interfering with dlPFC activity via TMS decreases the tendency to adopt a cooperative choice when individuals have to choose between a mutual or a selfish strategy in an iterative version of the prisoner's dilemma [72]. Similarly, perturbing the right dlPFC via rTMS alters the effective definition of decision-making strategy during an Ultimatum Game task [80]. And again, inhibition of the right inferior frontal gyrus resulted in a notable increase of irrationally aggressive (though winning) strategy in a repeated economic predator-prey task, while the enhancement of right inferior frontal gyrus activity resulted in selection of more rational and strategic choices [29]. Considering other cortical sites, it was found that the stimulation of dorsomedial PFC through TMS impaired belief updating and strategic behavioral adjustments after negative prediction errors in a prisoner's dilemma game Christian et al. [24]. In another study exploiting the Ultimatum Game, interfering with the right temporoparietal junction via a train of TMS pulses had an impact on responders' capacity to adopt proposers' viewpoints, leading to increased rates of unfair offer rejection when proposals are viewed as apart from responders' deliberate inequity [62].

Stimulation of prefrontal cortex highlighted their contribution even when individuals receive, appraise and decide whether to accept or refuse economic offers, instead of actively proposing them. Liu and colleagues [52] recently observed that enhancing the activity within prefrontal areas lowers the acceptance rate of unfair offers at an economic decision-making task, while inhibiting such areas resulted in increased acceptance rates. Such findings once more confirm the role of prefrontal structures as core elements in a decision-making network. Interestingly, it was also reported that by inhibiting right dlPFC it is possible to reduce willingness to reject unfair offers, while not affecting explicit judgment of unfairness [48].

Focusing on risky decision-making, risk tolerance, and impulsivity, TMS-based research once again consistently underlined the crucial role of prefrontal structures. In particular, the inhibition of the right dorsolateral prefrontal cortex via low-frequency repetitive TMS resulted in both the increase of risky behavior and overestimation errors during a ball-throwing task [79] and in significantly riskier decision-making strategies in a gambling task [48]. Nonetheless, it has to be acknowledged that perturbation of the right dlPFC via continuous theta burst stimulation (cTBS) contrastingly resulted in a decrease of impulsivity at a delay discounting task [21, 22]. In line with this evidence, a more recent study utilizing cTBS demonstrated that the right (but not left) dlPFC stimulation increasing sensitivity to reinforcers and led to avoiding hazardous decisions and encouraging favorable ones during the Iowa Gambling Task [60].

In the clinical field of addiction, according to a meta-analysis based on 26 research, the application of rTMS over the left DLPFC reduces craving and the bilateral dlPFC reduces the substance consumption, compared to sham stimulation [82]. For substance use disorders, the dlPFC appears to be the most accurate area to target, and generally, rTMS targets the left DLPFC at a low frequency to inhibit it [78], or the right DLPFC with a high frequency to excite the area [1], however the effects derived from the lateralization are still debated [32, 54, 71]. Interestingly, the insula and mPFC, which are implicated in the maladaptive response in a stressful or rewarding situation and impact decision-making, may also be viable TMS targets [33]. To date, only one double-blinded, randomized study using bilateral stimulation over the dlPFC and insula was applied to cigarette smoker. Results showed that high-frequency rTMS, as opposed to low-frequency rTMS and sham rTMS, significantly reduced the number of cigarettes consumed and increased the abstinence rate, especially when applied in conjunction with smoking cue exposure [30].

Even inferior parietal areas seem to play a role in the decision-making process. Coutlee et al. [27], indeed, reported that, whereas inhibition of the inferior frontal area generally slowed decision times, the inhibition of the intraparietal sulcus area quite specifically reduced risk-taking and biased participants' choices toward safer and certain options (low risk-low reward). Lohse et al. [53] demonstrated that the implementation of general impulsive inclinations (trait impulsivity) into actual conduct (state impulsivity) is causally related to the pre-supplementary motor area (pre-SMA). Findings showed an overall decoupling of trait impulsivity and state impulsivity during gambling after real-rTMS, with low-impulsivity individuals showing a rise in preference for hazardous options and high-impulsivity participants showing the reverse pattern. Pre-SMA activation changed in a way that was consistent with the behavioral change induced by rTMS.

Finally, TMS-based research on moral decision-making rather equally underlined the involvement of prefrontal and inferior parietal areas—namely, the temporoparietal junction (TPJ). By using a novel donation-decision task (i.e., a decision-making task in which decision also has moral connotations, e.g., donating to a good cause or sending money to a bad cause), Obeso et al. [59] showed that inhibiting the right TPJ via repetitive TMS did not modulate overall motivation to donate or reactions to social reputation cues but affected how moral and economic pieces of information are integrated to make the final choice to donate or not. Again, by comparing individual responses to personal and impersonal moral dilemmas, Jeurissen et al. [44] observed that inhibition of right dlPFC during the decision-making process influences how participants answer to personal dilemmas and reduce the ratings of regret about the decision, whereas inhibition of the right TPJ influences only impersonal dilemmas. Such findings confirm previous empirical observations concerning increased utilitarian decisions during evaluation of moral dilemmas following the inhibition of right dlPFC, which hinted at the contribution of such structure in regulating and integrating emotional contextual information during the decision-making process [74]. Notably, dlPFC was also implicated into norm-based judgments and determination of blameworthiness and punishment to wrong behaviors. Buckholtz et al. [18] reported that perturbing dlPFC via rTMS reduced the amount of punishment for wrongful acts without affecting blameworthiness ratings. Conversely, in a recent review [23] it was displayed that TMS stimulation over rDLPFC seem does not directly enhance or reduce norm-guided behavior; rather, contextual factors dictate the involvement of the rDLPFC in social interactions The authors specifically noted that when decision makers must weigh the intentions and expectations of their interaction partners for fairness against their own self-interests, rDLPFC primarily encourages norm-guided behavior (reactive fairness); by contrast, prosocial giving (proactive fairness) appears to be unaffected by rDLPFC TMS. These outcomes pave the way for a more thorough investigation of the moral decision-making process that considers both intervening contextual factors and social dynamics.

2.4 Social Processes: Cooperation and More

Thought neuroscience research tradition has initially focused mainly on core elements of cognition—besides sensory, perceptual and motor functions—the interest in neural correlates of human ability to feel and understand affective states as well as to interact and create social bonds grew more and more, leading to the formal definition of the social neuroscience discipline.

In this latter field, the TMS technique has been firstly and most systematically used to complement clinical, psychophysiological, electrophysiological, and neuroimaging research and investigate the contribution of specific cortical structures as primary hubs or as elements of broad networks mediating the ability to recognize and regulate emotions and affective states. Then, a second trend implied the use of such magnetic NIBS technique to explore the neural circuitry that supports human ability to perceive and process social signals, as well as to interact and exert prosocial behaviors.

Thus, in addition to the references to social decision-making and social factors already mentioned in the previous paragraphs, TMS can be employed for determining the contribution of specific neural regions in cooperative or competitive behaviors.

Recent neuroimaging studies highlighted the contribution of dlPFC in prosocial behaviors [13] and PFC lateralization in cooperative behaviors, showing a significant increased left PFC activity in cooperative contexts [12]. In a previous study, high-frequency rTMS was administered to the dlPFC while participants were asked to decide whether or not to directly intervene in emotionally charged social scenarios, including cooperative, non-cooperative, conflictual, and neutral contexts [6]. The findings revealed that, relative to both the sham and control conditions, rTMS stimulation elicited heightened prosocial behavior across all emotional contexts, such as empathic responsiveness. Moreover, the impact of rTMS was particularly pronounced in negative situations, as evidenced by both behavioral responses and electromyography (EMG) measures. Notably, the stimulation of the dlPFC resulted in an augmentation of zygomatic muscle activity during positive scenarios and corrugator muscle activity during negative scenarios, indicating a "facilitation effect" of the emotional response. Regarding lateralization, another study investigated the engagement of the left dlPFC by using low-frequency rTMS to reduce reactions to facial emotional expressions while participants engaged in tasks necessitating empathy toward emotional stimuli [7].

While most research on TMS and social behavior has been conducted in healthy populations, it could be of interest to apply TMS interventions to clinical populations characterized by impaired social functioning. By targeting specific brain regions implicated in social cognition, TMS could have the potential to modulate social behavior and improve interpersonal functioning in these populations.

3 To Conclude: Some Critical Considerations

The present work has summarized the most consistent findings concerning TMS studies on attention and executive functions, memory, language, emotion processing, and social perception and behavior. Without claiming to be exhaustive, this tentative systematization also allowed to point out specific open questions and to outline novel research topic to be better explored via TMS techniques.

As a final remark, the analysis of the evidence base concerning TMS basic research in cognitive and affective neuroscience highlighted even a few critical issues.

Indeed NIBS, in all its various forms, is considered a safe technique, characterized only by mild adverse effects (e.g., tingling or itching sensations, sensations of heat, moderate fatigue or headache). For this reason, this form of NIBS stands out for its versatility and usability and has become a technique of choice in the construction of innovative neuroscientific intervention programs, both in the neurorehabilitation field and, albeit still to an extent minor, of neuroenhancement [2]. Nonetheless, requiring the use of an electromedical instrument and involving the exogenous stimulation of nervous tissues, the management of NIBS protocols requires technical-methodological training and compliance with shared standards of practice, for which the interested reader is referred to the specific literature [17, 66].

More specifically, it has to be acknowledged that research in cognitive-affective and, in particular, social neuroscience typically targets complex and highly interdependent processes. It is often rather difficult, then, to model and parcel out mechanisms or processing steps that clearly and solely identify a specific cognitive function. As a consequence, implementing TMS-based investigation of higher cognitive-affective-social processes requires strict and reasoned theoretical and methodological assumptions, especially if there is a lack of consolidated reference models for the explanation of complex processes such as higher level ones. Implications of such first point could be accounted for by assuming a multi-method investigation approach—i.e. by integrating behavioral outcome measures with hemodynamic and electrophysiological measures to better outline the effects of TMS-induced perturbations on global neural dynamics and functional connections within the network of structures supporting the investigated functions. An additional critical issue is the well-known limitation concerning the depth of TMS-induced perturbations. It is, indeed, impossible to selectively reach and modulate the activity of deep structures by using such technique. It is then not possible to exploit the explanatory potential of the technique to investigate the causal role of many relevant subcortical nodes of networks mediating higher functions, especially affective and social ones. Finally, much still needs to be done to properly devise and implement valid NIBS intervention protocols targeting many higher cognitive, affective and social skills. Further basic research on long-term behavioral and functional effects induced by TMS protocols in both healthy participants and neurology-psychiatry patients are indeed needed to accurately assess the potential of non-invasive magnetic stimulation as a supportive tool for early intervention at prodromic or initial clinical phases and, in particular, for neurocognitive empowerment, besides neurorehabilitation. A last critical point is, then, represented by the ethical implications of using TMS—and, more generally, NIBS—as a tool to pre-emptively induce the modulation of neural activity and of individual cognitive-affective-social profile, especially in healthy individuals. Possible implications in terms of undesired secondary effects of NIBS protocols, as well as of threats to the core concepts of human authenticity and inter-individual equity are yet to be fully explored. Between the other the equity critical issue may be related to the fact that the access to treatment is not guaranteed by default to all individuals, given cultural, economic, social differences.

References

1. Amiaz R, Levy D, Vainiger D, et al. Repeated high-frequency transcranial magnetic stimulation over the dorsolateral prefrontal cortex reduces cigarette craving and consumption. Addiction. 2009;104:653–60. https://doi.org/10.1111/J.1360-0443.2008.02448.X.
2. Antal A, Luber B, Brem AK, et al. Non-invasive brain stimulation and neuroenhancement. Clin Neurophysiol Pract. 2022;7:146–65. https://doi.org/10.1016/J.CNP.2022.05.002.
3. Balconi M. Dorsolateral prefrontal cortex, working memory and episodic memory processes: insight through transcranial magnetic stimulation techniques. Neurosci Bull. 2013;29:381–9. https://doi.org/10.1007/S12264-013-1309-Z.
4. Balconi M, Bortolotti A. Emotional face recognition, empathic trait (BEES), and cortical contribution in response to positive and negative cues. The effect of rTMS on dorsal medial prefrontal cortex. Cogn Neurodyn. 2013;7:13–21. https://doi.org/10.1007/s11571-012-9210-4.
5. Balconi M, Canavesio Y. High-frequency rTMS improves facial mimicry and detection responses in an empathic emotional task. Neuroscience. 2013;236:12–20. https://doi.org/10.1016/j.neuroscience.2012.12.059.
6. Balconi M, Canavesio Y. High-frequency rTMS on DLPFC increases prosocial attitude in case of decision to support people. Soc Neurosci. 2014;9:82–93. https://doi.org/10.1080/17470919.2013.861361.
7. Balconi M, Canavesio Y. Empathy, approach attitude, and rTMs on left DLPFC affect emotional face recognition and facial feedback (EMG). J Psychophysiol. 2016;30:17–28. https://doi.org/10.1027/0269-8803/A000150.
8. Balconi M, Cobelli C. Motivational mechanisms (BAS) and prefrontal cortical activation contribute to recognition memory for emotional words. rTMS effect on performance and EEG (alpha band) measures. Brain Lang. 2014;137:77–85. https://doi.org/10.1016/j.bandl.2014.08.006.
9. Balconi M, Crivelli D. Transcranial magnetic stimulation for cognitive neurosciences: applications and open questions. In: Non invasive brain stimulation in psychiatry and clinical neurosciences. Cham: Springer; 2020. p. 191–208. https://doi.org/10.1007/978-3-030-43356-7_14.
10. Balconi M, Ferrari C. rTMS stimulation on left DLPFC increases the correct recognition of memories for emotional target and distractor words. Cogn Affect Behav Neurosci. 2012;12:589–98. https://doi.org/10.3758/s13415-012-0090-1.
11. Balconi M, Ferrari C. Repeated transcranial magnetic stimulation on dorsolateral prefrontal cortex improves performance in emotional memory retrieval as a function of level of anxiety and stimulus valence. Psychiatry Clin Neurosci. 2013;67:210–8. https://doi.org/10.1111/PCN.12041.
12. Balconi M, Crivelli D, Vanutelli ME. Why to cooperate is better than to compete: brain and personality components. BMC Neurosci. 2017;18:68. https://doi.org/10.1186/S12868-017-0386-8.
13. Balconi M, Fronda G, Vanutelli ME. Donate or receive? Social hyperscanning application with fNIRS. Curr Psychol. 2019;38:991–1002. https://doi.org/10.1007/S12144-019-00247-4/METRICS.
14. Balconi M, Angioletti L, Cassioli F, Crivelli D. Neurocognitive empowerment in healthy aging: a pilot study on the effect of non-invasive brain stimulation on executive functions. J Cogn Enhanc. 2021;5:343–50. https://doi.org/10.1007/S41465-020-00203-2/METRICS.
15. Ballard IC, Aydogan G, Kim B, McClure SM. Causal evidence for the dependence of the magnitude effect on dorsolateral prefrontal cortex. Sci Rep. 2018;8(1):16545. https://doi.org/10.1038/s41598-018-34900-y.
16. Berger C, Domes G, Balschat J, et al. Effects of prefrontal rTMS on autonomic reactions to affective pictures. J Neural Transm. 2017;124:139–52. https://doi.org/10.1007/s00702-015-1491-4.
17. Bikson M, Grossman P, Thomas C, et al. Safety of transcranial direct current stimulation: evidence based update 2016. Brain Stimul. 2016;9:641–61. https://doi.org/10.1016/J.BRS.2016.06.004.

18. Buckholtz JW, Martin JW, Treadway MT, et al. From blame to punishment: disrupting prefrontal cortex activity reveals norm enforcement mechanisms. Neuron. 2015;87:1369–80. https://doi.org/10.1016/J.NEURON.2015.08.023.
19. Camus M, Halelamien N, Plassmann H, et al. Repetitive transcranial magnetic stimulation over the right dorsolateral prefrontal cortex decreases valuations during food choices. Eur J Neurosci. 2009;30:1980–8. https://doi.org/10.1111/J.1460-9568.2009.06991.X.
20. Candidi M, Stienen BMC, Aglioti SM, de Gelder B. Event-related repetitive transcranial magnetic stimulation of posterior superior temporal sulcus improves the detection of threatening postural changes in human bodies. J Neurosci. 2011;31:17547–54. https://doi.org/10.1523/JNEUROSCI.0697-11.2011.
21. Cho SS, Ko JH, Pellecchia G, et al. Continuous theta burst stimulation of right dorsolateral prefrontal cortex induces changes in impulsivity level. Brain Stimul. 2010;3:170–6. https://doi.org/10.1016/J.BRS.2009.10.002.
22. Cho SS, Pellecchia G, Ko JH, et al. Effect of continuous theta burst stimulation of the right dorsolateral prefrontal cortex on cerebral blood flow changes during decision making. Brain Stimul. 2012;5:116–23. https://doi.org/10.1016/J.BRS.2012.03.007.
23. Christian P, Soutschek A. Causal role of right dorsolateral prefrontal cortex for norm-guided social decision making: a meta-analysis of TMS studies. Neuropsychologia. 2022;176:108393. https://doi.org/10.1016/J.NEUROPSYCHOLOGIA.2022.108393.
24. Christian P, Kaiser J, Taylor PC, et al. Belief updating during social interactions: neural dynamics and causal role of dorsomedial prefrontal cortex. J Neurosci. 2024;44:e1669232024. https://doi.org/10.1523/JNEUROSCI.1669-23.2024.
25. Chung HK, Sjöström T, Lee HJ, et al. Why do irrelevant alternatives matter? An fMRI-TMS study of context-dependent preferences. J Neurosci. 2017;37:11647–61. https://doi.org/10.1523/JNEUROSCI.2307-16.2017.
26. Civai C, Miniussi C, Rumiati RI. Medial prefrontal cortex reacts to unfairness if this damages the self: a tDCS study. Soc Cogn Affect Neurosci. 2015;10:1054–60. https://doi.org/10.1093/SCAN/NSU154.
27. Coutlee CG, Kiyonaga A, Korb FM, et al. Reduced risk-taking following disruption of the intraparietal sulcus. Front Neurosci. 2016;10:214756. https://doi.org/10.3389/FNINS.2016.00588/BIBTEX.
28. Crivelli D, Balconi M. The agent brain: a review of non-invasive brain stimulation studies on sensing agency. Front Behav Neurosci. 2017;11:229. https://doi.org/10.3389/fnbeh.2017.00229.
29. de Dreu CKW, Kret ME, Sligte IG. Modulating prefrontal control in humans reveals distinct pathways to competitive success and collective waste. Soc Cogn Affect Neurosci. 2016;11:1236–44. https://doi.org/10.1093/SCAN/NSW045.
30. Dinur-Klein L, Dannon P, Hadar A, et al. Smoking cessation induced by deep repetitive transcranial magnetic stimulation of the prefrontal and insular cortices: a prospective, randomized controlled trial. Biol Psychiatry. 2014;76:742–9. https://doi.org/10.1016/J.BIOPSYCH.2014.05.020.
31. Donhauser PW, Belin P, Grosbras M-H. Biasing the perception of ambiguous vocal affect: a TMS study on frontal asymmetry. Soc Cogn Affect Neurosci. 2014;9:1046–51. https://doi.org/10.1093/scan/nst080.
32. Dubuson M, Dousset C, Noël X, Campanella S. Neuromodulation techniques in the treatment of addictions. In: Balconi M, Campanella S, editors. Advances in substance and behavioral addiction. Advances in mental health and addiction. Cham: Springer; 2021. p. 169–96. https://doi.org/10.1007/978-3-030-82408-2_7.
33. Euston DR, Gruber AJ, McNaughton BL. The role of medial prefrontal cortex in memory and decision making. Neuron. 2012;76:1057–70. https://doi.org/10.1016/J.NEURON.2012.12.002.
34. Ferrari C, Gamond L, Gallucci M, et al. An exploratory TMS study on prefrontal lateralization in valence categorization of facial expressions. Exp Psychol. 2017a;64:282–9. https://doi.org/10.1027/1618-3169/a000363.

35. Ferrari C, Nadal M, Schiavi S, et al. The dorsomedial prefrontal cortex mediates the interaction between moral and aesthetic valuation: a TMS study on the beauty-is-good stereotype. Soc Cogn Affect Neurosci. 2017b;12:707–17. https://doi.org/10.1093/scan/nsx002.
36. Ferrari C, Oldrati V, Gallucci M, et al. The role of the cerebellum in explicit and incidental processing of facial emotional expressions: a study with transcranial magnetic stimulation. NeuroImage. 2018;169:256–64. https://doi.org/10.1016/j.neuroimage.2017.12.026.
37. Fregni F, Pascual-Leone A. Technology insight: noninvasive brain stimulation in neurology-perspectives on the therapeutic potential of rTMS and tDCS. Nat Clin Pract Neurol. 2007;3:383–93. https://doi.org/10.1038/NCPNEURO0530.
38. Georgiev D, Rocchi L, Tocco P, et al. Continuous theta burst stimulation over the dorsolateral prefrontal cortex and the pre-SMA Alter drift rate and response thresholds respectively during perceptual decision-making. Brain Stimul. 2016;9:601–8. https://doi.org/10.1016/J.BRS.2016.04.004.
39. Gomes-Osman J, Indahlastari A, Fried PJ, et al. Non-invasive brain stimulation: probing Intracortical circuits and improving cognition in the aging brain. Front Aging Neurosci. 2018;10 https://doi.org/10.3389/FNAGI.2018.00177.
40. Harmon-Jones E. Asymmetrical frontal cortical activity, affective valence, and motivation direction. In: Harmon-Jones E, Winkielman P, editors. Social neuroscience. Integrating biological and psychological explanations of social behavior. New York: The Guildford Press; 2007. p. 137–56.
41. Heinisch C, Krüger MC, Brüne M. Repetitive transcranial magnetic stimulation over the temporoparietal junction influences distinction of self from famous but not unfamiliar others. Behav Neurosci. 2012;126:792–6. https://doi.org/10.1037/a0030581.
42. Hill CA, Suzuki S, Polania R, et al. A causal account of the brain network computations underlying strategic social behavior. Nat Neurosci. 2017;20:1142–9. https://doi.org/10.1038/nn.4602.
43. Hoekert M, Vingerhoets G, Aleman A. Results of a pilot study on the involvement of bilateral inferior frontal gyri in emotional prosody perception: an rTMS study. BMC Neurosci. 2010;11:93. https://doi.org/10.1186/1471-2202-11-93.
44. Jeurissen D, Sack AT, Roebroeck A, et al. TMS affects moral judgment, showing the role of DLPFC and TPJ in cognitive and emotional processing. Front Neurosci. 2014;8:62910. https://doi.org/10.3389/FNINS.2014.00018/BIBTEX.
45. Klein PA, Olivier E, Duque J. Influence of reward on corticospinal excitability during movement preparation. J Neurosci. 2012;32:18124–36. https://doi.org/10.1523/JNEUROSCI.1701-12.2012.
46. Klein-Flügge MC, Bestmann S. Time-dependent changes in human corticospinal excitability reveal value-based competition for action during decision processing. J Neurosci. 2012;32:8373–82. https://doi.org/10.1523/JNEUROSCI.0270-12.2012.
47. Klucharev V, Munneke MAM, Smidts A, Fernández G. Downregulation of the posterior medial frontal cortex prevents social conformity. J Neurosci. 2011;31:11934–40. https://doi.org/10.1523/JNEUROSCI.1869-11.2011.
48. Knoch D, Pascual-Leone A, Meyer K, et al. Diminishing reciprocal fairness by disrupting the right prefrontal cortex. Science. 2006;314:829–32. https://doi.org/10.1126/science.1129156.
49. Lawson L, Spivak S, Webber H, et al. Alterations in brain activity induced by transcranial magnetic stimulation and their relation to decision making. Biology. 2023;12 https://doi.org/10.3390/BIOLOGY12111366.
50. Lefaucheur JP, Antal A, Ayache SS, et al. Evidence-based guidelines on the therapeutic use of transcranial direct current stimulation (tDCS). Clin Neurophysiol. 2017;128:56–92. https://doi.org/10.1016/J.CLINPH.2016.10.087.
51. Lefaucheur JP, Aleman A, Baeken C, et al. Evidence-based guidelines on the therapeutic use of repetitive transcranial magnetic stimulation (rTMS): an update (2014-2018). Clin Neurophysiol. 2020;131:474–528. https://doi.org/10.1016/J.CLINPH.2019.11.002.
52. Liu S, Liu Y, Ni Y, et al. A review of mindfulness improves decision making and future prospects. Psychology. 2018;9:229–48. https://doi.org/10.4236/PSYCH.2018.92015.

53. Lohse A, Løkkegaard A, Siebner HR, Meder D. Linking impulsivity to activity levels in pre-supplementary motor area during sequential gambling. J Neurosci. 2023;43:1414–21. https://doi.org/10.1523/JNEUROSCI.1287-22.2023.
54. Maiti R, Mishra BR, Hota D. Effect of high-frequency transcranial magnetic stimulation on craving in substance use disorder: a meta-analysis. J Neuropsychiatry Clin Neurosci. 2017;29:160–71. https://doi.org/10.1176/APPI.NEUROPSYCH.16040065.
55. Mazzoni N, Jacobs C, Venuti P, et al. State-dependent TMS reveals representation of affective body movements in the anterior intraparietal cortex. J Neurosci. 2017;37:7231–9. https://doi.org/10.1523/JNEUROSCI.0913-17.2017.
56. Möbius M, Lacomblé L, Meyer T, et al. Repetitive transcranial magnetic stimulation modulates the impact of a negative mood induction. Soc Cogn Affect Neurosci. 2017;12:526–33. https://doi.org/10.1093/scan/nsw180.
57. Müller-Leinß J-M, Enzi B, Flasbeck V, Brüne M. Retaliation or selfishness? An rTMS investigation of the role of the dorsolateral prefrontal cortex in prosocial motives. Soc Neurosci. 2018;13:701–9. https://doi.org/10.1080/17470919.2017.1411828.
58. Notzon S, Steinberg C, Zwanzger P, Junghöfer M. Modulating emotion perception: opposing effects of inhibitory and excitatory prefrontal cortex stimulation. Biol Psych Cogn Neurosci Neuroimag. 2018;3:329–36. https://doi.org/10.1016/j.bpsc.2017.12.007.
59. Obeso I, Moisa M, Ruff CC, Dreher JC. A causal role for right temporo-parietal junction in signaling moral conflict. elife. 2018;7 https://doi.org/10.7554/ELIFE.40671.
60. Obeso I, Herrero MT, Ligneul R, et al. A causal role for the right dorsolateral prefrontal cortex in avoidance of risky choices and making advantageous selections. Neuroscience. 2021;458:166–79. https://doi.org/10.1016/J.NEUROSCIENCE.2020.12.035.
61. Palm U, Segmiller FM, Epple AN, et al. Transcranial direct current stimulation in children and adolescents: a comprehensive review. J Neural Transm (Vienna). 2016;123:1219–34. https://doi.org/10.1007/S00702-016-1572-Z.
62. Panico F, Ferrara A, Sagliano L, Trojano L. The involvement of rTPJ in intention attribution during social decision making: a TMS study. Cogn Affect Behav Neurosci. 2024;24:755. https://doi.org/10.3758/S13415-024-01188-7.
63. Paracampo R, Pirruccio M, Costa M, et al. Visual, sensorimotor and cognitive routes to understanding others' enjoyment: an individual differences rTMS approach to empathic accuracy. Neuropsychologia. 2018;116:86–98. https://doi.org/10.1016/j.neuropsychologia.2018.01.043.
64. Philiastides MG, Auksztulewicz R, Heekeren HR, Blankenburg F. Causal role of dorsolateral prefrontal cortex in human perceptual decision making. Curr Biol. 2011;21:980–3. https://doi.org/10.1016/j.cub.2011.04.034.
65. Roesmann K, Dellert T, Junghoefer M, et al. The causal role of prefrontal hemispheric asymmetry in valence processing of words - insights from a combined cTBS-MEG study. NeuroImage. 2019;191:367–79. https://doi.org/10.1016/j.neuroimage.2019.01.057.
66. Rossini PM, Burke D, Chen R, et al. Non-invasive electrical and magnetic stimulation of the brain, spinal cord, roots and peripheral nerves: basic principles and procedures for routine clinical and research application. An updated report from an I.F.C.N. Committee. Clin Neurophysiol. 2015;126:1071–107. https://doi.org/10.1016/J.CLINPH.2015.02.001.
67. Sagliano L, D'Olimpio F, Panico F, et al. The role of the dorsolateral prefrontal cortex in early threat processing: a TMS study. Soc Cogn Affect Neurosci. 2016;11:1992–8. https://doi.org/10.1093/scan/nsw105.
68. Sanfey AG. Social decision-making: insights from game theory and neuroscience. Science. 2007;318:598–602. https://doi.org/10.1126/SCIENCE.1142996.
69. Schutter DJLG, van Honk J. Increased positive emotional memory after repetitive transcranial magnetic stimulation over the orbitofrontal cortex. J Psychiatry Neurosci. 2006;31:101–4.
70. Sliwinska MW, Pitcher D. TMS demonstrates that both right and left superior temporal sulci are important for facial expression recognition. NeuroImage. 2018;183:394–400. https://doi.org/10.1016/j.neuroimage.2018.08.025.

71. Song S, Zilverstand A, Gui W, et al. Effects of single-session versus multi-session non-invasive brain stimulation on craving and consumption in individuals with drug addiction, eating disorders or obesity: a meta-analysis. Brain Stimul. 2019;12:606–18. https://doi.org/10.1016/J.BRS.2018.12.975.
72. Soutschek A, Sauter M, Schubert T. The importance of the lateral prefrontal cortex for strategic decision making in the prisoner's dilemma. Cogn Affect Behav Neurosci. 2015;15:854–60. https://doi.org/10.3758/S13415-015-0372-5/TABLES/1.
73. Soutschek A, Ruff CC, Strombach T, et al. Brain stimulation reveals crucial role of overcoming self-centeredness in self-control. Sci Adv. 2016;2:e1600992. https://doi.org/10.1126/sciadv.1600992.
74. Tassy S, Oullier O, Duclos Y, et al. Disrupting the right prefrontal cortex alters moral judgement. Soc Cogn Affect Neurosci. 2012;7:282–8. https://doi.org/10.1093/SCAN/NSR008.
75. Tatti E, Rossi S, Innocenti I, et al. Non-invasive brain stimulation of the aging brain: state of the art and future perspectives. Ageing Res Rev. 2016;29:66–89. https://doi.org/10.1016/J.ARR.2016.05.006.
76. Tecilla M, Guerra A, Rocchi L, et al. Action selection and motor decision making: insights from transcranial magnetic stimulation. Brain Sci. 2022;12 https://doi.org/10.3390/BRAINSCI12050639.
77. Tosun T, Berkay D, Sack AT, et al. Inhibition of pre-supplementary motor area by continuous theta burst stimulation leads to more cautious decision-making and more efficient sensory evidence integration. J Cogn Neurosci. 2017;29:1433–44. https://doi.org/10.1162/JOCN_A_01134.
78. Trojak B, Meille V, Achab S, et al. Transcranial magnetic stimulation combined with nicotine replacement therapy for smoking cessation: a randomized controlled trial. Brain Stimul. 2015;8:1168–74. https://doi.org/10.1016/J.BRS.2015.06.004.
79. Tulviste J, Bachmann T. Diminished risk-aversion after right DLPFC stimulation: effects of rTMS on a risky ball throwing task. J Int Neuropsychol Soc. 2019;25:72–8. https://doi.org/10.1017/S1355617718000930.
80. van't Wout M, Kahn RS, Sanfey AG, Aleman A. Affective state and decision-making in the ultimatum game. Exp Brain Res. 2006;169:564–8. https://doi.org/10.1007/S00221-006-0346-5.
81. Willacker L, Roccato M, Can DN, et al. Reducing variability of perceptual decision making with offline theta-burst TMS of dorsal medial frontal cortex. Brain Stimul. 2020;13:1689–96. https://doi.org/10.1016/J.BRS.2020.09.011.
82. Zhang JJQ, Fong KNK, Ouyang R, et al. Effects of repetitive transcranial magnetic stimulation (rTMS) on craving and substance consumption in patients with substance dependence: a systematic review and meta-analysis. Addiction. 2019;114:2137–49. https://doi.org/10.1111/ADD.14753.

Bridging Network Neuroscience and Neuroethics: Can TMS and tACS Solve the Creativity Crisis?

Daniel Moseley and Flavio Fröhlich

1 Introduction

There is an ongoing decline in standardized measurements of creativity. This phenomenon has been dubbed a "creativity crisis" by some educational psychologists. There is interest in exploring novel methods to enhance creativity in order to effectively respond to this challenge. Non-invasive brain stimulation (NIBS) technologies, such as transcranial magnetic stimulation (TMS), transcranial alternating current stimulation (tACS), and transcranial direct current stimulation (tDCS) have the potential to provide a direct, technological solution to this decline in creativity. This chapter examines the potential of NIBS to solve the creativity crisis. We review empirical studies suggesting that these interventions can produce modest improvements in creative thinking and contribute to a deeper neuroscientific understanding of the neural underpinnings of creativity. By targeting specific brain regions implicated in creative cognition, NIBS may facilitate the neural mechanisms that underlie creative thought. However, the application of NIBS for creativity enhancement raises complex ethical considerations. We explore central ethical considerations in this context: the ethical significance of the distinction between treatment and enhancement, debates between bioconservatives and bioliberals, concerns about authenticity, respect for autonomy and cognitive liberty, justice in the allocation of resources, and global health concerns. While NIBS holds promise as a research tool to advance our understanding of creativity, we argue that careful ethical deliberation

D. Moseley (✉)
East Carolina University, Greenville, NC, USA
e-mail: moseleyd24@ecu.ed

F. Fröhlich
University of North Carolina at Chapel Hill, Chapel Hill, NC, USA
e-mail: flavio_frohlich@med.unc.edu

V. Dubljević, J. R. Young (eds.), *TMS and Neuroethics*, Advances in Neuroethics,
https://doi.org/10.1007/978-3-031-92401-9_8

is necessary before pursuing widespread use for enhancement purposes. Interdisciplinary collaboration among neuroscientists, ethicists, and policymakers will be necessary to navigate these challenges effectively and responsibly.

2 Creativity and the Creativity Crisis

Our capacity for creative thinking is the basis for progress in art, science, technology, mathematics, and culture. One worrying trend suggested by recent studies is that there is a significant decline in creativity among students at all educational levels [1]. This phenomenon has been described as a "creativity crisis" and has raised concerns about the future of human innovation [2].

What is creativity? Literally, the term "creative" refers to the ability to create. Ordinary usage of the term has positive and end-related connotations. There has been some recent philosophical debate about whether creativity is a valuable capacity at all and whether it is always directed at valuable ends [3]. Those issues are discussed below in Sect. 6.

Psychologists have identified two cognitive styles frequently associated with creative thinking [4, 5]. *Divergent thinking* involves flexibility, imagination and usually involves the generation of novel ideas. It is a style of thinking that searches for multiple possible answers to questions, and is commonly associated with creativity given its link to innovative thought. *Convergent thinking* involves analytical focus and narrowing down the possibilities to arrive at the best answer (or answers) to a problem, with attention to accuracy and efficiency. It is often associated with general intelligence. However, creative thinking has been found to involve proficiency at alternating between convergent and divergent thinking [4–6].

The Torrance Test of Creative Thinking (TTCT) is the most widely used instrument for evaluating creativity. The TTCT was created in 1966 by the psychologist Ellis Torrance [7]. It was developed to measure creativity in individuals ranging from kindergarten through adulthood, with a focus on evaluating divergent thinking abilities. The TTCT consists of verbal and non-verbal tasks that assess different aspects of divergent thinking: fluency, flexibility, originality, and elaboration in both verbal and non-verbal tasks. The TTCT is a key source of extensive data on creative thinking abilities across diverse populations that have been the basis for longitudinal analysis of creativity trends.

Kyung Hee Kim, an Educational Psychologist, conducted a comprehensive analysis of TTCT scores from American K-12 students from 1966 to 2008 that identifies a noticeable decline in scores starting in the 1990s [2]. The decreases were particularly striking in the dimensions of originality and elaboration. Kim describes the decline as *the creativity crisis, concluding* that the data suggests that students are becoming less adept at generating unique ideas and developing them in detail, which are abilities that are essential for innovation and adaptability.

Several factors have been proposed to explain this decline in creativity. The increasing emphasis on standardized testing and a narrow curriculum that often prioritizes memorization and conformity over exploration and innovative thinking are salient educational factors [8]. Budget cuts and policy shifts have led to diminished support for arts education in schools, and arts education fosters creativity by encouraging self-expression and divergent thinking. The pervasive use of technology and digital media has transformed how individuals engage with information and each other. While technology can be a tool for creativity, excessive passive consumption, such as endless scrolling through social media, may reduce opportunities for imaginative play and independent thought [1]. The rise of AI technologies raises concerns that dependence on algorithm-generated content could undermine creative skills. Some argue that while AI can assist in creative activities, over-reliance on it may lead to a reduction in the human capacity to think creatively without the aid of technology [9].

If human creative capacities are on the decline, there are significant implications for individuals and society. A less creative populace may struggle to develop novel and effective solutions to complex problems such as climate change, public health crises and socioeconomic disparities. Industries also rely on creative thinking for product development, marketing and adapting to market changes. A reduction in creativity could have a negative impact on global economies. Creative thinking also contributes to personal fulfillment and psychological well-being. It enhances problem-solving skills, resilience and the ability to cope with uncertainty. Artistic and cultural advancements also rely on creative expression. The creativity crisis could have negative implications on all of these dimensions of human life.

Network neuroscience may offer insights into addressing the creativity crisis. Electric activity patterns in brain networks, especially synchronized rhythmic activity across different brain areas, have recently been found to represent the biological substrate of cognition and emotion [10]. By studying the neuronal networks and their role in creative thinking, we are likely to improve our understanding of the underlying mechanisms of creativity at a biological level. Non-invasive brain stimulation technologies, such as transcranial magnetic stimulation (TMS) and transcranial alternating current stimulation (tACS) have emerged as tools to modulate rhythmic neural activity associated with creativity [11–13]. These technologies have the potential to enhance creative capacities by targeting specific brain regions and networks involved in divergent and convergent thinking. While still an emerging field, the application of network neuroscience may offer innovative approaches to fostering creativity in educational and professional settings.

3 Non-Invasive Brain Stimulation Technologies

Non-invasive brain stimulation (NIBS) technologies have garnered significant attention for their potential to modulate neural activity without the need for surgical intervention. Standardly, in neuroscience, techniques such as TMS and tACS are considered less invasive in comparison to electroconvulsive therapy (ECT) and deep brain stimulation (DBS). The term "non-invasive" traditionally refers to procedures that do not require incisions into the body or removal of tissue. In the context of NIBS, TMS and tACS are applied externally, delivering electromagnetic or electrical stimulation through the scalp to influence cortical activity [14].

tACS involves applying weak oscillating electrical currents across the scalp at specific frequencies. The low-amplitude electric current generates an exogenous electric field in the brain that interacts with endogenous rhythmic activity patterns [15]. The low stimulation amplitude makes tACS a safe method to perturb rhythmic brain activity and to explore the functional role of targeted activity in cognition and behavior. By synchronizing endogenous neural oscillations, tACS can entrain and influence rhythmic brain activity associated with various cognitive functions, including attention, memory, and creativity. This entrainment allows for targeted modulation of neural networks implicated in creative thinking.

TMS and tACS are generally considered safe when used within established guidelines. The U.S. Food and Drug Administration (FDA) has approved TMS for the treatment of medication-resistant major depressive disorder, obsessive-compulsive disorder and migraine [16, 17]. However, tACS remains an investigational modality that has yet to receive FDA approval or clearance for a clinical indication. Dubljević emphasizes that the regulatory landscape for NIBS technologies is fragmented and lacks comprehensive policies addressing their use for treatment and enhancement purposes [18].

4 Network Neuroscience and Creative Thinking

Network neuroscience emphasizes that cognitive functions arise from dynamic interactions within complex neural networks [14]. Neural oscillations, or rhythmic patterns of neuronal activity, are fundamental to these interactions and facilitate communication across different brain regions. Oscillations at specific frequency bands are associated with distinct cognitive processes. For instance, alpha oscillations (8–12 Hz) are linked to attentional control and information processing. Fröhlich highlights the importance of these oscillations in coordinating neural network activity [14]. By modulating oscillatory patterns, it is possible to influence creative thinking by intervening on the functional connectivity of neural circuits involved in higher-order cognitive functions.

Lustenberger and colleagues examined the role of frontal alpha oscillations in creative thinking [13]. The study hypothesized that enhancing frontal alpha activity could facilitate creative cognition by modulating neural networks associated with divergent thinking. In a double-blind, sham-controlled experiment, 20 healthy participants underwent two sessions involving creative thinking tasks from TTCT. During the active session, individualized alpha-frequency tACS was applied over the frontal cortex. In the sham session, participants received a brief period of stimulation to mimic the sensory experience of tACS without delivering an effective dose of stimulation ("active sham"). The use of individualized frequencies was crucial to align the tACS with each participant's natural alpha rhythm, maximizing the potential for neural entrainment. The findings revealed that participants exhibited significantly higher originality scores during the active tACS session compared to the sham session [13]. The study concludes that this enhancement in creative performance implies that frontal alpha oscillations play a functional role in creative thinking. By synchronizing neuronal activity in the frontal cortex, tACS may facilitate the neural processes underlying the generation of novel ideas. These results support the conclusion that modulating specific neural oscillations can influence cognitive functions. The study provides causal evidence linking frontal alpha activity to creativity, which advances our understanding of the neural mechanisms involved. The main strength of this study is the sham-controlled device and the use of a rigorous, gold-standard assessment of divergent thinking. The main limitations are the small sample size and the lack of direct demonstration that the stimulation modulated alpha oscillations since no EEG recordings were included due to the technical challenges of recording brain activity with EEG during stimulation.

In another study of the relationship between neural activity and cognition, Riddle and Fröhlich discuss how mental activity serves as a bridge between neural biomarkers and psychiatric symptoms [19]. They contend that neural oscillations reflect cognitive and emotional states and can be targeted to modulate these states therapeutically. By identifying neural biomarkers associated with specific mental activities, interventions like tACS can be designed to normalize aberrant neural patterns. Although that study focuses on psychiatric symptoms, the principles apply to cognitive functions such as creativity.

5 TMS and tDCS as Creativity Enhancement Technologies

Studies also suggest that TMS and Transcranial Direct Current Stimulation (tDCS) have the potential to produce modest enhancements in creative thinking and to promote neuroscientific understanding of its basis. Peña and colleagues examine whether combined tDCS and random noise stimulation of the left dorsolateral prefrontal cortex and inferior frontal gyrus would enhance creativity [20]. They report that their intervention induced robust effects on the creativity domains of convergent thinking, divergent thinking and insight problem-solving tasks. Lucchiari and colleagues present a critical review of the literature on the use of tDCS for creativity enhancement, and conclude that tDCS does appear to stimulate certain types of creative thinking [21].

The findings from these studies suggest that NIBS holds promise to address the creativity crisis by directly enhancing the creative capacities of individuals. By targeting neural oscillations linked to creative thinking, it may be possible to facilitate the neural processes that underlie creativity. However, there are feasibility constraints on this proposal. First, there is individual variability. Responses to NIBS can vary widely among individuals due to differences in brain anatomy, baseline neural activity, and susceptibility to neural entrainment [14]. Second, the sustainability of creativity enhancement through NIBS remains uncertain. Further research is needed to determine the duration of effects and whether repeated sessions could lead to lasting changes or unintended consequences. Third, while studies provide evidence of effect, the precise neural mechanisms by which NIBS influence creativity are not fully understood. Advancing our knowledge of the underlying neurophysiology of neural networks is essential for refining interventions and predicting outcomes.

Continued interdisciplinary research integrating network neuroscience and ethical considerations will be crucial in harnessing the benefits of NIBS while addressing feasibility concerns.

6 Creative and Moral Capacities

Before discussing the ethical issues involved in using NIBS to solve the creativity crisis, it is helpful to clarify the relation between our creative and ethical capacities. While creativity is often associated with positive innovation and problem-solving, it is important to recognize that enhancing creative abilities does not inherently lead to moral improvement. The concept of *malevolent creativity, which involves the use of creative thinking for unethical purposes* has gained attention in creativity science [22]. For instance, a bank robber could display a great deal of innovation and divergent thinking in successfully robbing a bank, oran assassin might exhibit creativity in developing methods of killing targets and eluding capture. Some studies suggest that individuals with high creative potential may engage in unethical behavior if they possess certain personality traits or if the environment encourages such actions [23]. So, creativity enhancement technologies could be used for nefarious purposes. On the other hand, enhancing the creative capacities of individuals and groups to prevent crime could lead to innovative strategies for individual and public safety [22]. These scenarios illustrate that creativity is a morally neutral tool: it can be used for praiseworthy or blameworthy pursuits. The potential for NIBS technologies to enhance creativity raises ethical concerns about unintended consequences. If creative capacities are amplified without a corresponding enhancement of moral capacities, there is a risk of facilitating harmful behavior. This underscores the importance of considering the ethical dimensions of creativity enhancement interventions.

There is a vast literature on the relation between cognitive and moral enhancement. Cognitive enhancement involves improving mental processes such as memory, attention, and creativity through the use of biomedical technology. Moral enhancement aims to improve moral action and ethical decision-making, potentially through interventions that influence empathy, altruism, or aggression [24]. Defining "morality" is important in this context, because different philosophical frameworks offer varying conceptions of what constitutes moral behavior [25]. For instance, a utilitarian perspective on moral enhancement would regard improvements that increase the net utility/happiness in the world as moral enhancements. A deontological theory that posits moral duties and rules might consider improvements to creativity that lead people to have innovative forms of moral engagement with the world might count as moral enhancements. These differing perspectives on morality lead to substantially different accounts of what is involved in enhancing our moral capacities. Joshua May argues that moral enhancement is complex due to the multifaceted nature of morality, which encompasses cognitive, emotional, and social components [26]. Enhancing moral capacities may involve improving a wide set of psychological capacities, such as our capacities for moral reasoning, empathy, and self-control. However, there is debate about whether biomedical interventions can truly enhance morality or if they risk infringing on individual autonomy and authenticity [27].

In sum, enhancing our creative capacities through NIBS technologies does not entail that we would also enhance our moral capacities via that process. As we explore the possibilities of NIBS interventions, integrating ethical considerations becomes imperative to ensure that advancements in creativity contribute positively to society.

7 Ethical Considerations in Using NIBS for Creativity Enhancement

7.1 Treatment vs. Enhancement

In medical care and research, it is common for people to make a distinction between using a medical treatment or device with the goal of therapy or enhancement. Therapeutic interventions aim to restore a baseline of health or normal functioning, whereas enhancement uses seek to improve human capacities above a baseline of health or normal functioning [28, 29]. In the case of using NIBS to improve creativity, there would be a difference between using it therapeutically to improve a person's diminished creative capacities to a baseline of health and enhancement uses, which would improve an already healthy and normal range of capacities to a higher-level of performance. Juengst and Moseley provide a detailed overview of the ethical issues involved in making the treatment/enhancement distinction and some of the common objections to using medical resources for enhancement purposes [28]. In the context of creativity enhancement, it is worthwhile exploring to what extent reduced creativity could be considered a facet of psychopathology. For example, could cognitive rigidity in obsessive-compulsive disorder, eating disorders, and depression be related to a loss in creativity? Strikingly, the same stimulation paradigm that was used in the above discussed tACS study has emerged as a promising stimulation modality for the treatment of depression [30, 31].

7.2 Bioconservatives vs. Bioliberals

The conservatives and liberals involved in this debate is not to be confused with political conservatives and liberals. In the context of debate about human enhancement, bioconservatism is the view that we must reject or ban the use of enhancement technologies because they undermine something intrinsically valuable about being human. Fukuyama warns of the potential loss of human dignity and the erosion of essential human qualities [32]. Sandel argues that the pursuit of enhancement undermines the appreciation of natural talents and the acceptance of human limitations, which he refers to as the "giftedness of life." [33] Bioliberals reject these prohibitive stances and contend that it is permissible to use enhancement technologies to improve our human capacities above the baseline of health or normal functioning.

This perspective is defended by Nicholas Agar, John Harris and Julian Savulescu [34–36]. Some bioliberals argue for the mere permissibility of using enhancement technologies, while others argue that there is a duty to enhance our capacities to face the complex problems within human survival. Savulescu argues that we have a moral duty to enhance our physical and cognitive abilities in order to improve human well-being [36]. Harris also presents a utilitarian argument for the bioliberal position that enhancing human capacities is a continuation of humanity's long-standing endeavor to overcome limitations and we must do so with biomedical technologies in order to improve our quality of life [35].

Dubljević contends that in the context of enhancement uses of NIBS there needs to be a thorough assessment of the risks [18]. Since enhancement interventions are not medically necessary, the threshold for acceptable risk may be lower, necessitating stringent safety evaluations. This type of position could open the door to a bioliberal stance if the technology surpasses the threshold of acceptable risk. This type of consumer-protection model is not straightforwardly bioliberal or bioconservative.

7.3 Moral Concerns about Authenticity

Enhancing creativity through NIBS raises questions about the authenticity of creative thought and action. Authenticity pertains to being genuine and true to oneself. Critics argue that artificially augmenting creativity may undermine the authenticity of creative processes and outputs. Elliott explores how biomedical enhancements can affect personal identity and the meaning derived from accomplishments [37]. If creativity results from technological intervention rather than individual effort and innate ability, it may diminish the perceived value and authenticity of creative works. Bioliberals counter that enhancement does not necessarily compromise authenticity. As Douglas suggests, if a person's enhanced capacities are integrated into one's self-concept and contribute to personal development, then that person's accomplishments would be authentic [24]. The debate centers on whether the method by which accomplishments are achieved impacts the value of those endeavors.

7.4 Autonomy and Cognitive Liberty

Respecting individual autonomy is a cornerstone of ethical practice, particularly in medical and research settings. The value of respecting patient autonomy is the basis for laws and policies governing informed consent. Securing the informed consent of patients and research subjects requires that individuals understand the risks, benefits, and alternatives associated with a medical treatment (or enhancement). There are three challenges to autonomy that could arise from using NIBS for the purpose of creativity enhancement. First, there are concerns about securing informed consent to the use of NIBS from vulnerable populations, such as children or persons with cognitive impairments. Second, there are risks associated with the availability of DIY brain stimulation devices. Individuals who attempt to assemble these devices or use them at home may lack the expertise to use them safely, potentially leading to significant harm. Third, there are concerns about coercive pressures to use the enhancement technologies if they become widely available, such as in competitive environments where individuals may feel compelled to use NIBS to enhance performance.

Another type of autonomy-concern that has been gaining traction in the neuroethics literature is a concern with protecting cognitive liberty, which is the right to self-determination over one's own mental processes. Farahany emphasizes the importance of protecting individuals' rights to mental privacy and freedom from unwarranted intrusion [38]. Balancing cognitive liberty involves ensuring that individuals can make informed choices about using NIBS while safeguarding against coercion and undue influence.

7.5 Justice in Distribution

There are also ethical considerations about the fair distribution of benefits and burdens associated with NIBS technologies. If NIBS are expensive, access may be limited to the wealthy and the benefits of their access could exacerbate economic disparities. Enhanced creativity could confer competitive advantages in education and employment and deepen existing inequalities. Moreover, there are concerns about the representation of racialized minorities in research. Studies on NIBS may not adequately represent racialized minorities and generate findings that are not generalizable. Underrepresentation of minority groups can perpetuate health disparities and limit the equitable application of benefits. Policies that promote equitable access and address potential disparities in the development and deployment of NIBS technologies may be developed to address these distributive justice concerns.

7.6 Global Health Concerns

Applying NIBS technologies globally introduces complex ethical and policy challenges. There is a lot of variation in cultural understandings of creativity and its value. The promotion of NIBS for enhancing creativity on the global scale may conflict with conceptions of creativity and its value in these cultures and impose the conceptual frameworks and values of high-income countries in areas that do not share those conceptions. Moreover, in low-income countries, the prioritizing of enhancement technologies could divert resources from essential healthcare needs.

Limited infrastructure and access to technology in some regions make the implementation of NIBS for creativity enhancement impractical. These economic limitations could exacerbate wealth disparities on the global scale.

8 Concluding Remarks

The use of NIBS for creativity enhancement presents a multifaceted ethical landscape. The exploration of NIBS technologies as a means to enhance creativity presents intriguing possibilities and significant challenges. While the allure of a technological solution to the creativity crisis is compelling, we must critically assess whether public resources should be allocated toward making NIBS widely accessible for creative enhancement, especially when traditional methods of fostering creativity face resource constraints.

We have examined the central ethical considerations: the ethical significance of the distinction between treatment and enhancement, concerns about undermining fundamental features of human nature, concerns about authenticity, respecting autonomy and cognitive liberty, promoting justice and equity, and global health issues.

NIBS holds significant promise as a research tool in network neuroscience. Utilizing NIBS to probe the neural underpinnings of creative thinking can advance our understanding of the complex neural networks and oscillatory dynamics that give rise to creativity. This knowledge has the potential to inform not only neuroscience but also educational strategies and interventions that cultivate creativity through evidence-based practices.

Interdisciplinary collaboration is essential as we navigate the intersection of neuroscience, neuroethics, and public policy. The emerging discourse on cognitive liberty and "neurorights" is particularly pertinent. As we develop technologies that can influence cognitive functions, safeguarding individual rights to mental privacy and autonomy becomes paramount. Ethicists, neuroscientists, legal scholars, and policymakers must work together to establish frameworks that promote innovation while protecting individual freedoms.

In conclusion, while NIBS may not be the panacea for the creativity crisis, its role in advancing our understanding of the neurological basis of creativity is invaluable. Prioritizing research that elucidates the neural mechanisms of creative thinking, rather than immediate application for enhancement, is a prudent path forward. Such an approach balances the pursuit of scientific knowledge with ethical responsibility, fostering advancements that can ultimately benefit society in meaningful and equitable ways.

Bibliography

1. Christakis E. Should we worry that American children are becoming less creative? The Washington Post. 2019 Dec 9 [cited 2024 Oct 16]; Available from: https://www.washingtonpost.com/education/2019/12/09/should-we-worry-that-american-children-are-becoming-less-creative/
2. Kim KH. The creativity crisis: the decrease in creative thinking scores on the Torrance tests of creative thinking. Creat Res J. 2011;23(4):285–95. https://doi.org/10.1080/10400419.2011.627805.
3. Brainard L. What is creativity? Philos Q. 2024; https://doi.org/10.1093/pq/pqae075.
4. Ciccarelli SK, White JN. Psychology. 5th ed. Boston, MA: Pearson Education; 2018.

5. Feldman RS. Understanding psychology. 14th ed. New York: McGraw-Hill Education; 2020.
6. Shear J. The creative process. In: The inner dimension: philosophy and the experience of consciousness. New York: Peter Lang; 1990. p. 119–40.
7. Torrance EP. Torrance tests of creative thinking: norms-technical manual. Princeton, NJ: Personnel Press; 1966.
8. Robinson K. Out of our minds: learning to be creative. Chichester: Capstone Publishing; 2011.
9. Uzzi B. Will AI kill human creativity? Forbes. March 29, 2023.
10. Fries P. Rhythms for cognition. Neuron. 2015;88(1):220–35. https://doi.org/10.1016/j.neuron.2015.09.034.
11. Grabner RH, Krenn J, Fink A, Arendasy M, Benedek M. Effects of alpha and gamma transcranial alternating current stimulation (tACS) on verbal creativity and intelligence test performance. Neuropsychologia. 2018;118(Pt A):91–8. https://doi.org/10.1016/j.neuropsychologia.2017.10.035.
12. Santarnecchi E, Sprugnoli G, Bricolo E, Costanzo F, Bonetti L, Ferrone S, et al. Gamma tACS over the temporal lobe increases the occurrence of Eureka! moments. Sci Rep. 2019;9(1):5778. https://doi.org/10.1038/s41598-019-42192-z.
13. Lustenberger C, Boyle MR, Foulser AA, Mellin JM, Fröhlich F. Functional role of frontal alpha oscillations in creativity. Cortex. 2015;67:74–82. https://doi.org/10.1016/j.cortex.2015.03.012.
14. Fröhlich F. Network neuroscience. London: Academic Press (Elsevier); 2016.
15. Huang WA, Stitt IM, Negahbani E, Passey DJ, Luftig D, Baruah V, et al. Transcranial alternating current stimulation entrains alpha oscillations by preferential phase synchronization of fast-spiking cortical neurons to stimulation waveform. Nat Commun. 2021;12(1):3151. https://doi.org/10.1038/s41467-021-23021-2.
16. U.S. Food and Drug Administration. FDA clears first transcranial magnetic stimulation device for treatment of depression. Silver Spring, MD: FDA; 2008.
17. Food US, Administration D. FDA allows marketing of first device to relieve migraine headache pain. Silver Spring, MD: FDA; 2013.
18. Dubljević V. Neurostimulation devices for cognitive enhancement: toward a comprehensive regulatory framework. Neuroethics. 2015;8(2):115–26. https://doi.org/10.1007/s12152-014-9225-0.
19. Riddle J, Fröhlich F. Mental activity as the bridge between neural biomarkers and symptoms of psychiatric illness. Clin EEG Neurosci. 2023;54(4):399–408. https://doi.org/10.1177/15500594231154012.
20. Peña J, Muthalib M, Sampedro A, Cardoso-Botelho M, Zabala O, Ibarretxe-Bilbao N, et al. Enhancing creativity with combined transcranial direct current and random noise stimulation of the left dorsolateral prefrontal cortex and inferior frontal gyrus. J Creat Behav. 2022;57(1):65–81. https://doi.org/10.1002/jocb.562.
21. Lucchiari C, Sala PM, Vanutelli ME. Promoting creativity through transcranial direct current stimulation (tDCS): a critical review. Front Behav Neurosci. 2018;12:167. https://doi.org/10.3389/fnbeh.2018.00167.
22. Cropley DH, Kaufman JC, Cropley AJ. Malevolent creativity: a functional model of creativity in terrorism and crime. Creat Res J. 2008;20(2):105–15. https://doi.org/10.1080/10400410802059424.
23. Gino F, Ariely D. The dark side of creativity: original thinkers can be more dishonest. J Pers Soc Psychol. 2012;102(3):445–59. https://doi.org/10.1037/a0026406.
24. Douglas T. Moral enhancement. J Appl Philos. 2008;25(3):228–45. https://doi.org/10.1111/j.1468-5930.2008.00412.x.
25. Rowlands M. What is moral enhancement? Royal Inst Philos Suppl. 2018;83:5–18. https://doi.org/10.1017/S1358246118000036.
26. May J. Regard for reason in the moral mind. Oxford: Oxford University Press; 2018.
27. Harris J. How to be good: the possibility of moral enhancement. Oxford: Oxford University Press; 2016.

28. Juengst ET, Moseley D. Human enhancement [Internet]. In: Zalta EN, editor. The Stanford encyclopedia of philosophy. Stanford, CA: Metaphysics Research Lab, Stanford University; 2019. [cited 2024 Oct 16]. Available from: https://plato.stanford.edu/archives/spr2019/entries/enhancement/.
29. President's Council on Bioethics. Beyond therapy: biotechnology and the pursuit of happiness. Washington, DC: Government Printing Office; 2003.
30. Alexander ML, Alagapan S, Lugo CE, Mellin JM, Lustenberger C, Rubinow DR, et al. Double-blind, randomized pilot clinical trial targeting alpha oscillations with transcranial alternating current stimulation (tACS) for the treatment of major depressive disorder (MDD). Transl Psychiatry. 2019;9(1):106. https://doi.org/10.1038/s41398-019-0439-0.
31. Schwippel T, Pupillo F, Feldman Z, Walker C, Townsend L, Rubinow D, et al. Closed-loop transcranial alternating current stimulation for the treatment of major depressive disorder: an open-label pilot study. Am J Psychiatry. 2024;181(9):777–845. https://doi.org/10.1176/appi.ajp.20230838.
32. Fukuyama F. Our posthuman future: consequences of the biotechnology revolution. New York: Farrar, Straus and Giroux; 2002.
33. Sandel MJ. The case against perfection: ethics in the age of genetic engineering. Cambridge, MA: Harvard University Press; 2007.
34. Agar N. Truly human enhancement: a philosophical defense of limits. Cambridge, MA: MIT Press; 2014.
35. Harris J. Enhancing evolution: the ethical case for making better people. Princeton, NJ: Princeton University Press; 2007.
36. Savulescu J. New breeds of humans: the moral obligation to enhance. Ethics Law Moral Philos Reprod Biomed. 2005;1(1):36–9.
37. Elliott C. Better than well: American medicine meets the American dream. New York: W. W. Norton & Company; 2004.
38. Farahany NA. The battle for your brain: defending the right to think freely in the age of neurotechnology. New York: St. Martin's Press; 2023.

New Directions and Ethical Issues in Clinical TMS Research

Subgrouping "Treatment Responders": Ethical Implications for those Seeking Participation in TMS Treatment and/or Research

Stephanie M. Hare and Daniel Roche

1 Introduction

In recent years, the notion of precision medicine has taken a stronghold in many medical subdisciplines, including psychiatry [3]. This is born out of the idea that the *same* medical condition will manifest *differently* from individual to individual. Those that are diagnosed with the same type of cancer, for instance, may share some of the same daily ailments and symptoms, but the etiology, site, and severity of underlying pathology as well as the duration of illness may differ between individuals. The same holds true for mental disorders such as depression, where specific symptoms and biological underpinnings may vary from person to person. Since each person's symptoms will fluctuate over time, it stands to reason that treatment efficacies will also dynamically change over time. Thus, there is growing consensus that medical conditions, including psychiatric disorders, are heterogeneous, and a one-size-fits-all approach is poorly equipped to treat *diverse* symptoms and health outcomes impacting *diverse* populations.

The central goal of precision medicine is to tailor treatments to the individual to improve outcomes. A close cousin of precision medicine is stratified medicine, where the goal is to subgroup individuals into different classes to aid clinicians in making more accurate diagnoses and treatment decisions [3]. Both approaches—precision medicine and stratified medicine—require the results of clinical research to feed forward into clinical practice. This begins with clinical research teams developing statistical models that can predict things like treatment outcomes, clinical risk and/or diagnosis with a high degree of accuracy. The hope is that these models might later be used in clinical practice to guide clinical decision-making.

An emerging trend in psychiatric research is to use these models to subgroup the treatment population into those likely to respond to treatment (e.g., "treatment

S. M. Hare (✉) · D. Roche
Maryland Psychiatric Research Center, University of Maryland School of Medicine, Baltimore, MD, USA

V. Dubljević, J. R. Young (eds.), *TMS and Neuroethics*, Advances in Neuroethics,
https://doi.org/10.1007/978-3-031-92401-9_9

responders") and those likely to not show a response to the treatment (e.g., "treatment non-responders"). When it comes to the landscape of available treatments for many psychiatric disorders, transcranial magnetic stimulation (TMS) still lies on the fringes—it is not FDA-approved for treatment of most conditions, and even for those with FDA-approval (like major depression), it is not a front-line treatment. It is used to treat depression in those whom multiple pharmacological interventions have failed. This means that most individuals that pursue TMS treatment have been battling disabling symptoms for years and may have never found a therapy that provides relief; if they did, the positive benefits of the therapy may have waned over time and/or the side effects may have become too much of a burden.

Notably, this emerging trend in *clinical research* to use biological and other demographic and/or health data to predict who will (not) respond to treatment has not yet translated into *clinical practice.* However, there is hope that 1 day this research will translate into clinical practice, where pre-treatment consultations may occur with clinicians performing tests (scans, blood draws, administer questionnaires) and analyzing data to determine the likelihood of success of treatment. If clinical practice and treatment adopts similar methods and terminology of those employed in clinical research, a likely scenario is that these data will be used to subtype probable "responders" and "non-responders" prior to treatment initiation. At this stage, we need to ask: *What does this mean for patients seeking TMS treatment? For these patients, what are the ethical implications of being labeled a "responder" or "non-responder" prior to initiating TMS treatment?* The purpose of this chapter is to explore the ethical implications of clinical research aiming to subtype treatment responders/non-responders as it translates into clinical practice and the impact that it may have on those seeking to undergo TMS treatment.

We begin by providing an overview of how the terms "treatment response" and "treatment resistance" have classically been conceptualized. Drawing from this conceptual analysis, we argue that the methods used to make the treatment responder/resistant subgrouping are somewhat arbitrary, overly reductive, and poorly operationalized. We then proceed to discuss the potential impact of this subgrouping on individuals seeking TMS treatment. For instance, those labeled "treatment responsive" may develop unrealistic expectations about the success of TMS treatment. More ethical concerns arise when we consider how the "treatment resistant" label may perpetuate or amplify stigma of one's illness and be used to rule out TMS as a treatment option for those that have been chronically battling disabling symptoms.

2 Conceptualizing "Treatment (Non) Response"

Modern conceptualizations of *treatment response* are centered around the idea of whether there is an adequate change in symptoms when a certain treatment is administered. Meanwhile, conceptualizations of *treatment resistance* often describe cases where symptoms show inadequate response [13]. Here, we see how the notion of adequate *symptom response* lies at the heart of notions of both treatment response and treatment resistance.

The term "adequate" here is vague, so two questions immediately arise: What is considered an *adequate* treatment response and how was this criterion derived? Anyone that had conducted clinical research and/or clinical trials knows that study criteria and outcome measures must be painstakingly operationalized to clearly demarcate classes of research participants (e.g., who is considered eligible or ineligible to participate in the study, who is a so-called "treatment responder," and many more). Surely, any attempt to bifurcate treatment response into two categories as consequential and potentially stigmatizing as *treatment resistant* vs. *responsive* must have also painstakingly operationalized the supporting criterion.

In a recent review examining brain biomarkers of treatment response in those with major depressive disorder, the authors explored different depression treatments including pharmacological trials [22]. Here, we will focus on their review of brain stimulation studies that analyzed treatment response to TMS. Across the eight TMS studies included in the review [2, 4, 5, 9, 17, 19, 23, 24], it is worth noting that there was substantial heterogeneity in the clinical scales used to assess depression symptoms over the course of the trials. Commonly used scales included the Beck Depression Inventory (BDI) [6], the Hamilton Depression Rating Scale [11], and the Montgomery-Asperg Depression Rating Scale [25]. There was also heterogeneity in how symptom change/improvement was operationalized and analyzed. Three of the eight studies included depression score(s) as a continuous variable in correlation and/or regression analyses [9, 19, 23]. Meanwhile, the remaining five studies adopted similar definitions of *adequate* or *clinically significant* treatment response defined as a ≥ 50% reduction in total depression scores [2, 4, 5, 17, 24]. If this benchmark is met, a patient is classified as a "treatment responder"; all others were classified as "non-responders."

Several issues arise as we consider the conceptualizations of treatment "response" vs. "non-response" in these TMS studies. First, this seems like an arbitrary way to designate who is not a responder to treatment. It is unclear exactly how this cutoff point and definition were derived (e.g., demonstrating a ≥ 50% reduction in total depression scores). According to a 2001 article, it was "widely known (and accepted) … that a response is defined as a ≥ 50% decrease from baseline depression scale scores to trial endpoint [18]." In support of this claim, the authors cite a 1991 report from a task force that was convened to discuss "ways in which change points in the course of depressive illness had been described" and develop consensus in the wake of mounting inconsistencies in these descriptions [10]. Relying on this timeline, we see it took roughly a decade for this "≥ 50% decrease from baseline" definition of treatment response to become the modus operandi in psychiatric research. Although this definition strikes one as arbitrary, it seems there has been little pushback in the field to adopt alternative definitions backed by a more well-rounded scientific approach. For instance, a more reasonable approach may be to model, at the individual level, the percent reduction in symptoms that corresponds to an equivalent percent increase in quality of life, or reduction in suicide attempts, etc.

Second, this binary definition of treatment response is overly reductive in the sense that a 50% reduction in the total depression score oversimplifies the complex symptom presentation of a disorder into one lump score. For instance, the BDI has questions that probe one's motivation, energy/fatigue, sadness, guilt, pleasure, suicidality, and somatic effects such as appetite and weight gain. When speaking about the efficacy of a depression treatment, should each of these items carry equal weight? Is suicidality equivalent with a loss of motivation? Physicians and patients may have different perspectives and opinions on these questions. When considering the efficacy of a treatment like TMS, this "$\geq$ 50% reduction in total score" definition may lead to some problematic cases that defy logic and challenge our commonsense intuitions. Consider the case of an individual that undergoes TMS, and TMS treatment fully resolves suicidality, but doesn't move anhedonia, motivation, appetite, energy/fatigue, etc. In this case, the TMS treatment may be deemed ineffective (and the person labeled a non-responder), because the TMS didn't yield a 50% change in total symptom score. But, do we think the TMS treatment *should* be labeled as ineffective and the patient labeled as a non-responder? It seems a bit absurd to answer yes since an important facet of their illness and daily struggles—suicidality—was effectively treated and resolved with TMS. This is an especially salient point given the risk for fatality and injury that comes with suicidality. Since all that matters is a sizable reduction in the total depression score, this seemingly *clinically significant* outcome is essentially treated as insignificant. We are of the mind that treatment response is not a simple topic and should not be treated as such. Even modest improvements in certain domains can be *clinically significant* if they elevate one's quality of life and/or keep the person safe.

Third, this way of defining treatment response seems to set a very high bar for what is considered clinically significant symptom improvement. This classification implies that someone whose total depressive symptoms improved 43% did not respond to treatment. However, for the patient, this 43% reduction in depressive symptoms may make a world of difference; the patient might be shocked if they found out they were categorized as "treatment non-responsive."

In addition to questioning the validity of such a reductive, seemingly arbitrary conceptualization, we must consider the *quality* of the TMS treatment response studies that might be used to guide clinical decision-making. In their review, Strafella and coauthors [22] provided metrics on the quality of the eight TMS studies using the Grading of Recommendations Assessment, Development, and Evaluation (GRADE) working group methodology [21] to rank the study quality as high, moderate, low, and very low. High-quality studies included randomized, double-blinded, and placebo-controlled studies, whereas participants in low-quality studies were non-randomized with a placebo or control group. Very low studies lacked a placebo group and were excluded from the review to focus on higher quality, and reliable designs. Of the eight TMS studies, three were of high quality [9, 19, 23], three were of moderate quality [5, 17, 24] and two were of low quality [2, 4]. Here, we can see the considerable heterogeneity in the adopted quality standards of the TMS clinical trials. Furthermore, over 60% of the studies lacked important criteria (participant randomization, double-blind, placebo-control) to meet the highest accepted standards for clinical trial design. Together, these considerations suggest that we presently lack high-quality empirical support to serve as a solid foundation to guide clinical decision-making on who should and should not get TMS as a treatment.

To summarize, the methods used to make the treatment responder/non-responder subgrouping are somewhat arbitrary, oversimplified, and may set too high of a bar for what is considered "adequate" or "clinically significant" improvement. There is also considerable heterogeneity in the standards and methods of TMS clinical trials, with only a small portion of high-quality studies (using the GRADE criteria). Given these considerations, we caution against using the results of existing TMS clinical trials to make clinical decisions about who should and should not get TMS as a treatment. It is therefore imperative that we explore the ethical implications of applying this responder/non-responder subgrouping for identifying which participants are eligible to undergo TMS treatment.

3 Ethical Implications for Treatment Responders

As mentioned in the Introduction, the clinical research on TMS treatment response has not *yet* been translated into clinical practice, but there is hope that it will down the line. The hope is that this research will help us identify generalizable biomarkers of treatment response, such that people entering clinics for TMS treatment can undergo preliminary tests and screening, and be given some idea of their likelihood for treatment response *before* they receive treatment: some individuals may be told that they are likely to be "treatment responders" while others may be told that they will likely not respond to TMS treatment. Several ethical challenges arise when we consider the prospective labeling of someone seeking TMS treatment as a likely "treatment responder."

First, those labeled "treatment responders" may develop unrealistic expectations about the success of TMS treatment. Most people will find this news encouraging and will feel optimistic about the treatment. Many would argue this is a good thing, or at least it can't hurt. However, some may place heavy weight on this news and think TMS is the one treatment that might finally work for providing substantial relief of symptoms.

One reason to take special care in counseling patients to help mitigate against inflated expectations of TMS treatment is that there is evidence that the media's messages of TMS may already present a biased view of the promise of TMS treatment. A recent review article examined the scope and tone of academic articles on TMS vs. media portrayals of TMS [20]. The authors reported that 98% of the academic articles reviewed tended to present a balanced tone with respect to TMS (e.g., presenting both "pros" and "cons"/risks of TMS). Meanwhile, 63% of media prints of TMS presented a strongly optimistic tone promoting the many successes of TMS, with only 34% providing a balanced tone [20]. This analysis demonstrates the abundance of media articles hyping the positive benefits of TMS, which may lead many folks to have an inflated, overly optimistic view of TMS treatment when they come to the clinic. If this hyped view is the norm, we can easily see how being identified as a "treatment responder" may springboard off existing biases, leading to *unrealistically* high expectations for TMS treatment response.

There are additional reasons why someone may have overly inflated, unrealistic expectations for the success of TMS. As we've discussed, most people seeking out TMS are there because they have been non-responsive or showed resistance to other treatments. Now, we have a scenario where some of the people in this "previously non-responsive" group might be labeled as likely "responders for TMS." Those labeled as "TMS responders" will likely have inflated levels of excitement tied to the thought, "Finally, there is a treatment that will work for me." So, what if TMS doesn't work?

We know from our survey of clinical trials in Sect. 2 that there are going to be people that are labeled likely TMS treatment responders that won't respond (at least not using the strict "$\geq$ 50% decrease from baseline" criteria). However, we don't know how many likely treatment responders won't respond, and what the consequences of non-response would be in these individuals. At a minimum, we expect that failing to respond to TMS when a patient is told they have a strong likelihood of responding will be met with disappointment. But for someone with depression (or another disabling chronic mental disorder), the outcomes could be catastrophic. For instance, the *New York Times* recently published an article describing a case where a man with terminal cancer enrolled in a placebo-controlled psilocybin clinical trial [7]. He was disappointed because he didn't experience the anticipated effects. After angrily leaving the trial on the first treatment visit, he asked to return, but was told he could not. A week later, he died by suicide [7].

We are not saying such catastrophic consequences will abound in TMS treatment trials, but this story provides a cautionary note that they may occur (hopefully on very rare occasions). The man with cancer in this NYT story happened to receive placebo in the clinical trial, which led to him not seeing the effects he desired. But what about those TMS seekers that are told they are (likely) "responders," undergo active TMS treatment, and still feel terrible even if they happen to meet the "$\geq$ 50% decrease in total depression score" criterion? Since they met this criterion, they will be labeled a successful "responder" after the trial is over, but they still may feel miserable in many aspects of their life. They may still be plagued by certain symptoms, feel that their quality of life is still very low (despite there being a substantial reduction in total symptom score on paper), and be left wondering: "Wow this is really as good as its going to get?". For these people, the whole experience could be extremely confusing and demoralizing.

To be clear, we are not arguing that individuals seeking TMS treatment should be devoid of any optimism. Rather, our concern is that *unrealistic* expectations for the success of TMS are likely to be met with disappointment and other potential negative consequences. As we previously discussed, the TMS treatment response clinical trials show high heterogeneity of symptomatic responses in both groups (responders, non-responders). This should be discussed with patients seeking treatment to guide and ground their expectations of TMS treatment. It is possible to provide consultation to those seeking TMS treatment that supports an optimistic, yet realistic attitude toward the outcomes of TMS. Such consultations with clinicians and clinical staff should include discussion of the heterogeneity of TMS outcomes, side effects of TMS, and how TMS performs compared to other neuromodulatory treatments like electroconvulsive therapy (ECT).

4 Ethical Implications for Non-Responders

4.1 A Valid Reason to Exclude People from TMS Treatment?

There are several adverse outcomes that might be tied to being prospectively labeled a "treatment non-responder" when seeking TMS treatment. First, this classification may be used to rule out TMS as a treatment option for those that have been battling chronically disabling symptoms. There is a lot on the line for people seeking TMS; they want a therapy that may provide some relief for burdensome symptoms that have been resistant to other treatments. If we are to exclude or rule out certain people from receiving a potentially helpful treatment, there should be solid rationale and justification for doing so.

Different attitudes and reactions might accompany receiving the news that one is likely to be resistant to TMS treatment ranging from outright pessimism to confusion: "What does this mean? Does this mean I will not respond to TMS treatment? How do you know?" Given our discussion of the heterogeneity of findings and the historically reductive and seemingly arbitrary operationalization of "treatment response" in TMS clinical research in Sect. 2, it becomes evident that these pressing questions lack straightforward answers. It is outside the scope of this chapter to wrangle with the difficult ethical question of how clinicians *should* respond to these questions and provide counsel to their patients in these scenarios. Nonetheless, it is important to highlight that clinical teams will need to be ready to address these difficult questions as research findings on TMS treatment response translate into clinical practice.

Another ethical question arises concerning the weight and significance that we might place on the results of pre-treatment clinical assessments that suggest an individual will likely not respond to TMS treatment: Should these results be used as a valid reason for excluding a person from TMS treatment? Drawing from the considerations discussed in Sect. 2 (e.g., heterogeneity of responses, poor operationalization of "treatment response" categories, and many more), we think there is no solid foundation from empirical research to merit categorically ruling out someone for TMS treatment based on results of pre-treatment assessments and analyses. Doing so may run the risk of denying a potentially helpful treatment modality to someone suffering based only on a flawed, imprecise scientific basis.

4.2 The Potential for Health Insurance Companies to Deny Coverage

Even if the categorization as a "treatment non-responder" is not used by the clinician or clinical staff to overtly block or deny TMS treatment for a given individual, there are other ways that one's access to TMS treatment may be limited. For countries with healthcare systems that require healthcare costs to be covered and reimbursed by insurance companies, a scenario that will likely emerge is that insurance coverage may be declined for those that are labeled "treatment non-responders." For most individuals, the impact of this outcome is essentially the same as having your clinician categorically deny TMS treatment. In the USA, the current cost of a single session of TMS for depression treatment varies from $300 to $500 without insurance [26, 27]. With most TMS protocols for depression specifying at least 20 sessions over 4 weeks [27], the total comes to ~$8000 in out-of-pocket costs. This precludes anybody, except those with the highest levels of wealth and financial means, to be able to receive TMS.

4.3 Amplification and Perpetuation of Stigma

In addition, being labeled a treatment non-responder will likely perpetuate and/or amplify stigma of one's mental illness. Stigma of mental illness is a complex concept with many layers [1, 8] including *self-stigma* (e.g., stigmatizing attitudes projected by single individual onto oneself), *public stigma* (e.g., shared attitudes about mental disorder shared by those in local cultures), and *structural stigma* (e.g., where negative attitudes or stereotypes are enacted in the form of laws or public regulations) [1, 8]. Given its complexity, researchers have operationalized stigma of mental illness into four major dimensions: (1) desire for social distance from the person, (2) blame that we may place on the person for their actions, (3) perceived dangerousness of the person, and (4) pessimistic attitudes that we may hold toward their recovery (e.g., prognostic pessimism) [15, 16].

Individual and public attitudes regarding mental illness can be shaped by a variety of factors including existing racial and/or gender stereotypes or the type of mental illness. For example, there may be different public perceptions of schizophrenia and depression. Our understanding of the causes of these illnesses also plays a role in shaping stigmatizing attitudes. In this regard, biological explanations of mental illnesses (e.g., bad genetics, dysfunctional brains, and many more) have been shown to amplify public perceptions of perceived dangerousness [14, 15], desire for social distance [14] and prognostic pessimism [15], in comparison to psychosocial explanations.

Being labeled "treatment resistant" to TMS may further amplify stigma. It is rather obvious that being labeled as "treatment resistant" will fuel prognostic pessimism and ideas like "I am (or they are) a lost cause," "I (or they) will not get better (at least not with this therapy)." As we have discussed, TMS is often a down-the-line treatment that is only considered after multiple other treatments have failed. Being told that you will likely not respond to TMS treatment will likely diminish hope that one's symptoms can be successfully managed and treated, or even that recovery is possible.

It is less clear how being labeled as a non-responder or "resistant" to TMS treatment may impact other aspects of stigma such as desire for social distance or perceived dangerousness. It has been theorized that the reason biogenetic explanations of mental illness fuel many aspects of stigma is they convey the idea that the person has a pathological *essence* that is *essentially* different from what is normal; this creates a rift between "us" (e.g., the normal ones) and "them" (e.g., those that are marked by defect or pathology at some essential level) [12]. Being told that you're "resistant" to TMS treatment before even initiating the treatment may further perpetuate essentialist thinking that deepens the divide between "us" vs. "them."

5 Conclusions

This chapter began by providing an overview of how "treatment response" and "treatment responder" have been traditionally defined in TMS clinical trials, focusing on those for major depressive disorder. In Sect. 2, we argued that the common method of subdividing "responders" and "non-responders" into two groups based on the "$\geq$ 50% decrease in total symptom score" criterion is somewhat arbitrary, oversimplified, and may set too high of a bar for what is considered "adequate" or "clinically significant" improvement. We noted the considerable heterogeneity in the adopted quality standards of TMS clinical trials, with only a small portion of high-quality studies being identified in a recent meta-analysis of depression studies [22]. We therefore cautioned against using the results of existing TMS clinical trials to make clinical decisions about who should and should not get TMS as a treatment.

Next, we examined the societal impact and ethical implications of applying the reductive responder/non-responder subgrouping for identifying which participants are eligible to undergo TMS. In Sect. 3, we discussed how those prospectively labeled "responders" may have inflated, unrealistic expectations for the success of TMS treatment. Given how things pan out with their TMS treatment, this hyped outlook could be met with disappointment, or even more severe, adverse consequences. In Sect. 4, we discussed ways in which being labeled "treatment resistant" may perpetuate and amplify different aspects of stigma of mental disorders. We also returned to the question of whether results of pre-treatment tests indicating that one is a (likely) non-responder to TMS treatment should be used to rule out TMS as viable treatment? Drawing from our earlier discussions in Sect. 2, we concluded that the answer is no; we argued that using those clinical tests to rule out TMS as a treatment runs the risk of denying a potentially helpful treatment modality to someone suffering based only on a flawed, imprecise scientific basis.

Overall, the purpose of this chapter was to draw attention to the trend in clinical research to subgroup patients into treatment "response" vs. "non-response" groups and consider the ethical implications for future patients seeking TMS treatment if these rigid and oversimplified categories are applied before treatment is initiated. Admittedly, we focused on various ethical challenges that may emerge from this subgrouping including the potential to overly inflate patients' expectations, amplify stigma, and hamper patients' abilities to receive TMS treatment. However, we acknowledge that, *in theory*, there are also advantages or positive outcomes that may emerge from performing this subgrouping. One primary advantage of this subgrouping is that it relies on a simple, dual-category system, making it easy for clinical staff to make a concrete determination of who is likely to respond and who (likely) won't respond. However, this theoretical advantage or benefit may directly conflict with various ethical challenges that arise *in practice*. While a dual-category system may be useful in assisting clinical teams in making treatment decisions (and who, potentially should or should not receive a treatment), we argued in this chapter that this oversimplified categorization fails to capture and explain heterogeneous and complex treatment trajectories and should not serve as the cornerstone for making decisions about who should receive TMS treatment.

An additional theoretical advantage is that we can save patients' time and financial burden if we know ahead of time that TMS likely won't treat their symptoms; we can also avoid wasting valuable (expensive) resources and tools on those that will likely not respond to TMS treatment. However, this theoretical claim rests heavily on the practical assumption that we have reliable biomarkers and metrics that can predict diverse treatment trajectories. It remains uncertain whether those in the clinical research space will deliver on their promise to develop highly reliable biomarkers and metrics. Although it seems like we might be on a promising path in this regard, clinical researchers have historically failed to identify robust and reproducible biomarkers and metrics that can predict diverse treatment trajectories. Instead, researchers have prioritized simplifying clinical trial designs and the interpretation of their findings (e.g. subgrouping "responders" vs. "non-responders). While this approach is understandable, it is hard to be hopeful that these same researchers will be able to use findings from their future studies to meaningfully predict diverse treatment trajectories and improve upon the treatment "responder"/"non-responder" dichotomy.

Notably, our discussion here might be labeled as "too early" or "unnecessary" given that the emerging trend in clinical research to use data to predict who will (not) respond to treatment has not yet translated into clinical practice. But we know that the goal of this research is not just to let scientists make fancy predictions or publish in high impact academic journals. It is to develop individualized health profiles and perform assays to assist clinicians on how to better treat patients (rather than taking a blind "guess-and-check" approach with existing treatments). Several ethical questions and concerns arise in this context, and we think it is important to be ahead of the curve in considering the societal impact of the translation of clinical research on TMS treatment response into clinical practice.

References

1. American Psychiatric Association Stigma, Prejudice and Discrimination Against People with Mental Illness. https://www.psychiatry.org/patients-families/stigma-and-discrimination. Accessed 29 Apr 2024.
2. Arns M, Cerquera A, Gutiérrez RM, et al. Non-linear EEG analyses predict non-response to rTMS treatment in major depressive disorder. Clin Neurophysiol. 2014;125:1392–9. https://doi.org/10.1016/j.clinph.2013.11.022.
3. Arns M, van Dijk H, Luykx JJ, et al. Stratified psychiatry: tomorrow's precision psychiatry? Eur Neuropsychopharmacol. 2022;55:14–9. https://doi.org/10.1016/j.euroneuro.2021.10.863.
4. Bailey NW, Hoy KE, Rogasch NC, et al. Differentiating responders and non-responders to rTMS treatment for depression after one week using resting EEG connectivity measures. J Affect Disord. 2019;242:68–79. https://doi.org/10.1016/j.jad.2018.08.058.
5. Bares M, Novak T, Kopecek M, et al. The effectiveness of prefrontal theta cordance and early reduction of depressive symptoms in the prediction of antidepressant treatment outcome in patients with resistant depression: analysis of naturalistic data. Eur Arch Psychiatry Clin Neurosci. 2015;265:73–82. https://doi.org/10.1007/s00406-014-0506-8.
6. Beck AT, Steer RA, Brown GK. Beck depression inventory II. Psychological Association; 1996.
7. Borrell B. The psychedelic evangelist. The New York Times (2024)
8. Corrigan PW, Druss BG, Perlick DA. The impact of mental illness stigma on seeking and participating in mental health care. Psychol Sci Public Interest. 2014;15:37–70. https://doi.org/10.1177/1529100614531398.
9. Eshel N, Keller CJ, Wu W, et al. Global connectivity and local excitability changes underlie antidepressant effects of repetitive transcranial magnetic stimulation. Neuropsychopharmacology. 2020;45:1018–25. https://doi.org/10.1038/s41386-020-0633-z.
10. Frank E, Prien RF, Jarrett RB, et al. Conceptualization and rationale for consensus definitions of terms in major depressive disorder. Remission, recovery, relapse, and recurrence. Arch Gen Psychiatry. 1991;48:851–5. https://doi.org/10.1001/archpsyc.1991.01810330075011.
11. Hamilton M. A rating scale for depression. J Neurol Neurosurg Psychiatry. 1960;23(1):56–62. https://doi.org/10.1136/jnnp.23.1.56.
12. Haslam N, Kvaale EP. Biogenetic explanations of mental disorder: the mixed-blessings model. Curr Dir Psychol Sci. 2015;24:399–404. https://doi.org/10.1177/0963721415588082.
13. Howes OD, Thase ME, Pillinger T. Treatment resistance in psychiatry: state of the art and new directions. Mol Psychiatry. 2022;27:58–72. https://doi.org/10.1038/s41380-021-01200-3.
14. Kvaale EP, Gottdiener WH, Haslam N. Biogenetic explanations and stigma: a meta-analytic review of associations among laypeople. Soc Sci Med. 2013a;96:95–103. https://doi.org/10.1016/j.socscimed.2013.07.017.
15. Kvaale EP, Haslam N, Gottdiener WH. The 'side effects' of medicalization: a meta-analytic review of how biogenetic explanations affect stigma. Clin Psychol Rev. 2013b;33:782–94. https://doi.org/10.1016/j.cpr.2013.06.002.
16. Loughman A, Haslam N. Neuroscientific explanations and the stigma of mental disorder: a meta-analytic study. Cogn Res Princ Implic. 2018;3:43. https://doi.org/10.1186/s41235-018-0136-1.
17. Narushima K, McCormick LM, Yamada T, et al. Subgenual cingulate theta activity predicts treatment response of repetitive transcranial magnetic stimulation in participants with vascular depression. J Neuropsychiatry Clin Neurosci. 2010;22:75–84. https://doi.org/10.1176/jnp.2010.22.1.75.
18. Nierenberg AA, DeCecco LM. Definitions of antidepressant treatment response, remission, nonresponse, partial response, and other relevant outcomes: a focus on treatment-resistant depression. J Clin Psychiatry. 2001;62(Suppl 16):5–9.
19. Philip NS, Leuchter AF, Cook IA, et al. Predictors of response to synchronized transcranial magnetic stimulation for major depressive disorder. Depress Anxiety. 2019;36:278–85. https://doi.org/10.1002/da.22862.

20. Scheper A, Rosenfeld C, Dubljević V. The public impact of academic and print media portrayals of TMS: shining a spotlight on discrepancies in the literature. BMC Med Ethics. 2022;23:25. https://doi.org/10.1186/s12910-022-00760-5.
21. Schünemann HJ, Oxman AD, Brozek J, et al. Grading quality of evidence and strength of recommendations for diagnostic tests and strategies. BMJ. 2008;336:1106–10. https://doi.org/10.1136/bmj.39500.677199.AE.
22. Strafella R, Chen R, Rajji TK, et al. Resting and TMS-EEG markers of treatment response in major depressive disorder: a systematic review. Front Hum Neurosci. 2022;16:940759. https://doi.org/10.3389/fnhum.2022.940759.
23. Voineskos D, Blumberger DM, Rogasch NC, et al. Neurophysiological effects of repetitive transcranial magnetic stimulation (rTMS) in treatment resistant depression. Clin Neurophysiol. 2021;132:2306–16. https://doi.org/10.1016/j.clinph.2021.05.008.
24. Widge AS, Avery DH, Zarkowski P. Baseline and treatment-emergent EEG biomarkers of antidepressant medication response do not predict response to repetitive transcranial magnetic stimulation. Brain Stimul. 2013;6:929–31. https://doi.org/10.1016/j.brs.2013.05.001.
25. Williams JBW, Kobak KA. Development and reliability of a structured interview guide for the Montgomery Asberg depression rating scale (SIGMA). Br J Psychiatry. 2008;192:52–8. https://doi.org/10.1192/bjp.bp.106.032532.
26. TMS Therapy Cost And Effectiveness | BetterHelp. https://www.betterhelp.com. Accessed 5 Sep 2024.
27. TMS Therapy Costs | Mental Health & Wellness Clinics located in Aurora, Louisville and Westminster, CO | Axis Integrated Mental Health. https://www.axismh.com/content/tms-therapy-costs. Accessed 5 Sep 2024.

TMS for Mood Disorders: An Ethical Perspective

Mohsen Poorganji, Emma C. Boyd, Deepinder S. Nagra, and Lawrence G. Appelbaum

1 Guiding Questions

We begin this chapter by posing a series of thought-provoking questions, which we will explore and address through a neuroscientific lens throughout the manuscript. Does the human brain change over the course of a lifetime? If so, how do mood disorders, and particularly depression-influence these changes? Do these changes imply that an individual's personhood is also altered? Finally, does society bear responsibility for shaping or reshaping an individual's personhood in response to these changes? Through this framework, we aim to provide a deeper understanding of the interplay between brain function, mental health, and the societal dimensions of personhood.

2 Plastic Changes in the Brain Throughout Life as the Results of Lived Experiences

The human brain goes through several stages of development throughout one's lifespan. Within the first 3–4 years of life, children undergo considerable synaptogenesis wherein large numbers of synaptic connections are formed. Later in development, through pruning and consolidation, these connections start to specialize and decrease in number [31, 48]. These changes happen as the result of both activity-dependent and activity-independent processes and their interactions [48]. Activity-independent processes refer to development based on genetic and molecular factors

M. Poorganji · E. C. Boyd · L. G. Appelbaum (✉)
Department of Psychiatry, University of California, San Diego, CA, USA
e-mail: lappelbaum@health.ucsd.edu

D. S. Nagra
Department of Psychiatry and Behavioral Sciences, MIND Institute, University of California Davis Health Center, Sacramento, CA, USA

V. Dubljević, J. R. Young (eds.), *TMS and Neuroethics*, Advances in Neuroethics,
https://doi.org/10.1007/978-3-031-92401-9_10

of neurons, such as pathfinding in axons and selection of target neurons, whereas activity-dependent processes, such as neuroplasticity, are influenced by neuronal activity to external and internal stimuli. Both processes and their interactions play crucial roles in the development of the nervous system.

Scientists now consider biology as the origin of human behavior, personality, emotion, and cognition, making the brain the most important organ in the human body toward the expression of personhood. The processes of development can also be viewed as a representation of the interaction between nature and nurture, where no single process solely determines the formation of personality traits, emotions, cognition, and overall personhood. Plastic changes in the brain can occur at different stages of one's life span as the result of nature/nurture interactions, and as such, the human brain is subject to changes throughout life as the result of various experiences. Evidence for this comes in the form of higher prevalences of developing psychiatric disorders in those who have experienced trauma in their childhood, but that trauma alone does not solely explain the development of trauma disorders [38]. Changes in the brain can also be observed in survivors of trauma who develop post-traumatic stress disorder (PTSD) [62]. Accordingly, while not deterministic, life experiences have a dramatic effect on the way their brain develops and the way the plastic changes manifest.

3 Mood Disorders and Dysregulated Brain

Mood disorders such as major depressive disorder (MDD) and bipolar disorder (BD), are known for their negative impact on an individual's quality-of-life as well as the impact on family members and society [49]. MDD is the second leading cause of disability worldwide [22] with a 16% lifetime prevalence [34], while BD is also highly prevalent with a 5% lifetime prevalence [39]. Other mood disorders include seasonal affective disorder, premenstrual dysphoric disorder, depression related to a medical condition, and several other categories of mental health illnesses characterized by disturbances in a person's emotional state or mood. These disorders can be debilitating, disabling, and carry an increased risk of suicide and emotional and economical burden on individuals, families, and society. MDD is defined by the presence of symptoms which include depressed mood, loss of interest or pleasure, significant weight gain or loss in a short period of time, and sleep disturbances, among other symptoms [1]. Bipolar disorder shares many of these symptoms but is also characterized by periodic shifts in emotional states from depressive to manic episodes, and vice versa.

Several theories regarding the underlying mechanism of mood disorders have been proposed, many of which are supported by replicated neurophysiological studies, offering a clearer understanding of physiological basis of mood disorders. While ultimately, the neuropathology of depression and other mood disorders involves complex interactions between genetic, biochemical, structural, and functional factors in the brain, dysregulation in each of these systems has been identified. A central theory of the underlying dysregulation stems from neurochemical imbalances in neurotransmitters like serotonin, norepinephrine, and dopamine wherein low levels of these neurotransmitters in the synaptic cleft are thought to contribute to depressive symptoms [29]. This mechanistic theory, known as the "Monoamine Hypothesis" is also accompanied by evidence of dysregulation in both excitatory glutamate transmission [53] and inhibitory GABA neurotransmission [59] that also plays a role in depression.

The presence of structural and functional abnormalities in the brain have also been widely reported in mood disorders with reduced volume and atrophy in the hippocampus [54], and pathophysiology in prefrontal cortex [18] and the amygdala [17]. These areas, involved in decision-making, emotional regulation, and executive function, often show structural abnormalities in patients with mood disorders. Moreover, the function of these areas is frequently maladaptive, with aberrant functional connectivity in the default mode network (DMN) that is typically active during rest and self-referential thought [40]. Increased connectivity within the DMN is thought to contribute to rumination, a common symptom of depression.

Other common functional anomalies observed in patients with MDD are elevated activity in the subgenual cingulate cortex (SGC) and increased connectivity between SGC and the dorsolateral prefrontal cortex (DLPFC) [19, 23, 27]. This increased connectivity between the SGC and DLPFC in depression suggests a dysfunctional communication pathway between emotional and cognitive brain regions that may contribute to the persistence of negative thought patterns and emotional dysregulation characteristic of depressive episodes. The DLPFC's role in modulating cognitive control over emotions might be compromised due to its overactive linkage with the SGC, leading to difficulty in regulating mood and maintaining balanced emotional responses. Understanding this connectivity provides a rationale for therapeutic interventions targeting these brain regions [36]. Transcranial magnetic stimulation (TMS) can be used to modulate the activity of the DLPFC, aiming to restore balanced connectivity and improve emotional regulation. By targeting the DLPFC, TMS may help normalize the communication between the DLPFC and SGC, potentially alleviating depressive symptoms.

In summary, mood disorders are multifaceted with roots in neurotransmitter imbalances, structural and functional brain changes, genetic influences, and impaired neuroplasticity. Considering the neurophysiological characteristics of mood disorders on one hand, and the changes they cause in an individual's mood, energy, emotions, and physical state on the other hand, one can argue that mood disorders alter one's personhood.

4 How TMS Can Modulate Mood and Emotion?

Several treatment approaches are widely available for mood disorders including pharmacotherapy and psychotherapy. Pharmacotherapy primarily involves the use of antidepressant medications that act to alter the balance of certain neurotransmitters in the brain, such as serotonin and norepinephrine, while psychotherapy involves working with a trained therapist to identify and change troubling emotions, thoughts and behaviors. Patients not responding to two or more of these treatments are referred to as treatment resistant and are eligible for other therapies such as TMS, electroconvulsive therapy (ECT), and ketamine treatment.

TMS is a non-invasive form of brain stimulation that works based on principles of electromagnetic induction [5]. According to Faraday's law, the magnetic field produced by a TMS coil induces electric currents that when placed over the cortex induce currents that depolarize neurons and induce action potentials [60]. Based on the direct synaptic and trans-synaptic connections of the neurons, this change in neuronal firing can propagate to other connected cortical and subcortical areas of the brain [45]. When the stimulation is repeated several times, long-term potentiation (LTP) and long-term depression (LTD) can be induced [28, 30]. It is also important to note that due to the magnetic nature of TMS, the induced field reduces exponentially with distance from the coil, such that TMS mainly directly affects the superficial cortex within 2–3 cm of the coil [16]. Considering that the prefrontal cortex, anterior cingulate cortex, insula, and amygdala are some of the principal areas of the brain responsible for emotion regulation, therapeutic effects of TMS are attributed to induced plastic changes to these areas [9, 23, 26]. DLPFC and DMPFC are two of the common target areas of stimulation using TMS for treatment of mood disorders. Both of these areas are superficial parts of the cortex which can be directly stimulated by TMS. As mentioned earlier, through synaptic and trans-synaptic connection, the effect of superficial stimulation spreads to other cortical (insula and SGC) subcortical areas (amygdala) of the brain which will not be directly stimulated by magnetic pulse [14, 15, 26]. In other words, by stimulating the mood regulator part of the brain (PFC), rTMS tries to take control/ modulate the activities of the mood generator (insula, SGC, amygdala) part of the brain using various treatment protocols of rTMS including high-frequency rTMS and intermittent theta burst stimulation (iTBS) which are now FDA approved/cleared for treatment of treatment resistant depression (TRD) [8–10, 44].

5 Personhood and the Effects of TMS

Throughout history, philosophers, scientists, clergy, and people of different faiths have tried to answer the fundamental question of what defines us as humans? The answer to this question has sparked different schools of thoughts. Immanuel Kant considered that which is aware of its identity as a person [33, 37]. John Locke affiliated personhood to consciousness [25, 43] and Renes Descartes attributed personhood to solely "Res Cogitans" referring to the mind in mind-body dualism [32]. Since we are living in the era of artificial intelligence, we asked three of the common AI assistants, ChatGPT, MetaAI, and Copilot to "define personhood" (Appendix A). As one may expect, they all acknowledged that personhood is complicated and controversial. The answers also included attributes like emotions, consciousness, self-awareness, and responsibility. Considering these personal traits, it can be asked if a person is responsible for shaping these traits, or are they only inherited? How much of personhood is shaped by environment and experiences?

Different fields of science have tried answering to these questions. From studies on identical twins separated at birth [11] to studying the effects of starvation of one generation of mothers (imposed by Dutch famine in the 1940s) on the well-being of next generations [7], to the claims of pure behaviorists like Watson who view personhood as the product of reinforcement and consequences that are entirely outside the organism, scientists have tried to tackle this issue for decades. At each end of the spectrum are scientists considering humans as the pure production of their genetic makeup, and those who believe in pure behaviorism, which says that humans are solely formed by their environment. In the middle stand those who believe humans are the product of their inherited genes combined with their environment. Epigenetics tries to tackle this issue by emphasizing that not all the genes that one inherits shape their fate, as the activation/deactivation of certain genes are based on exposure to environmental factors including childhood trauma, immigration, environmental enrichment, and complications during pregnancy and birth. Strong evidence can emerge from studies demonstrating that individuals with presence of risk factors for developing schizophrenia and other psychiatric disorders do not develop these conditions as the result of their environment [47, 61]. From biological perspective alone, a deterministic approach toward human will is inevitable [55, 56]. Though determinism must be approached with caution, like any other scientific argument, it is modulated by interaction between inheritance and environment, considering the effect that different layers of environment (core and extended family, educational system, society, and global interactions to name a few) have on individuals. Accordingly, the extensive combinations of these interactions provide a balance between determinism and free will. This on its own, intensifies the important role of decision makers to provide a nourishing atmosphere where individuals can live a more fulfilling life with less burden and higher achievement.

In sum, personhood is not a static phenomenon. The dynamics of it come from the fact that individuals' emotions, cognition, knowledge, the sense of self, and judgment can be altered as the result of life experiences and interactions between environment and genes. By viewing mood disorders through the lens of disorders of brain network dysregulation, any treatments targeting the dysregulations can have therapeutic effects and accordingly change the already altered personhood as the result of the disorder. As such, considering that emotion, mood, and executive functioning are different aspects of personhood, and given the plastic changes that TMS can have on the centers/networks responsible for these human functions, one can conclude that TMS has the potential to alter an individual's personhood. The same rationale is true about the effects of psychoactive compounds, such as traditional antidepressant medications, prescription drugs, illicit substances, as well as ketamine, psilocybin, and other hallucinogens on an individual's personhood.

Multiple therapies including pharmacotherapy, TMS, and ECT are available for the treatment of depression, each with different therapeutic and risk profiles. Though ECT is an effective form of treatment for severe cases of treatment resistant depression, it comes at the cost of short-term memory loss and invasiveness [35, 46]. With pharmacotherapy, including serotonin re-uptake inhibitors, come other side effects such as sleep disturbance, loss of libido, and possible weight gain [20, 21]. All these side effects would essentially point toward changes in one's personhood and not necessarily in the desirable way that a patient would seek. However, it is not wrong to claim that TMS conveys among the fewest negative side effects for patients with short-term headaches the most frequently reported side effect and seizures the most severe, but very rare side effect [28, 51, 58]. Accordingly, TMS is one of the safer treatments through which a patient can achieve desirable outcomes with minimal burdens, leading to changes in both brain activity and the sense of personhood which comes from these changes.

6 Ethical Considerations of Not Treating Psychiatric Disorders

Society has made significant progress toward understanding psychiatric illness, recognizing that depression and other mood disorders are brain conditions requiring treatment to reduce the heightened risk of suicide and the substantial social and economic burdens on individuals, families, and communities. The severe impact of MDD and bipolar disorder on daily functioning highlights the crucial need for comprehensive treatment [24].

Although some individuals may have a genetic predisposition for mood disorders, most individuals are not born with fully developed psychiatric illnesses. These disorders result from the interplay between genetic factors and environmental influences [57]. Accepting a harmful environment as unchangeable and viewing the resulting disorders as an unalterable aspect of personhood is misguided. A civilized society has the responsibility to create supportive environments that enable individuals to thrive. Addressing psychiatric illness is an essential part of this supportive framework. Given the demonstrated effectiveness of TMS in treating mood disorders, it is imperative for a responsible society to ensure access to this treatment for all of its members. Providing access to effective treatments like TMS is both an ethical obligation and a practical strategy to alleviate the extensive impact of mood disorders on individuals and society.

7 Freedom of Choice in Choosing Treatment Option for Mood Disorders

Although in the broadest sense, there may not ever be complete freedom of choice, within the current context patients are typically provided with standard care subject to limitations of accessibility, affordability, and feasibility. Primary treatments for mood disorders (psychotherapy and pharmacotherapy) are typically covered by most insurance providers. If a patient fails to respond to at least two courses of these treatments, alternative therapies like TMS may then be offered. Given the demonstrated efficacy and minimal side effects of TMS in treating psychiatric disorders, there is now opportunity to move TMS from a second line treatment, implemented after psychotherapy and pharmacotherapy have failed, to a first line treatment that is more readily available to patients in need. In fact, there is emerging arguments that TMS applied in the initial phases of MDD might reduce the chance of developing TRD [50]. Considering the increasing number of treatments available to those with mood disorders, patients should have more of a choice of which line of treatment they would like to pursue, which can be achieved through providing accessible treatments.

8 Democratizing TMS Accessibility

Given the importance of accessible treatments for depression, ensuring broad access to TMS is imperative for achieving equitable healthcare. As a device-based therapy, TMS is less scalable than pharmacotherapy or psychotherapy due to equipment costs, additional training, and lower volumes of patients that can be seen. Despite this, TMS is becoming increasingly available with more medical clinics offering this treatment and many insurance companies now providing coverage for acute courses of treatment. However, most clinics providing this service to patients are in large cities and affluent communities. Despite the continued growth in clinics and medical centers providing TMS, socioeconomic status, geographic location, and insurance limitations have continued to hinder equitable access to TMS, raising ethical concerns about justice and fairness in mental healthcare treatment. Addressing these disparities in access to TMS treatment is essential for ensuring all individuals can benefit from it.

Disparities in access to TMS stem from socioeconomic, geographic, racial, ethnic, and gender factors [41]. Racial and ethnic minorities, particularly Black, Indigenous, and People of Color (BIPOC), encounter disproportionate obstacles in accessing TMS compared to their white counterparts. Socioeconomic barriers exacerbate the issue, with minorities from lower socioeconomic backgrounds often burdened with out-of-pocket expenses due to insufficient insurance coverage for TMS treatment [13]. Geographic disparities further hinder access, with TMS clinics primarily concentrated in urban areas, leaving rural or underserved communities to face logistical challenges such as transportation [4]. Gender and LGBTQIA+ disparities compound the issue, with women, transgender, and non-binary individuals encountering discrimination in accessing TMS [52]. Addressing these disparities is crucial for promoting justice and fairness within the mental healthcare landscape. Inclusive policies and practices must be implemented to ensure equitable access for individuals, regardless of socioeconomic status, race, ethnicity, gender, or sexual orientation.

Navigating the ethical considerations of addressing disparities in access to TMS requires a close look at fundamental principles such as justice, autonomy, beneficence, and nonmaleficence in mental healthcare[63]. The principle of justice underscores the moral obligation to mitigate healthcare disparities and ensure the equitable distribution of effective treatments like TMS. Timely and accessible alternatives are imperative, particularly considering the delayed effectiveness often observed with traditional treatments. Additionally, the principle of autonomy highlights the importance of empowering patients to make informed decisions about their healthcare. Limited access to TMS treatment restricts patients' autonomy by limiting their range of healthcare options. Providing patients with comprehensive information about TMS enables them to make autonomous decisions aligned with their values, preferences, and treatment goals. Furthermore, healthcare providers are ethically bound by the principles of beneficence and nonmaleficence to maximize benefits and minimize harm. Disparities in access to TMS raise concerns about potential harm from untreated conditions and less effective or riskier treatments. By addressing these disparities, healthcare providers uphold ethical principles and contribute to improved mental health outcomes for all individuals.

Reducing disparities in access to TMS requires a multifaceted approach grounded in policy improvements, educational initiatives, and inclusive practices. Policy improvements, such as enhancing insurance coverage, subsidizing treatment costs, and increasing the number of TMS clinics in underserved communities, are necessary to bridge the access gap [64]. These changes would not only make TMS more financially accessible, but also geographically available to a wider population. Additionally, educational initiatives raise awareness and provide comprehensive information about TMS to individuals from all backgrounds, allowing them to make informed decisions about their healthcare. Moreover, the development of culturally competent care models is vital for addressing the diverse needs of all individuals. Inclusive practices can help to eliminate the barriers faced by marginalized populations, ensuring that all individuals, regardless of their background, have equitable access to TMS treatment. By implementing these strategies, stakeholders can work toward an equitable and inclusive mental healthcare landscape.

As TMS emerges as a frontrunner for improving outcomes for patients grappling with MDD, it will offer a promising alternative to traditional treatments with fewer side effects, thereby enhancing overall patient health. Moreover, the integration of TMS provides additional treatment options and allows for personalized care, leading to improved patient satisfaction and adherence to treatment regimens. As disparities in access to TMS are minimized, healthcare providers transition toward a more patient-centered care model, fostering an environment where every individual is seen, heard, and valued. Furthermore, with more equitable access to TMS, enhanced productivity, reduced healthcare costs, and more inclusive society can all be anticipated. By reducing the overall burden of depression, TMS can lower healthcare costs associated with prolonged and repeated hospitalizations, emergency care, and the long-term use of medications, contributing to a more sustainable and efficient healthcare system while benefiting individual patients.

Democratizing access to TMS is necessary for fostering an equitable and inclusive healthcare system that respects individual personhood. Addressing the socioeconomic, geographic, racial, ethnic, and gender disparities in access to TMS ensures that all individuals can benefit from this treatment. Policy reforms aimed at enhancing insurance coverage and expanding clinic availability, alongside educational initiatives and culturally competent care, are imperative to reduce these barriers. By prioritizing access to TMS, we uphold the ethical principles of justice, autonomy, beneficence, and nonmaleficence, empowering patients to make informed decisions about their mental health. This commitment to accessibility not only improves individual health outcomes but also supports the ongoing efforts to personalize TMS treatments. Ultimately, by ensuring that every individual can benefit from TMS, we move toward a mental healthcare landscape that values the personhood of all individuals.

9 Conclusion

TMS has the potential to change the neurocircuitry of emotions, and thus personhood. With TMS, we may be able to induce longer-lasting plastic changes to the benefit of patients, families, and societies by ameliorating the burden of psychiatric disorders. TMS can reduce the burden of mental illness and society should continue to advance TMS as an available treatment option to all who may benefit. There are several directions in which neuromodulation may advance to increase the effectiveness of treatments for mood disorders. One important area of improvement with promising initial results is that the field of brain stimulation has started to move away from "one size fits all" standard treatments to a more individualized approach [3]. This individualization is reflected in various aspects of delivering treatment including incorporating anatomical and functional variability between individuals. Personalizing the intensity, frequency, and number of treatment pulses is another active area of research. Furthermore, incorporating the ongoing dynamics of the brain into therapy through closed-loop approaches [65, 66] and delivering TMS concurrently with behavioral therapies [42] are important emerging approaches to improve therapeutic outcomes. Such future technological innovations, in conjunction with improved access, will continue to propel neuromodulation as an important technique to improve patients' outcomes. Furthermore, considering the mechanism of action of TMS and the nature of plastic changes in the brain, it is essential to acknowledge the difficulty of sustaining the induced changes as the result of TMS acute course of treatment. Several studies have demonstrated the possibility of having prolonged therapeutic effect for MDD patients by incorporating additional maintenance treatment, which significantly improves the possibility of maintaining response and remission for the patients [2, 6, 12]. In this regard, it is crucial to note that relapse of depression imposes significant emotional, economical, and social costs to the patients, their families, as well as society which all can be prevented by making maintenance treatments more accessible for patients.

Appendix A: AI Responses Generated (on August 10, 2024) Regarding the Definition of Personhood and Whether It Is Limited to Humans?

Inquiry	Define personhood	Is it limited to humans?
Copilot	Personhood refers to the status of being a person. It is a concept that is controversial and has significant implications in philosophy and law. Let's explore it further: 1. Legal and political context: In legal and political contexts, personhood is closely tied to concepts of citizenship, equality, and liberty. Rights, protections, privileges, responsibilities, and legal liability are associated with legal persons, which can be either natural (individuals) or juridical (entities like corporations) 2. Philosophical definitions: Philosophically, defining personhood is challenging due to its historical, cultural, and contextual variability. Common attributes include: Human nature. Agency. Self-awareness. Notion of past and future. Possession of rights and duties. Boethius, a philosopher from the 6th century, defined a person as an "individual substance of a rational nature" The naturalist view considers continuous consciousness over time as a criterion for personhood 3. Cross-cultural perspectives: Social and legal recognition of personhood varies across cultures. For example: Among the Wari' people in Brazil, personhood is tied to social relations The Gebusi people in Papua New Guinea incrementally become persons through social interactions Western and Indigenous legal systems intersect in discussions about corporate personhood and extending personhood to non-human entities In summary, personhood is a multifaceted concept that encompasses legal, philosophical, and cultural dimensions. It remains a topic of ongoing debate and has implications in various domains of human existence.	Personhood is not limited to humans alone. It is a concept that extends beyond our species. Let's explore this further: 1. Human personhood: Traditionally, personhood has been associated with human beings. Legal systems recognize individuals as legal persons, granting them rights, responsibilities, and protections. Philosophically, personhood attributes include self-awareness, agency, and the capacity for rational thought. 2. Beyond humans: Non-human animals, corporations, and even environmental entities have been subjects of personhood discussions: Non-human animals: Animal rights activists argue for recognizing their personhood based on sentience, cognition, and emotional experiences. Corporations: The concept of corporate personhood allows legal entities (like companies) to have certain rights and responsibilities. Environmental entities: Some propose extending personhood to natural features like rivers or ecosystems. 3. Cross-cultural perspectives: Notions of personhood vary across cultures. For example: Among the Wari' people in Brazil, personhood is tied to social relations. The Gebusi people in Papua New Guinea gradually become persons through social interactions. 4. Debates and challenges: Personhood debates arise in contexts such as abortion, fetal rights, and reproductive rights. The personhood movement seeks to bridge Western and indigenous legal systems by granting non-human entities personhood. In summary, personhood transcends species boundaries, encompassing legal, philosophical, and cultural dimensions

(continued)

Inquiry	Define personhood	Is it limited to humans?
GPT	Personhood refers to the state or condition of being an individual human being with legal and moral rights, typically including attributes such as consciousness, self-awareness, rationality, and the capacity to make decisions and interact with others in society. It encompasses the inherent dignity and worth of an individual as recognized within legal, ethical, philosophical, and social contexts.	The concept of personhood traditionally pertains to humans, as it relates to the unique attributes and capacities that distinguish human beings from other entities. However, there are ongoing philosophical and ethical discussions about extending personhood to certain non-human entities, such as animals or artificially intelligent systems, particularly when considering their cognitive abilities, consciousness, and capacity for suffering. This extension of personhood to non-human entities is a subject of debate and varies depending on cultural, legal, and philosophical perspectives.
Meta AI	Personhood is a complex and multifaceted concept that has been debated by philosophers, legal scholars, and medical professionals. At its core, personhood refers to the state of being a person, with all the rights, dignity, and characteristics that come with it. Some common attributes associated with personhood include: 1. Consciousness and self-awareness 2. Rationality and the ability to make decisions 3. Autonomy and agency 4. Capacity for emotions, thoughts, and experiences 5. Individuality and uniqueness 6. Social and relational capabilities The concept of personhood is often tied to questions of human rights, dignity, and the moral and ethical treatment of individuals. It is a central issue in debates around: 1. Abortion and fetal rights 2. Euthanasia and end-of-life care 3. Artificial intelligence and machine consciousness 4. Animal rights and non-human personhood 5. Disability and vulnerability The definition of personhood can vary depending on cultural, religious, and philosophical perspectives, leading to ongoing discussions and disagreements on its nature and scope.	The concept of personhood is often associated with humans, but it can be extended to other entities, depending on the context and perspective. Some examples include: 1. Non-human animals: Some argue that certain animals, like primates, dolphins, or elephants, possess personhood due to their advanced cognitive abilities and emotional experiences. 2. Artificial intelligence: As AI becomes increasingly sophisticated, questions arise about its potential personhood, including rights and responsibilities. 3. Legal personhood: Corporations, organizations, and even some natural entities like rivers or forests have been granted legal personhood in certain jurisdictions, allowing them to hold rights and liabilities. 4. Fictional entities: In literature, myth, and fantasy, personhood can be attributed to fictional beings like characters, spirits, or supernatural entities. 5. Theological or spiritual perspectives: In some belief systems, personhood can be attributed to divine beings, spirits, or souls, which may be seen as distinct from human persons. These examples illustrate how personhood can be conceived and extended beyond humans, sparking debates about the boundaries and implications of personhood in various contexts.

References

1. American Psychiatric Association. Diagnostic and statistical manual of mental disorders. 5th ed. American Psychiatric Association; 2013. https://doi.org/10.1176/appi.books.9780890425596.
2. Andrea G, Mancusi G, Santovito MC, Marrangone C, Martino F, Santorelli M, Miuli A, et al. Investigating the role of maintenance TMS protocols for major depression: systematic review and future perspectives for personalized interventions. J Person Med. 2023;13(4):697. https://doi.org/10.3390/jpm13040697.

3. Appelbaum LG, Shenasa MA, Stolz L, Daskalakis Z. Synaptic plasticity and mental health: methods, challenges and opportunities. Neuropsychopharmacology. 2023;48(1):113–20. https://doi.org/10.1038/s41386-022-01370-w.
4. Bailey RK, Mokonogho J, Kumar A. Racial and ethnic differences in depression: current perspectives. Neuropsychiatr Dis Treat. 2019;15:603–9. https://doi.org/10.2147/NDT.S128584.
5. Barker AT, Jalinous R, Freeston IL. Non-invasive magnetic stimulation of human motor cortex. Lancet. 1985;325(8437):1106–7. https://doi.org/10.1016/S0140-6736(85)92413-4. Originally published as Volume 1, Issue 8437
6. Benadhira R, Thomas F, Bouaziz N, Braha S, Andrianisaina PS-K, Isaac C, Moulier V, Januel D. A randomized, sham-controlled study of maintenance rTMS for treatment-resistant depression (TRD). Psychiatry Res. 2017;258(December):226–33. https://doi.org/10.1016/j.psychres.2017.08.029.
7. Bleker LS, de Rooij SR, Painter RC, Ravelli ACJ, Roseboom TJ. Cohort profile: the Dutch famine birth cohort (DFBC)— a prospective birth cohort study in The Netherlands. BMJ Open. 2021;11(3):e042078. https://doi.org/10.1136/bmjopen-2020-042078.
8. Blumberger DM, Maller JJ, Thomson L, Mulsant BH, Rajji TK, Maher M, Brown PE, et al. Unilateral and bilateral MRI-targeted repetitive transcranial magnetic stimulation for treatment-resistant depression: a randomized controlled study. J Psych Neurosci. 2016;41(4):E58–66. https://doi.org/10.1503/jpn.150265.
9. Blumberger DM, Vila-Rodriguez F, Thorpe KE, Feffer K, Noda Y, Giacobbe P, Knyahnytska Y, et al. Effectiveness of theta burst versus high-frequency repetitive transcranial magnetic stimulation in patients with depression (THREE-D): a randomised non-inferiority trial. Lancet. 2018;391(10131):1683–92. https://doi.org/10.1016/S0140-6736(18)30295-2.
10. Blumberger DM, Vila-Rodriguez F, Wang W, Knyahnytska Y, Butterfield M, Noda Y, Yariv S, et al. A randomized sham controlled comparison of once vs twice-daily intermittent theta burst stimulation in depression: a Canadian rTMS treatment and biomarker network in depression (CARTBIND) study. Brain Stimul. 2021;14(6):1447–55. https://doi.org/10.1016/j.brs.2021.09.003.
11. Bouchard TJ, Lykken DT, McGue M, Segal NL, Tellegen A. Sources of human psychological differences: the Minnesota study of twins reared apart. Science. 1990;250(4978):223–8. https://doi.org/10.1126/science.2218526.
12. Chang J, Chu Y, Ren Y, Li C, Wang Y, Chu X-P. Maintenance treatment of transcranial magnetic stimulation (TMS) for treatment-resistant depression patients responding to acute TMS treatment. Int J Physiol Pathophysiol Pharmacol. 2020;12(5):128–33.
13. Cohen RA, Cha AE. Health insurance coverage: early release of estimates from the National Health Interview Survey, 2022. National Center for Health Statistics; 2023. https://www.cdc.gov/nchs/data/nhis/earlyrelease/insur202305_1.pdf
14. Cole EJ, Phillips AL, Bentzley BS, Stimpson KH, Nejad R, Barmak F, Veerapal C, et al. Stanford neuromodulation therapy (SNT): a double-blind randomized controlled trial. Am J Psychiatry. 2022;179(2):132–41. https://doi.org/10.1176/appi.ajp.2021.20101429.
15. Cole EJ, Stimpson KH, Bentzley BS, Gulser M, Cherian K, Tischler C, Nejad R, et al. Stanford accelerated intelligent neuromodulation therapy for treatment-resistant depression. Am J Psychiatry. 2020;177(8):716–26. https://doi.org/10.1176/appi.ajp.2019.19070720.
16. Deng Z-D, Lisanby SH, Peterchev AV. Electric field depth–focality tradeoff in transcranial magnetic stimulation: simulation comparison of 50 coil designs. Brain Stimul. 2013;6(1):1–13. https://doi.org/10.1016/j.brs.2012.02.005.
17. Drevets WC. Neuroimaging abnormalities in the amygdala in mood disorders. Ann N Y Acad Sci. 2003;985(1):420–44. https://doi.org/10.1111/j.1749-6632.2003.tb07098.x.
18. Drevets WC, Price JL, Simpson JR, Todd RD, Reich T, Vannier M, Raichle ME. Subgenual prefrontal cortex abnormalities in mood disorders. Nature. 1997;386(6627):824–7. https://doi.org/10.1038/386824a0.
19. Drevets WC, Savitz J, Trimble M. The Subgenual anterior cingulate cortex in mood disorders. CNS Spectr. 2008;13(8):663–81.

20. Edinoff AN, Akuly HA, Hanna TA, Ochoa CO, Patti SJ, Ghaffar YA, Kaye AD, et al. Selective serotonin reuptake inhibitors and adverse effects: a narrative review. Neurol Int. 2021;13(3):387–401. https://doi.org/10.3390/neurolint13030038.
21. Ferguson JM. SSRI antidepressant medications: adverse effects and tolerability. Prim Care Compan J Clin Psych. 2001;3(1):22–7.
22. Ferrari AJ, Charlson FJ, Norman RE, Patten SB, Freedman G, Murray CJL, Vos T, Whiteford HA. Burden of depressive disorders by country, sex, age, and year: findings from the global burden of disease study 2010. PLoS Med. 2013;10(11):e1001547. https://doi.org/10.1371/journal.pmed.1001547.
23. Fox MD, Buckner RL, White MP, Greicius MD, Pascual-Leone A. Efficacy of transcranial magnetic stimulation targets for depression is related to intrinsic functional connectivity with the Subgenual cingulate. Biol Psychiatry. 2012;72(7):595–603. https://doi.org/10.1016/j.biopsych.2012.04.028.
24. GBD 2019 Mental Disorders Collaborators. Global, Regional, and National Burden of 12 Mental Disorders in 204 Countries and Territories, 1990–2019: A Systematic Analysis for the Global Burden of Disease Study 2019. Lancet Psychiatry. 2022;9(2):137–50. https://doi.org/10.1016/S2215-0366(21)00395-3.
25. Gordon-Roth J. Locke on personal identity. In: Zalta EN, editor. The Stanford encyclopedia of philosophy. Metaphysics Research Lab, Stanford University; 2020. https://plato.stanford.edu/archives/spr2020/entries/locke-personal-identity/.
26. Hadas I, Sun Y, Lioumis P, Zomorrodi R, Jones B, Voineskos D, Downar J, Fitzgerald PB, Blumberger DM, Daskalakis ZJ. Association of Repetitive Transcranial Magnetic Stimulation Treatment with Subgenual Cingulate Hyperactivity in patients with major depressive disorder: a secondary analysis of a randomized clinical trial. JAMA Netw Open. 2019;2(6):e195578. https://doi.org/10.1001/jamanetworkopen.2019.5578.
27. Hajek T, Gunde E, Bernier D, Slaney C, Propper L, Grof P, MacQueen G, Duffy A, Alda M. Subgenual cingulate volumes in affected and unaffected offspring of bipolar parents. J Affect Disord. 2008;108(3):263–9. https://doi.org/10.1016/j.jad.2007.10.024.
28. Hallett M. Transcranial Magnetic Stimulation: A Primer. Neuron. 2007;55(2):187–99. https://doi.org/10.1016/j.neuron.2007.06.026.
29. Hirschfeld RMA. History and evolution of the monoamine hypothesis of depression. J Clin Psychiatry. 2000;61(suppl 6):8272.
30. Huang Y-Z, Edwards MJ, Rounis E, Bhatia KP, Rothwell JC. Theta burst stimulation of the human Motor cortex. Neuron. 2005;45(2):201–6. https://doi.org/10.1016/j.neuron.2004.12.033.
31. Huttenlocher PR, Dabholkar AS. Regional differences in synaptogenesis in human cerebral cortex. J Comp Neurol. 1997;387(2):167–78. https://doi.org/10.1002/(SICI)1096-9861(19971020)387:2<167::AID-CNE1>3.0.CO;2-Z.
32. Jorgensen LM. Seventeenth-century theories of consciousness. In: Zalta EN, editor. *The Stanford Encyclopedia of Philosophy*. Metaphysics Research Lab, Stanford University; 2020. https://plato.stanford.edu/archives/spr2020/entries/consciousness-17th/.
33. Kerstein S. Treating persons as means. In: Zalta EN, editor. The Stanford Encyclopedia of Philosophy. Metaphysics Research Lab, Stanford University; 2024. https://plato.stanford.edu/archives/sum2024/entries/persons-means/.
34. Kessler RC, Berglund P, Demler O, Jin R, Koretz D, Merikangas KR, John Rush A, Walters EE, Wang PS. The epidemiology of major depressive DisorderResults from the National Comorbidity Survey Replication (NCS-R). JAMA. 2003;289(23):3095–105. https://doi.org/10.1001/jama.289.23.3095.
35. Lisanby SH. Electroconvulsive therapy for depression. N Engl J Med. 2007;357(19):1939–45. https://doi.org/10.1056/NEJMct075234.
36. Liston C, Chen AC, Zebley BD, Drysdale AT, Gordon R, Leuchter B, Voss HU, Casey BJ, Etkin A, Dubin MJ. Default mode network mechanisms of transcranial magnetic stimulation in depression. Biol Psychiatry. 2014;76(7):517–26. https://doi.org/10.1016/j.biopsych.2014.01.023.

37. Longuenesse B. Kant's multiple concepts of person*. In: Watkins E, editor. Kant on persons and agency. Cambridge: Cambridge University Press; 2017. p. 155–73. https://doi.org/10.1017/9781316856529.010.
38. McKay MT, Cannon M, Chambers D, Conroy RM, Coughlan H, Dodd P, Healy C, O'Donnell L, Clarke MC. Childhood trauma and adult mental disorder: a systematic review and meta-analysis of longitudinal cohort studies. Acta Psychiatr Scand. 2021;143(3):189–205. https://doi.org/10.1111/acps.13268.
39. Merikangas KR, Akiskal HS, Angst J, Greenberg PE, Hirschfeld RMA, Petukhova M, Kessler RC. Lifetime and 12-month prevalence of bipolar Spectrum disorder in the National Comorbidity Survey Replication. Arch Gen Psychiatry. 2007;64(5):543–52. https://doi.org/10.1001/archpsyc.64.5.543.
40. Mulders PC, van Eijndhoven PF, Schene AH, Beckmann CF, Tendolkar I. Resting-state functional connectivity in major depressive disorder: a review. Neurosci Biobehav Rev. 2015;56(September):330–44. https://doi.org/10.1016/j.neubiorev.2015.07.014.
41. Nagra DS, Stolz LA, Weissman CR, Appelbaum LG. Access to interventional psychiatric treatments in the United States: disparities and proposed solutions. GHES. 2024;2(1):10.36922/ghes.2456.
42. Neacsiu AD, Beynel L, Powers JP, Szabo ST, Appelbaum LG, Lisanby SH, LaBar KS. Enhancing cognitive restructuring with concurrent repetitive transcranial magnetic stimulation: a transdiagnostic randomized controlled trial. Psychother Psychosom. 2022;91(2):94–106. https://doi.org/10.1159/000518957.
43. Nimbalkar N. John Locke on personal identity. Mens Sana Monog. 2011;9(1):268–75. https://doi.org/10.4103/0973-1229.77443.
44. O'Reardon JP, Brent Solvason H, Janicak PG, Sampson S, Isenberg KE, Nahas Z, McDonald WM, et al. Efficacy and safety of transcranial magnetic stimulation in the acute treatment of major depression: a multisite randomized controlled trial. Biol Psych Depress New Persp Treat Etiol. 2007;62(11):1208–16. https://doi.org/10.1016/j.biopsych.2007.01.018.
45. Paulus W, Peterchev AV, Ridding M. Transcranial electric and magnetic stimulation: technique and paradigms. Handb Clin Neurol. 2013;116:329–42. https://doi.org/10.1016/B978-0-444-53497-2.00027-9.
46. Payne NA, Prudic J. Electroconvulsive therapy part I: a perspective on the evolution and current practice of ECT. J Psychiatr Pract. 2009;15(5):346–68. https://doi.org/10.1097/01.pra.0000361277.65468.ef.
47. Pettersson E, Lichtenstein P, Larsson H, Song J, Agrawal A, Børglum AD, Bulik CM, et al. Genetic influences on eight psychiatric disorders based on family data of 4 408 646 full and half-siblings, and genetic data of 333 748 cases and controls. Psychol Med. 2019;49(7):1166–73. https://doi.org/10.1017/S0033291718002039.
48. Power, Jonathan D., and Bradley L. Schlaggar. 2017. "Neural plasticity across the lifespan." Wiley Interdiscip Rev Dev Biol 6 (1): https://doi.org/10.1002/wdev.216. 10.1002/wdev.216.
49. Rakofsky J, Rapaport M. Mood disorders. CONTINUUM Lifelong Learn Neurol. 2018;24(3):804. https://doi.org/10.1212/CON.0000000000000604.
50. van Rooij SJH, Arulpragasam AR, McDonald WM, Philip NS. Accelerated TMS - moving quickly into the future of depression treatment. Neuropsychopharmacology. 2024;49(1):128–37. https://doi.org/10.1038/s41386-023-01599-z.
51. Rossini PM, Burke D, Chen R, Cohen LG, Daskalakis Z, Di Iorio R, Di Lazzaro V, et al. Non-invasive electrical and magnetic stimulation of the brain, spinal cord, roots and peripheral nerves: basic principles and procedures for routine clinical and research application. An updated report from an I.F.C.N. Committee. Clin Neurophysiol. 2015;126(6):1071–107. https://doi.org/10.1016/j.clinph.2015.02.001.
52. Safer JD, Coleman E, Feldman J, Garofalo R, Hembree W, Radix A, Sevelius J. Barriers to healthcare for transgender individuals. Curr Opin Endocrinol Diabetes Obes. 2016;23(2):168–71. https://doi.org/10.1097/MED.0000000000000227.
53. Sanacora G, Treccani G, Popoli M. Towards a glutamate hypothesis of depression. Neuropharmacology. 2012;62(1):63–77. https://doi.org/10.1016/j.neuropharm.2011.07.036.

54. Sapolsky RM. Glucocorticoids and hippocampal atrophy in neuropsychiatric disorders. Arch Gen Psychiatry. 2000;57(10):925–35. https://doi.org/10.1001/archpsyc.57.10.925.
55. Sapolsky RM. Behave: the biology of humans at our best and worst. New York: Penguin Books; 2018.
56. Sapolsky RM. Determined: a science of life without free will. New York: Penguin Books; 2024.
57. Shadrina M, Bondarenko EA, Slominsky PA. Genetics factors in major depression disease. Front Psych. 2018;9(July):334. https://doi.org/10.3389/fpsyt.2018.00334.
58. Shafi MM. Seizures with TMS: much ado about (almost) nothing? Clin Neurophysiol. 2019;130(8):1397–8. https://doi.org/10.1016/j.clinph.2019.04.315.
59. Shiah I-S, Yatham LN. GABA function in mood disorders: an update and critical review. Life Sci. 1998;63(15):1289–303. https://doi.org/10.1016/S0024-3205(98)00241-0.
60. Siebner HR, Funke K, Aberra AS, Antal A, Bestmann S, Chen R, Classen J, et al. Transcranial magnetic stimulation of the brain: what is stimulated?—a consensus and critical position paper. Clin Neurophysiol. 2022;140:59. https://doi.org/10.1016/j.clinph.2022.04.022.
61. Stilo SA, Murray RM. Non-genetic factors in schizophrenia. Curr Psychiatry Rep. 2019;21(10):100. https://doi.org/10.1007/s11920-019-1091-3.
62. van Der Kolk B, editor. The Body Keeps the Score. PenguinRandomHouse; 2014. Accessed May 16, 2024. https://www.penguinrandomhouse.com/books/313183/the-body-keeps-the-score-by-bessel-van-der-kolk-md/
63. Varkey B. Principles of clinical ethics and their application to practice. Med Princ Pract. 2021;30(1):17–28. https://doi.org/10.1159/000509119.
64. Weissman CR, Bermudes RA, Voigt J, Liston C, Williams N, Blumberger DM, Fitzgerald PB, Daskalakis ZJ. Repetitive transcranial magnetic stimulation for treatment-resistant depression: mismatch of evidence and insurance coverage policies in the United States. J Clin Psychiatry. 2023;84(3):46813. https://doi.org/10.4088/JCP.22com14575.
65. Zrenner C, Desideri D, Belardinelli P, Ziemann U. Real-time EEG-defined excitability states determine efficacy of TMS-induced plasticity in human Motor cortex. Brain Stimul. 2018;11(2):374–89. https://doi.org/10.1016/j.brs.2017.11.016.
66. Zrenner C, Ziemann U. Closed-loop brain stimulation. Biol Psych. 2024;95(6):545–52. https://doi.org/10.1016/j.biopsych.2023.09.014.

"The Fewer Pills the Better": Attitudes Toward Transcranial Magnetic Stimulation for Depressed Adolescents

Patricia Henegan, Paul E. Croarkin,
Manisha Kodavatiganti, and Laura Y. Cabrera

1 Introduction

The prevalence of depression has been rapidly increasing in recent years particularly in young people (ages 12–25) and across sex, racial, income, and education groups [14]. For example, 15.1% of adolescents ages 12–17 years had a major depressive episode in 2018–2019 [1].

Considering that antidepressant medications are not always effective or even sought out by patients suffering from depression, other treatment modalities have entered the armamentarium of psychiatric tools. One such modality is transcranial magnetic stimulation (TMS)—a neurotechnology which is considered to be minimally invasive (from a physical perspective), to have a low-risk profile, and to produce a significant antidepressant effect (compared to sham) [6, 29]. TMS was first cleared by the U.S. Food and Drug Administration for treatment resistant depression in adults in 2008. While as of March 2024, TMS received FDA clearance for adolescents, there are unanswered questions related to efficacy, side effects, safety, and evidence based

P. Henegan
School of Public Health and Health Sciences, UMass Amherst, Amherst, MA, USA

P. E. Croarkin
Mayo Clinic, Rochester, MN, USA
e-mail: Croarkin.Paul@mayo.edu

M. Kodavatiganti
Neuroethics Lab, The Pennsylvania State University, University Park, PA, USA

L. Y. Cabrera (✉)
School of Public Health and Health Sciences, UMass Amherst, Amherst, MA, USA

Department of Engineering Science and Mechanics, and Rock Ethics Institute, The Pennsylvania State University, University Park, PA, USA

Center for Neural Engineering, The Pennsylvania State University, University Park, PA, USA
e-mail: lcabrera@psu.edu

V. Dubljević, J. R. Young (eds.), *TMS and Neuroethics*, Advances in Neuroethics,
https://doi.org/10.1007/978-3-031-92401-9_11

in its use for children or adolescents, with emerging studies examining the safety and efficacy of TMS for these patient populations [10, 11, 16, 23].

Priorly published literature on neuromodulation has highlighted a number of neuroethical considerations specific to the use of TMS for adult depression, including side effects, effectiveness, inconvenience of frequent sessions (typically 4–5 times per week for 4–6 weeks), and equitable access due to a shortage of equipment, a scarcity of healthcare providers who offer or refer patients, limited geographic availability, and issues with insurance coverage [2, 5, 6]. Moreover, with the more recent accelerated protocols incorporating theta burst stimulation, several safety considerations remain to be determined, including establishing guidelines for the total pulse count, the time gap between consecutive sessions, the level of stimulation intensity, and the cumulative frequency of daily or weekly applications [6]. When considering younger populations, like adolescents, ethical considerations become more salient, as there is even more limited evidence of safety and efficacy [19, 23, 25], and other ethical considerations appear, such as considerations related to long-term effects on brain development [6] and considerations related to informed consent, assent, and respecting the developing autonomy of children. Other key considerations are related to who determines what is in the best interest of the child, and how involved should adolescents be in shared decision-making [37]. For example, in the USA, the requirement for seeking assent (as opposed to legal consent) from children is set at a minimum age of 7 years old [17]. Wendler and Shah argue that the threshold for assent should be fixed at 14 years of age [36], while others argue that who meets the capacity to provide assent needs to be assessed on an individualized manner [20, 32] as some children might be capable of providing assent at a much earlier age compared to others [27]. Children of the same age may exhibit varying levels of maturity. Moreover, while some young children may demonstrate competence for decision-making in specific situations, their competence may be insufficient in other contexts.

A few studies have explored attitudes and ethical concerns in relation to the use of TMS among other electroceuticals [2, 5, 7], which have found a perceived lack of evidence on safety and efficacy, intrusiveness from a lifestyle perspective, and fear of side effects as key ethical concerns. Yet, to our knowledge attitudes and ethical concerns that adolescents and parents have in relation to the use of TMS for adolescent depression has not been previously studied in the USA.

This study aims to fill that gap by examining adolescent and parental attitudes toward the use of TMS as potential future treatments in adolescent depression. Inclusion of parental attitudes is key given the role of parents in initiating the process of seeking treatment and having responsibility for their adolescent's medical decisions (including in many cases consenting to treatment). Family dynamics and relationships can also impact an adolescent's willingness to participate in treatment. Moreover, adolescents often depend on their parents for transportation to appointments and financial support, and in the case of mental health treatment, parents often play a crucial role in the treatment process. In addition, better understanding of parental and adolescent perceptions of TMS and factors associated with their attitudes can allow for providers to offer curated information to parents considering TMS for their adolescents as well as to adolescents involved in deciding whether or not to undergo this therapy.

2 Methods

We conducted semi-structured interviews with adolescents (N = 5, aged 13–17 years) and parents of adolescents (N = 15) between November 2022 and February 2023.

2.1 Recruitment and Participants

We used a non-probability sampling design of purposive sampling. While we originally planned to achieve quotas on gender and race for all participants, our struggles to recruit participants caused us to realize that this goal was not feasible. We also initially were going to recruit dyads of parents and their adolescents with depression, but given recruitment difficulties we only ended up recruiting 2 dyads (Participants 1P/1A, and 9P/9A) and the rest were parents of adolescents and adolescents not related, with or without depression.

We distributed information about the study to mental health clinics associated with the Pennsylvania State University (PSU) and Mayo Clinic, later we expanded recruitment to depression disease support groups, clinician networks (through the National Network of Depression Centers), as well as in public places in Central Pennsylvania and social media posts. When scheduling each interview, we sent participants a letter that described the study and stated that proceeding with the interview indicated their voluntary agreement to participate. We also sent each participant a one page document with information about the intervention. The study was approved by PSU's Institutional Review Board (Exempt determination, STUDY00020208) whereby the formal recording of written consent was not deemed necessary. Participants were informed prior to the start of the interview about their right to decline to answer any questions, or to decide to stop or withdraw their involvement at any point during the interview.

Two team members conducted semi-structured interviews, a method used to facilitate exploration of unanticipated issues and in-depth understanding, while covering a core set of topics [22]. Interviews with participants were held via Zoom, an interview modality that has received extensive study in recent years [24]. We invited participants to ask any clarifying questions about the consent/assent sheets before the interview began.

We first asked both adults and adolescent participants general sociodemographic questions. Adult participants were asked questions focused on their experiences with their adolescent with depression and treatments (if they have an adolescent with depression); and we asked adolescents to share their experiences with depression and treatments (if applicable). For both groups, we continued the interviews with questions about their general knowledge, attitudes, and ethical concerns toward repetitive TMS (rTMS). We also discussed with parental participants how much their children had a voice in the decision-making process of choosing a treatment method, while balancing the parents' own decisions for their children. Toward the end of each interview, we asked participants if they felt differently when thinking about theta burst stimulation (TBS) compared to rTMS. Interviews were recorded

Table 1 Description of the study samples

	Parents of adolescents (n = 15)	Adolescents (n = 5)
Gender		
Male	5 (33.33%)	3 (60%)
Female	10 (66.67%)	2 (20%)
Age		
Median	N/A	16
Range	26–35 (n = 1, 6.67%)	13–17
	36–45 (n = 10, 66.57%)	
	46–55 (n = 4, 26.67%)	
Race/ethnicity		
Hispanic	2 (13.33%)	1 (20%)
Non-Latino white	5 (33.33%)	2 (40%)
African American	6 (40%)	2 (40.0%)
Other	2 (13.3%)	0 (0%)
Education		
Middle school		2 (40%)
High school		3 (60%)
College 4 years or more	5 (33.33%)	
Advanced degree	9 (60%)	
Professional certificate	1 (6.67%)	

by Zoom, anonymized, and transcribed verbatim. We used 3PlayMedia to transcribe the interviews from the de-identified Zoom audio files.

2.2 Participant Description

Within our parental participants, we had a slight majority of women compared to men. Most participants for both our parents and adolescents' groups self-identified as non-white or mixed race (Table 1).

2.3 Data Analysis

We used the web-based software Dedoose to analyze the text in the interview transcripts. Data analysis began immediately upon transcription to identify emergent issues that warranted further exploration in subsequent interviews. We used qualitative content analysis methods and a deliberative approach [18, 26]. We used the iteratively developed codebook to code the transcripts of the adolescent group, revising this where necessary in order to allow for the emergence of themes not found in the adults' group interviews. Team meetings provided further opportunities to revise the codebook as necessary and to reach a consensus on coding discrepancies. In our Results section below, we present relevant quotes to highlight the findings with non-content words and expressions removed for readability.

3 Results

Our results suggest that both parents and adolescents are enthusiastic about the potential for TMS in treating adolescents with depression, yet there is limited awareness about the intervention. Only a third of adult participants (n = 5) reported any knowledge about the intervention, and the rest reported no prior knowledge. In terms of our adolescent sample, three out of five adolescent participants self-identified as individuals with depression. Eleven out of fifteen of our interviewed parents reported having at least one adolescent with depression. Here, we describe our results pertaining to (a) attitudes about the treatment, (b) adolescent decision-making, and (c) barriers and facilitators around the use of TMS in this patient population.

3.1 Attitudes about TMS

The majority of our participants (n = 10) lacked prior awareness of TMS, as such several participants had questions about the mechanism of TMS treatment.

Participant 13P: "Yeah, I just want to know like this TMS therapy, exactly, does it work permanently?"

Participant 09A: "I don't know. I just think it's crazy, being able to do that basically. I'm just amazed by how that could treat it."

Some participants thought the mechanism behind TMS was interesting, while others felt initial hesitation by the "brain stimulation" component.

Participant 10P: "I think it's a little—I don't know, cartoonish. Almost like "Frankenstein" or something. I've had some different treatments for muscle things but it's not been on my head. It's been like on my back, or my leg, or whatever. It seems less scary than your actual head."

Participant 03P: "I don't want to say messing with the brain because it could be helpful, but that it's like right near the head, like right where the brain is that makes me a little nervous….if that makes sense, a little."

Participant 05A: "Electromagnetic waves going to the brain, and even if you research it people don't really understand the brain So that's the only thing, but that's what happens when you do stuff with your brain. It's a risk you have to take."

Several participants mention that a potential benefit of TMS is that it is less time consuming; explaining that it is not a treatment you need to do for your whole life (a view held regarding antidepressants). It was also perceived to be easier than having to remember to take pills every day, and potentially having less side effects than medications. It was also mentioned that it did not seem to be physically invasive.

Participant 09A: "Definitely time. I feel like that that's a big thing for me. Having to get up and go take my medicine and all of that. This is a one and done thing, right? And like—well, not one—is it one and done? I mean, no. But you know what I mean? You go do it and then you're done, right? It's not every night or you're not going and talking about it, talking about your problems. I don't know. I feel it could just be easier and less time consuming."

Participant 04P: "[…] the more I learned about this this doesn't really seem physically invasive."

Another benefit mentioned was that it provides alternatives to those who do not want to engage in counseling or psychotherapy, or who are not into taking pills. A number of parents seem to be more open to try TMS than to medicate their children, with a key concern being a perception that medications are addictive. While others mentioned that they would try cognitive-sociobehavioral therapy, and even medication before TMS.

Participant 04P: "Maybe this could be something that would be a better sell for those who are just totally not into taking pills […] the fewer pills the better."

Participant 03P: "I'm probably try to find a counselor […] Then, if that really didn't do anything, then maybe medication next, then TMS."

Participant 02P: "Talk therapy would be first I would try that first, and I don't know. I mean… I don't know other than medication. I would probably not put my adolescent on medication. […] I might be more willing to try TMS first".

Most participants agreed that TMS, if effective, would provide lasting results that would allow for their child to improve without daily medications. Another participant felt optimistic about an intervention that with 40 min (or 10 min in the accelerated protocols) could enable adolescents to live a normal life.

Participant 09P: "It's optimistic that if you can do this and after the initial 40 minutes, if it's just like 10 minutes a day, but just to be able to be like, OK, I can take 10 minutes out of my day to do this, and then I can live a normal life."

Participant 01A: "It sounds pretty good, like it helps people be back on track."

However, participants were cautiously optimistic about the results TMS could provide, saying that they would like to see results from people who have used it, but also acknowledging that is good to have more options.

Participant 06P: "We need to see results. But if it is working, it's good to have more options. Because like I said, there's no one size fits all when it comes to a lot of things, including depression."

Participant 05A: "I know I would want at least to see some people with who's done that long term, and how they turned out."

Despite the perceived benefits and reasons for optimism, parents and adolescents still had concerns about the safety and side effects associated with TMS, as well as risks, in particular age related or long-term effects of TMS.

Participant 16A: "OK, I just wanted to ask the side effects, how long can the side effect last."
Participant 04P: "I would say just the long-term effects. You get the people out there beginning their lives."
Participant 09A: "How it affects any comorbidities. [...] the long-term effects, what it does to your body."

3.2 Decision-Making for Adolescent TMS

Another important theme in our interviews deals with decision-making. Both parents and adolescents discussed the involvement of adolescents in decision-making related to their health care. Some parents brought up concerns about the maturity of their adolescents or them being too depressed to be "in their right mind" to have a say in such a decision. In cases like this, most of the parents agreed that it is ultimately the parents' decision to act in the best interest of their child. Though there were two parents who strongly held that the decision should be on them as parents.

Participant 01P: "We always try to have them take part and you know in the stuff that are happening [...] especially when there are things that really involves him directly. So we know that [he] needs to be part and we like to hear his ideas."
Participant 04P: "I don't think an adolescent while they should be consulted, I don't think they should be making the call as to what's best for them as far as [TMS] goes."
Participant 08P: "I would not be OK committing to a course of treatment without their consent. I would make it clear that I would want to have a conversation. But I would also work very hard to honor their preference as well. I mean, if I felt like they could not make a decision, then I would have another kind of discussion with my husband about how do we want to manage this, and how do we want to think about this. But it would be really critical for me that my kid was on board with the decision."

In the case of adolescents most of them said that they want to be asked about their opinion not just have someone decide for them, though a few also mentioned that they trust their parents to make the best decision for them. One adolescent, similar to some parents' views, mentioned that not all adolescents have the capacity or ability to decide what they want. This thought process complements another perspective under decision-making mentioned by both parents and adolescents, which was a reference to adolescent age as being relevant to whether and how involved adolescents should be in decision-making.

Participant 12P: "My child is around 17 years old, and being I know it's around 18 years, and considered that is he's man enough to make decisions. You know? So I was not really concerned, because I knew in a while, or in a couple of months, he is going to grow up and be an adult."

Participant 09P: "A 14-year-old or 16-year-old with no mental health concerns, in my opinion, does not have the capability to make rational thoughts and decision. They're just going through a lot. And so for them to be able to have complete 100% control by law boggles my mind. […] but he does have a say, I mean, primarily."

Participant 05A: [if I were younger than 12] I would be a little more hesitant just because I feel like it would be harder for me to have an understanding".

Participant 09A: "I think if you're over the age of 14, I mean, you have the right to therapy. You have the right to consent to therapy. So, I feel like that should be the same thing for this. If you're over the age of 14, you have every right to say yes or no without a parent's consent."

A few parents agree that involvement of adolescents in decision-making is important as it helps them build a sense of autonomy, whereas most older adolescents view this as important to their autonomy in deciding things that will affect them directly. One parent mentioned that as adolescents become older it can be a challenge to make adolescents attend treatment appointments, especially because of the time commitment.

Participant 02P: "[T]hey're developing their own agency at this age, and you can be very careful about respecting kind of that balance between respecting their autonomy and being the parent."

Participant 04P: "I'm wondering, especially with the time commitment how easier or challenge you would be to have someone who's maybe not cooperative or being depressed … to get them in a car 5 times a week, […] we're not talking about 2-year-old kids that they do what you say, right, and they go where you go, because that's all they know you can pick them up. But if you have a 16-year-old who doesn't wanna do it, I guess it's the same with medication if they don't want to take their pills, I guess, you can't make them, for the most part."

3.3 Barriers and Facilitators to Use of Adolescent TMS

The third main theme was access to TMS, including the considerations of cost, insurance coverage, and availability in the region.

Participant 03P: "There's like the factors that are like beyond our control like if it's covered by insurance, and how far we have to go and how many times we have to go."

Most parent participants had some concern about where the clinics would be located and how far they would have to drive to receive this treatment. Although there were concerns about the logistics associated with treatment, most parent participants agreed that they would find a way to "make it work" in order to give their child the best shot possible in improving.

However, this position was complicated if they have more than one child to take care and the frequency of the treatment as there would be other factors in making this family decision.

Participant 07P: "We go where we need to go when [our adolescent] need[s] help. I think maybe some parents who don't have the resource of a good reliable car—that might be a problem. […] When you're desperate for your kid to be well, you will do and go almost anywhere."

Participant 01P: "If it's going to be like the nearest facility is going to be […] more than 1 h drive just getting there. Going back is going to be like 2 h and a half every single day."

Another important discussed barrier was cost and insurance coverage. With several of our adolescent and parent participants mentioning that the treatment should be affordable for people in need to try it. Considerations of barriers often included acknowledgment of how barriers tend to be interconnected and compound each other.

Participant 16A: "I think it should be affordable and good so that a lot of people could try it."

Participant 08P: "If it's something that you have to do for a month, and it has to be five or six days a week, and it's a three-hour drive, then that means that maybe the family has to make a commitment to go live somewhere else for a month. And what does that mean? In that case, then it becomes sort of like is it prohibitively expensive now because we have to go and live in a hotel, or an apartment, or something like that and try to figure out how to make this work. So in the logistics way of thinking about things, I could imagine that would be an issue. That would have to be a consideration."

Treatment frequency of TMS was seen as a major barrier, and most participants were interested in novel protocols that would make it more likely for people to be able to join a course of treatment. Yet, as described above there was agreement that if there were no other options and this was the next thing to try, they would do their best to make it work. Importantly, as a parent raised, one needs to consider how difficult it can be to convince a depressed adolescent to get in a car 5 times a week. Another related consideration from the young adolescents not self-identified with depression was how this schedule of treatment would disrupt their after classes activities.

Participant 04P: "I'm wondering, especially with the time commitment how easier, or, challenge you would be to have someone who's maybe not cooperative or being depressed … to get them in a car 5 times a week."

While TMS was seen as having an impact on daily life, most participants saw it as something that could be managed.

Participant 08P: "So I think that 30 or 40 minutes a day for five or six weeks—I don't know—it might even be an opportunity to think about connecting with my kid, right, because this is a thing that we're doing together or something like that like."

Participant 09A: "I don't think it would be difficult for me to go personally because my mom would try to do her best. And I mean, I would schedule it around my free time so then I could be able to go."

4 Discussion

This study contributed to ongoing efforts examining adolescent and parental views related to TMS for adolescent depression. A prior Australian study of eight adolescents receiving investigational TMS suggests that TMS was well tolerated but perceived as not beneficial by adolescents and parents [21]. This diverged with views on antidepressant medications, as participants from that study viewed medication as a more helpful treatment [21]. A larger recent survey study on parents preferences for adolescent depression treatment suggested that past parental experiences with pharmacological treatment and parental perceptions of benefit of treatment influenced preference for antidepressants medications.

Similar to the finding from studies of views about the treatment for adult depression in the USA [2, 5, 7], our adolescent and parent participants did not experience the treatment as fear-provoking or too disturbing, and overall our findings suggest that parents are interested and open to this modality of treatment. Yet it is important to also emphasize that our findings also reveal a cautionary stance, as several perceived risks and barriers were mentioned.

4.1 Interested but with Caution

Similar to previous studies examining public and patient perceptions around the use of electroceuticals used for adult depression [2, 5, 7], our participants did not fear TMS. This is in contrast to technologies like electroconvulsive therapy (ECT) and deep brain stimulation (DBS), which previous studies have found to be perceived with fear by participants. Overall, there was interest on the possibility to offer this as an opportunity, which aligns with previous findings. For example, an interview study with 48 adult patients in Australia found that 65% of them believed TMS should be given to adolescents if indicated [34].

Our participants both recognized benefits and drawbacks of TMS for adolescents. Compared to other neuromodulation interventions, such as ECT and DBS, TMS has been found in our study and others to be perceived to be overall safe [2, 5, 21, 34], and minimally invasive (from a physical point of view) as it does not require anesthesia or surgery. However, while other studies have found that invasiveness is a much more complex concept that goes beyond physical invasiveness [3, 7], in this study, our participants minimally acknowledged the lifestyle invasiveness suggested in previous studies. This could to some extent be a result of the different populations, as in those previous studies, the focus was on the use of TMS for adults, whereas in this study, we were focused on TMS for adolescents. It could also be the result of having both participants with and without lived-experience.

Results on perceptions of effectiveness of TMS point to divergent views, with some studies finding participants perceiving TMS to be beneficial [30], others pointing that non-clinician groups perceived limited evidence of treatment effectiveness [9], and the only study of TMS in adolescents findings that parents are skeptical of effectiveness [21]. Yet as Wallman and colleagues point out [33], efficacy is not the only factor parents consider when evaluating possible treatments, with consideration such as side effects, long-term effects and degree of disruption to regular activities (e.g., school, extracurricular activities) playing an equally or more important role.

The present study showed worry about the side effects and long-term effects of TMS in adolescents, which is one of the reasons why several participants in our study and others seem to accept TMS but only as a last resort treatment, once other treatments failed. For example, in Mayer and colleagues' paper most TMS recipients (62%) mentioned they would agree to have TMS again if recommended to them, but only if medications and other treatments would not work [21].

Interestingly, similar to prior research, several parents in our sample, had negative views about medication [33], and welcome TMS as offering choice in selecting a different type of treatment of depression, in particular in light of a lack of alternative treatments for those adolescents were psychotherapy and medication does not work. For example, one parent with an adolescent with depression was excited to hear about an alternative treatment to therapy because of the difficultics connected to getting her adolescent to open-up about how they are feeling with a therapist, especially when the therapist is not always the same.

Participants in a study observing TMS versus fluoxetine treatment showed that participants were more likely to remember the side effects associated with the medication that with TMS [21]. A portion of this trend may be subject to recall bias; however, it may also mean that TMS side effects are relatively less than those of medication. Compared to views on TMS, parents perceived medications as having a greater impact on an adolescent's developing brain, viewing them as a lifelong commitment, and some as potentially addictive.

A key reason for a cautionary stand, is that children are considered to be more vulnerable than adults, which makes the ethical considerations more salient [6, 37]. As with other public perception studies around neuromodulation, the perceived uniqueness of the brain as well as the history of previous somatic interventions might explain a predominant cautionary attitude toward using neuromodulation interventions for depression in adolescents. However, in our study, most participants did not perceived TMS with stigma, as it is the case with other electrocuticals, such as ECT and even DBS. A second likely factor contributing to the cautionary stance is the very low treatment familiarity and knowledge observed around neuromodulation treatments for depression, including TMS [2, 5]. Moreover, while TMS has a relatively low risk profile, the potential for some rare but serious adverse effects might also contribute to the cautionary stance, and while recently approved, the study discussed the therapy as experimental for this age group [11], which likely also played a role in the cautionary stance. Another potential contributor to the cautionary stance is the lack of familiarity with this treatment modality. Similar to the lack of familiarity with TMS in our sample, in another study on parents preferences for adolescent depression treatment, only 16.1% of respondents indicated high familiarity with TMS [33]. Even in the case of TMS for adult depression, for which TMS has been FDA-cleared for over a decade, it is not a widely known treatment [31]. This is important, as research suggests that individuals who are not familiar with novel forms of neuromodulation, such as TMS, may default to their overall perceptions of more known treatments, such as ECTm which tends to be predominantly negative [31].

Considering the recent approval of TMS for this population, there is a need to create more educational opportunities with parents and their child to increased awareness of the treatment and promote more balanced views of the intervention [35].

4.2 Access Remains an Important Barrier

While overall our participants show interest moderated with caution, they pointed to an area that remains a key barrier to utilization for adults, access to the treatment [9]. A recent national survey with psychiatrists, caregivers, patients, and members of the public showed that cost and insurance were among top perceived barriers across all surveyed groups, and frequency of treatment being a top barrier for psychiatrists but not among the top barriers for the non-clinical groups. Another study acknowledges the intrusiveness of TMS given the time commitment for the standard protocols [3]; however, our results, with a particular focus on adolescent depression, suggest a much complex landscape, as concerns about availability and time commitment were mentioned, but ultimately participants agreed that if the treatment was deemed the right option for them or their adolescent they would do what they could to ensure that it is received. For example, some of our parent participants mentioned that they would be willing to travel great distances, pay high costs, and make long appointments to ensure that their child receives the proper care that they need to improve their symptoms. Further evolving TMS protocols may mitigate the time burden for participants [4, 8].

As found through the interviews from participants, parents seem willing to overcome discussed barriers or difficulties that may be encountered to get the proper treatment needed for their child. Weinzimmer et al. [35] explain that this finding was also relevant in their study as parents expressed that they would do everything within their power to ensure that their child is able to receive the care that they need to improve their conditions.

It is important to acknowledge that families who need to make these decisions are already in distress, so even if they are willing to overcome barriers, having to care for multiple family members, having just one car, and other considerations can make a difference for an adolescent to be able to complete treatment course. If this therapy is to help those in need, it is crucial to continue working on how to address TMS barriers to access, as having it be FDA approved is just part of the equation for utilization.

4.3 Decision-Making and Autonomy

Perspectives around the level of adolescents' involvement in decision-making varied greatly in our results. Each parent considered their role as a guardian of their child when responding to our questions. They shared their worries about making the correct decisions for their children, while also contemplating their child's ability to understand what the differences in treatments mean for their future lives. Previous published work by Sharp and Quigley supportively posited that adolescent assent should be determined on a case-by-case basis to see if the adolescents have the capacity with their life experiences to provide adequate assent for participation in a research study [27]. Sharp and Quigley [27] reject the age threshold and argue that an adolescent's human flourishing be taken into account in order to determine an adolescent's capacity for assent in the best way possible. Since each adolescent has different points at which they receive adequate life experience to provide assent, assent must be tailored to determine the maturity of an adolescent therefore an age threshold cannot be accurately defined. As an overall rule regarding assent, the majority of states agree that a child below the age of 18 requires parental consent and can provide independent assent on a case-by-case basis of capacity analysis.

Issues around informed consent and assent of children and adolescent in medical treatment and research is a sensitive topic [15]. In the context of neuromodulation, Davis pointed out that a major practical problem around the use of brain stimulation for children is developing a framework for establishing informed consent in children and their guardians [12]. Similarly Zuk and colleagues, in a review of neuroethics issues that arise with the development, translation and use of technologies for neuromodulation, including TMS, pointed to the issue around informed consent and agency [37]. In particular, they mentioned that it is not always clear who should determine what is in the best interest of a child, who should get to make the final decision about whether a neuromodulation intervention is to be used on a child or adolescent, and based on our own data, in the case of adolescents how much saying they should have [37]. In this regard, considering the goals of shared decision-making, assent and dissent of the adolescent involved should be sought and integrated in the final decision.

4.4 Study Limitations

Although not a novel limitation of including adolescent participatory research, one limitation of our study includes the relatively small number of adolescent participants This, in turn, limits the range of views and generalizability that may exist among this population and our study. In spite of our efforts to recruit adolescents via different platforms and venues, adolescent participants are notoriously difficult to attain, this may have skewed our results toward the views of parent/guardian participants. Another limitation is that while the original goal was to recruit parents and adolescent with TMS experience, recruitment challenges move us away from that goal to also include participants who did not necessarily have personal experience with the disorder in question or with TMS, such as having depression themselves, having tried TMS, or being a parent of an adolescent with depression or a child who has tried TMS. However, inclusion of these diverse perspectives is still a way to contribute to a more responsible and effective translation into clinical practice [13, 28] for this vulnerable population, even if the participant has not had direct experience with the intervention. In particular as these views enable to capture expectations, misconceptions, and potential structural and attitudinal barriers that might either limit uptake or blind acceptance.

5 Conclusion

As we were analyzing the data for this project, TMS received FDA clearance for depression in adolescents. While it is still early to truly say whether or not TMS will prove to be a clinically effective treatment for this patient population, our findings suggests that parents have an interest in the application of TMS as a treatment for adolescent depression is more salient now that they can access the treatment beyond enrolling their adolescents in a clinical trial. Our results also point to the continue need for educating the general public on what TMS is and how it works, as well as to continue working on practical considerations that remain to be addressed to make this treatment more accessible for this patient population.

Acknowledgments The authors are grateful to all the participants for their time and sharing their views. We also are thankful to the National Network of Depression Centers (NNDC) for funding this study as part of a Momentum Grant.

Appendix: TMS Information Page

Transcranial Magnetic Stimulation (TMS) is a minimally invasive brain stimulation technique approved by the U.S. Food and Drug Administration (FDA) since 2008 for the treatment of adults with Major Depressive Disorder who have failed prior antidepressant medication therapy. Therapy with TMS for children and adolescents with Major Depressive Disorder has been studied previously but is still considered investigational.

During a TMS session, which is often administered in an outpatient setting, a TMS-certified physician places an electromagnetic coil against the patient's scalp. Side effects of rTMS are generally mild to moderate and include headaches, pain or discomfort under the treatment coil, twitching of facial muscles, dizziness, and blurred vision. Not all patients experienced one of these side effects and, when present, they tend to decrease after the first weeks of treatment. Though rare, rTMS has the potential for serious side effects including seizures, induction of mania, worsening of depression and hearing loss. rTMS is not recommended for people with any metal devices or objects implanted in their bodies. In the case of adolescents, there may be other risks that are currently unknown.

Repetitive Transcranial Magnetic Stimulation (rTMS) involves passing an electrical current through a magnetic coil placed superficial to the scalp, producing a high-intensity magnetic field that passes through the scalp and skull to excite neuronal tissue. Treatment sessions typically last around 30–40 min. A typical course of TMS is done five times a week for 6 weeks. rTMS appears to be a relatively safe and reasonably well-tolerated treatment.

Theta Burst Stimulation (TBS) is a newer kind of rTMS that delivers the magnetic pulses in fast patterns, shortening the duration of rTMS treatment sessions from 37 min to 3 min with equivalent antidepressant response. Theta burst was FDA-cleared in 2018 after it was shown that both the efficacy and side-effect profiles were the same as standard 10 Hz rTMS treatment. TBS courses also involve daily stimulation sessions for 6 weeks.

Accelerated TMS A few studies have come out which have tried administering multiple sessions of TBS in 1 day (e.g., 3 sessions per day over 10 days or 10 session per day over five consecutive days), which is called Accelerated TMS. Although research and knowledge in this area is growing it is not known if accelerated TMS is more or less effective than the standard 6 week course.

References

1. Bitsko RH, Claussen AH, Lichstein J, Black LI, Jones SE, Danielson ML, Hoenig JM, et al. Mental health surveillance among children—United States, 2013–2019. MMWR Suppl. 2022;71:1–42. https://doi.org/10.15585/mmwr.su7102a1.
2. Bluhm R, Sipahi ED, Achtyes ED, McCright AM, Cabrera LY. Stakeholders' ethical concerns regarding psychiatric Electroceutical interventions: results from a US Nationwide survey. AJOB Emp Bioethic. 2024;15(1):11–21. https://doi.org/10.1080/23294515.2023.2224592.
3. Bluhm R, Cortright M, Achtyes ED, Cabrera LY. "They are invasive in different ways.": stakeholders' perceptions of the invasiveness of psychiatric electroceutical interventions. AJOB Neurosci. 2023;14(1):1–12. https://doi.org/10.1080/21507740.2021.1958098.

4. Blumberger DM, Vila-Rodriguez F, Thorpe KE, Feffer K, Noda Y, Giacobbe P, Knyahnytska Y, et al. Effectiveness of theta burst versus high-frequency repetitive transcranial magnetic stimulation in patients with depression (THREE-D): a randomised non-inferiority trial. Lancet. 2018;391:1683–92. https://doi.org/10.1016/s0140-6736(18)30295-2.
5. Cabrera LY, Gilbert MMC, McCright AM, Achtyes ED, Bluhm R. Beyond the cuckoo's Nest: patient and public attitudes about psychiatric Electroceutical interventions. Psychiatry Q. 2021;92:1425. https://doi.org/10.1007/s11126-021-09916-9.
6. Cabrera LY, van den Heuvel OA. Ethical considerations regarding the use of transcranial magnetic stimulation in mental health practice. Biol Psychiatry. 2023;95:491. https://doi.org/10.1016/j.biopsych.2023.05.003.
7. Cabrera LY, Achtyes ED, Bluhm R, McCright AM. Views about neuromodulation interventions for depression by stakeholder group, treatment modality, and depression severity. Compr Psychiatry. 2023;122:152365. https://doi.org/10.1016/j.comppsych.2023.152365.
8. Cole EJ, Phillips AL, Bentzley BS, Stimpson KH, Nejad R, Barmak F, Veerapal C, et al. Stanford neuromodulation therapy (SNT): a double-blind randomized controlled trial. Am J Psychiatry. 2022;179:132–41. https://doi.org/10.1176/appi.ajp.2021.20101429.
9. Cortright MK, Bluhm R, Achtyes ED, McCright AM, Cabrera LY. Perceived barriers to using neurostimulation: a national survey of psychiatrists, patients, caregivers, and the general public. J ECT. 2024;40:111. https://doi.org/10.1097/yct.0000000000000990.
10. Croarkin PE, Elmaadawi AZ, Aaronson ST, Randolph Schrodt G, Holbert RC, Verdoliva S, Heart KL, Demitrack MA, Strawn JR. Left prefrontal transcranial magnetic stimulation for treatment-resistant depression in adolescents: a double-blind, randomized, sham-controlled trial. Neuropsychopharmacology. 2021;46:462–9. https://doi.org/10.1038/s41386-020-00829-y.
11. Croarkin PE, MacMaster FP. Transcranial magnetic stimulation for adolescent depression. Child Adolesc Psychiatr Clin N Am. 2019;28:33–43. https://doi.org/10.1016/j.chc.2018.07.003.
12. Davis NJ. Transcranial stimulation of the developing brain: a plea for extreme caution. Front Hum Neurosci. 2014;8:600. https://doi.org/10.3389/fnhum.2014.00600.
13. Garden H, Bowman DM, Haesler S, Winickoff DE. Neurotechnology and society: strengthening responsible innovation in brain science. Neuron. 2016;92:642–6. https://doi.org/10.1016/j.neuron.2016.10.053.
14. Goodwin RD, Dierker LC, Melody W, Galea S, Hoven CW, Weinberger AH. Trends in U.S. depression prevalence from 2015 to 2020: the widening treatment gap. Am J Prev Med. 2022;63:726–33. https://doi.org/10.1016/j.amepre.2022.05.014.
15. Grootens-Wiegers P, Hein IM, van den Broek JM, de Vries MC. Medical decision-making in children and adolescents: developmental and neuroscientific aspects. BMC Pediatr. 2017;17:120. https://doi.org/10.1186/s12887-017-0869-x.
16. Hameed MQ, Dhamne SC, Gersner R, Kaye HL, Oberman LM, Pascual-Leone A, Rotenberg A. Transcranial magnetic and direct current stimulation in children. Curr Neurol Neurosci Rep. 2017;17:11. https://doi.org/10.1007/s11910-017-0719-0.
17. National Institutes of Health. Children's assent. 2023. https://www.cancer.gov/research/participate/clinical-trials/safety#childrens-assent. Accessed December 20.
18. Hsieh HF, Shannon SE. Three approaches to qualitative content analysis. Qual Health Res. 2005;15:1277–88. https://doi.org/10.1177/1049732305276687.
19. Illes J, Gallo M, Kirschen MP. An ethics perspective on transcranial magnetic stimulation (TMS) and human neuromodulation. Behav Neurol. 2006;17:149–57. https://doi.org/10.1155/2006/791072.
20. King CA, Kramer AC. Intervention research with youths at elevated risk for suicide: meeting the ethical and regulatory challenges of informed consent and assent. Suicide Life Threat Behav. 2008;38:486–97. https://doi.org/10.1521/suli.2008.38.5.486.
21. Mayer G, Faivel N, Aviram S, Walter G, Bloch Y. Repetitive transcranial magnetic stimulation in depressed adolescents. J ECT. 2012;28:104–7. https://doi.org/10.1097/yct.0b013e318250058c.
22. Miles MB, Huberman MA, Saldaña J. Qualitative data analysis: a methods sourcebook. SAGE; 2013.

23. Oberman LM, Hynd M, Nielson DM, Towbin KE, Lisanby SH, Stringaris A. Repetitive transcranial magnetic stimulation for adolescent major depressive disorder: a focus on neurodevelopment. Front Psych. 2021;12:642847. https://doi.org/10.3389/fpsyt.2021.642847.
24. Oliffe JL, Kelly MT, Montaner GG, Yu WF, Ko. Zoom interviews: benefits and concessions. Int J Qual Methods. 2021;20:16094069211053522. https://doi.org/10.1177/16094069211053522.
25. Rossi S, Antal A, Bestmann S, Bikson M, Brewer C, Brockmöller J, Carpenter LL, et al. Safety and recommendations for TMS use in healthy subjects and patient populations, with updates on training, ethical and regulatory issues: expert guidelines. Clin Neurophysiol. 2021;132(1):269–306. https://doi.org/10.1016/j.clinph.2020.10.003.
26. Saldana J. An intro to codes and coding. In: The coding manual for qualitative researchers. Sage; 2008. p. 1–31.
27. Sharp RR, Quigley RB. Knowing who you want to be when you grow up: implications for pediatric assent. Am J Bioeth. 2003;3:14–5. https://doi.org/10.1162/152651603322614427.
28. Sierawska A, Splittgerber M, Moliadze V, Siniatchkin M, Buyx A. Transcranial direct current stimulation (tDCS) in pediatric populations——voices from typically developing children and adolescents and their parents. Neuroethics. 2023;16:1–17. https://doi.org/10.1007/s12152-022-09507-w.
29. Somani A, Kar SK. Efficacy of repetitive transcranial magnetic stimulation in treatment-resistant depression: the evidence thus far. Gener Psych. 2019;32:e100074. https://doi.org/10.1136/gpsych-2019-100074.
30. Stillianesis G, Cavaleri R, Tang CY, Summers SJ. Exploring patient perceptions of noninvasive brain stimulation: a systematic review. Neuromodulation. 2021; https://doi.org/10.1111/ner.13461.
31. Tyron JM, Bluhm R, Achtyes ED, McCright AM, Cabrera LY. The influence of prior awareness on views about psychiatric electroceutical interventions among non-clinician stakeholders. NPJ Mental Heal Res. 2023;2:6. https://doi.org/10.1038/s44184-023-00028-9.
32. Waligora M, Dranseika V, Piasecki J. Child's assent in research: age threshold or personalisation? BMC Med Ethics. 2014;15:44. https://doi.org/10.1186/1472-6939-15-44.
33. Wallman EJ, Melvin GA. Parent preferences for adolescent depression treatment: the role of past treatment experience and biological etiological beliefs. J Affect Disord. 2022;316:17–25. https://doi.org/10.1016/j.jad.2022.07.057.
34. Walter G, Martin J, Kirkby K, Pridmore S. Transcranial magnetic stimulation: experience, knowledge and attitudes of recipients. Aust N Z J Psychiatry. 2001;35:58–61. https://doi.org/10.1046/j.1440-1614.2001.00852.x.
35. Weinzimmer SA, Schneider SC, Cepeda SL, Guzick AG, Lázaro-Muñoz G, McIngvale E, Goodman WK, Sheth SA, Storch EA. Perceptions of deep brain stimulation for adolescents with obsessive-compulsive disorder. J Child Adolesc Psychopharmacol. 2021;31:109–17. https://doi.org/10.1089/cap.2020.0166.
36. Wendler D, Shah S. Should children decide whether they are enrolled in nonbeneficial research? Am J Bioeth. 2010;3:1–7. https://doi.org/10.1162/152651603322614382.
37. Zuk P, Torgerson L, Sierra-Mercado D, Lázaro-Muñoz G. Neuroethics of neuromodulation: an update. Curr Opin Biomed Eng. 2018;8:45–50. https://doi.org/10.1016/j.cobme.2018.10.003.

Transcranial Magnetic Stimulation in Older Adults

Tajudeen O. Basiru, Joan A. Camprodon, and Michael D. Kritzer

1 Introduction

After the decades of sustained, multidisciplinary, research to improve therapeutic approaches for neuropsychiatric illness, transcranial magnetic stimulation (TMS) has emerged as one of the most exciting breakthroughs in recent years [1]. TMS is a non-invasive modality that uses an electromagnetic field to stimulate specific areas of the brain's cortical surface. Compared to existing psychopharmacological interventions, TMS is associated with fewer adverse effects and has been shown to be effective for treatment-resistant depression (TRD). TMS received FDA approval for TRD depression in 2008 and has subsequently been approved for treating many other conditions including obsessive compulsive disorder (OCD), smoking cessation, migraine, anxious depression, and neuropathic pain [2]. Research into novel approaches and targets of TMS will lead to additional indications for neuropsychiatric conditions and may become a first-line treatment for many common conditions like major depressive disorder (MDD) rather than a last resort after many treatment failures [3]. Given the rate of progress and the tremendous gains achieved in the field of neuromodulation thus far, there is little doubt that the future of TMS holds great promise for the treatment of neuropsychiatric illnesses.

In the 1980s, TMS emerged as a safe and effective means of modulating brain function and has now become an alternative to electroconvulsive therapy (ECT) that can be used when patients fail to respond to medication. Despite the recent growth

T. O. Basiru
Department of Behavioral Health, Community Health of South Florida, Miami, FL, USA

J. A. Camprodon · M. D. Kritzer (✉)
Division of Neuropsychiatry & Neuromodulation, Massachusetts General Hospital, Boston, MA, USA

Harvard Medical School, Boston, MA, USA
e-mail: jcamprodon@mgh.harvard.edu; mkritzer-cheren@mgh.harvard.edu

V. Dubljević, J. R. Young (eds.), *TMS and Neuroethics*, Advances in Neuroethics,
https://doi.org/10.1007/978-3-031-92401-9_12

of knowledge involving the application of TMS in neurophysiologic research and interventional psychiatry, much remains to be known about the safety and efficacy of TMS, particularly in specific populations such as the elderly and those with neurological conditions. TMS is clinically used for TRD, OCD, anxiety comorbid with MDD, migraine headaches with aura, smoking cessation, and for cortical mapping [4]. Neurostar was recently granted FDA approval for the treatment of MDD in people as young as 15 years old. While some research studies on the use of TMS in older populations have yielded promising results, many clinical trials exclude this population because of insufficient evidence regarding safety and efficacy [5]. For instance, the 2018 clinical trial by BrainsWay's deep TMS (dTMS) among 100 patients with OCD that led to the approval for OCD excluded patients older than 68 years [6]. Similarly, the RCT that studied dTMS in smokers that paved the way for FDA approval of TMS for smoking cessation excluded those aged over 70 years. This pattern of excluding the geriatric populations from clinical trials for the use of TMS is controversial as some authors opine that there is no scientific justification for such exclusion. This lack of evidence may create undue harm on older adults suffering from neuropsychiatric illness as providers and payers may be constrained to available evidence when offering these novel treatments. This substantially risks reducing equity for effective and even life-saving treatments. A review of available evidence is imperative to objectively justify the use of TMS in older adults.

2 Older Age and Neuropsychiatric Conditions

As the world's population ages, older adults, loosely defined as those aged 65 years and older, will become a substantial portion of the world's demographic [7]. Improved healthcare and public health systems in many countries have led to lower death rates and an increased portion of the elderly population at an unprecedented rate [8]. Countries with robust healthcare and proactive public health systems will have to deal with an increased burden of health conditions associated with older age. According to the U.S. Census, the proportion of the U.S. population aged 65 years and older grew from less than 1 in 20 in 1920 to about 1 in 6 in 2020, a growth rate that is about five times that of the entire U.S. population [9]. The decade between 2010 and 2020 saw the highest growth rate since 1880 and the highest rate compared to any other demographic [9]. Although common psychiatric conditions like mood disorders and anxiety are more common in younger people than the geriatric population, these disorders are associated with severe outcomes in older populations [10]. Meanwhile, stroke and neurodegenerative conditions like Alzheimer's disease and Parkinson's disease are more common in old age and can be associated with psychiatric symptoms [11]. The implication is that the health, societal, and economic burden posed by these disorders will increase as the proportion of the U.S. population that is elderly continues to rise.

As of 2019, based on the data from the Pan American Health Organization (PAHO), Alzheimer disease plus other dementias and depressive disorders are, respectively, the sixth and ninth leading causes of disability-adjusted life years

(DALYs) in the USA [12]. With the proportion of the U.S. population that is older than 65 set to become 22% in 2040 compared to 17% in 2020, these disorders commonly seen in the elderly will lead to significantly higher DALYs and economic loss for the country [13]. Importantly, neuropsychiatric symptoms are common, not only in older adults with neurocognitive disorders like Alzheimer's disease, but also in those without neurocognitive disorders. A study of close to 200 elderly patients in Paris revealed that while the prevalence of agitation was highest (about 37%) in the group with neurocognitive disorders, the group without these disorders had higher prevalence of major depressive disorder [14]. Neuropsychiatric symptoms are said to be present at some points in as high as three-fourth of patients with neurocognitive disorders, are a major cause of hospitalizations and death, and can be expected to constitute a high economic burden to the U.S. economy [15–18].

3 TMS and Geriatric Depression

Geriatric mood disorders are the most common psychiatric issues in old age and depression is associated with more adverse outcomes in older people than in younger people [10, 19]. The prevalence of geriatric depression is reported to be almost 40% with more than half of these being moderate to severe [20]. Meanwhile, psychopharmacological modalities that are efficacious in younger adults have been shown to be less effective in the geriatric population [19]. Physiological changes associated with old age, the increased incidence of medical comorbidities, and polypharmacy further complicate the use of psychopharmacological interventions in older people. For instance, first-line treatment for MDD and OCD such as selective serotonin reuptake inhibitors (SSRI) increases the risk of hyponatremia, a serious illness in the elderly that increases the risk of falls, confusion, and even death. TMS, on the other hand, offers a safe, nonpharmacological alternative to medications and has proven efficacy in the adult population in treating TRD [21]. While there is much less evidence for the safety and clinical efficacy of TMS in older adults, various studies have demonstrated the efficacy and safety of TMS in treating depression in the geriatric population and TMS has been approved for treating late-life depression in this population [22, 23]. Although some studies have associated older age with lower response following treatment of MDD with repetitive TMS (rTMS), a meta-analysis of TMS for depression in patients older than 50 years indicated that older age was associated with better improvement [21, 24]. There are several reasons for the lack of evidence for the efficacy and safety of TMS in geriatric populations including exclusion of people older than 65 or 70 years from most clinical trials involving TMS and potentially inadequate dosing of TMS in older adults when they are included in trials [5]. Furthermore, older age is associated with various factors that influence the pathophysiology of depression compared to that of younger age such as high prevalence of vascular diseases and amyloidopathy. These factors are associated with significant dysfunction of the frontal and subcortical networks which are important mediators of depressive disorders and may explain recurrence and chronicity of depression in this age group [21].

Therefore, there have been questions about the need to develop tailored TMS protocols for older adults since the existing protocols were developed based on trials in the younger adult population [21].

While there's still less evidence for the safety and efficacy of TMS in Alzheimer's disease, multiple studies have demonstrated improvements in cognition and depression in patients with Alzheimer's disease comorbid with depression treated with rTMS [25, 26]. For instance, Ahmed and colleagues demonstrated sustained improvement in multiple domains of cognition as well as in geriatric depression symptoms when high-frequency rTMS over bilateral dorsolateral prefrontal cortex (DLPFC) was applied to patients with Alzheimer's disease in comparison to low-frequency rTMS or sham TMS [27]. There is growing evidence that TMS can be directed trans-synaptically to other sites in the brain including the precuneus and hippocampal areas that are capable of modulating cognition [28, 29]. Meanwhile, improvement in cognition in patients with Alzheimers and depression has been suggested to be associated with improvement in depressive symptoms in these patients [30]. In a meta-analysis of 143 studies including 94 RCTs, most of the studies that included depression as comorbidity of Alzheimers showed improvement in cognition in the study participants, compared to sham TMS [31]. Drummond et al. (2015) found that rTMS among 34 patients with mild cognitive impairment led to improvement in episodic memory compared with patients in the sham TMS group [32]. Padala et al. also found improvements in apathy and multiple cognitive domains following rTMS application to the left DLPFC in patients with mild cognitive impairments [33]. Although available evidence remains weak, several meta-analyses have found rTMS can improve most cognitive dysfunction associated with many psychiatric disorders including schizophrenia, depression, as well as executive functioning in many dementias [24].

4 Depression with Comorbid Neurological Illness

Neurological disorders such as cerebrovascular diseases and neurodegenerative disorders are prevalent in old age. A 2019 Dutch study found that up to 50% of women and about a third of men will develop stroke, Alzheimer's or Parkinson's disease in their lifetime [34]. Depression has a bidirectional relationship with most neurological disorders in old age and has an estimated comorbidity rate of 25 to 40% with these disorders [35]. Synthesis of evidence in the Indian Psychiatric Society (IPS) clinical practice guidelines (CPG) indicates that rTMS is efficacious in post-stroke depression (1–5 pooled effect size) and MDD comorbid with Parkinson's disease (effect size less than 1), especially when rTMS is high frequency over the left DLPFC [24]. Evidence supporting the use of TMS in post-stroke depression is preliminary as some studies suggest that the use of TMS in clinical practice is not yet recommended [36]. However, a recent RCT demonstrated that rTMS led to more improvement of post-stroke depression compared to sham TMS although clinical improvement was demonstrated in both groups [37]. Similarly, while previous studies have yielded inconclusive results regarding the efficacy of TMS in depression comorbid with Parkinson's disease, recent studies have demonstrated improved symptoms of depression and anxiety in patients with Parkinson disease [38]. In this study, Jiang et al. demonstrated that patients with Parkinson disease and depression attained significant improvement in their depression ratings, anxiety symptoms, and motor symptoms immediately following the trial and up to 4 weeks afterwards [38].

5 OCD in Older Adults

Not only is there a lack efficacy and safety of FDA-approved treatments for OCD in older adults, there is a lack of the evaluation of symptoms of OCD in this population [39]. The prevalence of OCD in geriatric population varies substantially, with one study estimating up to 0.8% and another estimating up to 2.9% compared to a lifetime prevalence of 1.5% to 3.5% in the general population [40–42]. Studies have shown that OCD is less prevalent in older adults, rarely presents for the first time in people over 40 years old, and symptoms tend to improve with age [43]. However, MDD and other comorbidities have been shown to be higher in geriatric patients with OCD compared to younger people, although the severity of symptoms has not been shown to be different in older adults compared to younger adults [44, 45]. These findings may be related to the fact that geriatric OCD, like many geriatric psychopathologies, is poorly studied and therefore, further research in this population is necessary.

BrainsWay's dTMS was approved for the treatment of OCD in 2017, followed by MagVentures' TMS therapy in 2020. The pivotal study that led to the approval of TMS for the treatment of OCD included patients aged 22–68 years old, significantly limiting the involvement of the geriatric population in this study [6]. Therefore, although TMS has been shown to be safe and effective for treating OCD, studies are lacking in the evaluation of TMS for the treatment for geriatric OCD. A meta-analysis of published papers between 1997 and 2022 involving a total of 860 participants showed that rTMS has demonstrated significant clinical improvement in OCD symptoms compared to sham treatment and longer duration of stimulation but fewer sessions of TMS were associated with more improvement overall [46]. Another meta-analysis of RCTs involving 781 patients also showed that rTMS applied over both DLPFCs provide greater improvement in the Yale-Brown Obsessive Compulsive Scale (YBOCS) scores compared to sham TMS with follow-up studies displaying sustained symptom remission in the experimental groups compared to the controlled groups [47].

6 TMS and Neurological Conditions

There is less evidence for the use of TMS in neurological conditions compared to psychiatric disorders, but recent research has demonstrated the potential of TMS in the treatment of various neurological conditions including migraine, epilepsy, multiple sclerosis, motor and neurocognitive disorders. TMS has a variety of clinical utilities in many conditions due to its ability to modulate neuronal plasticity and attenuating inflammation in the nervous system [48]. There is also robust evidence suggesting that TMS is important in treating cognitive impairment associated with old age as well as tinnitus, disorders of consciousness, chronic pain, PTSD, and functional neurological disorders [29]. For instance, rTMS applied to bilateral dorsomedial prefrontal cortices (DMPFC) of adult males resulted in improved autism symptoms compared to sham in one RCT, while low-frequency rTMS applied to left DMPFC in children with autism resulted in improved symptoms compared to transcranial direct current stimulation (tDCS) [49, 50]. A recent systematic review of studies on TMS in autism spectrum disorders (ASD) reported only one RCT that actually involved comparison of rTMS with sham TMS in patients with ASD, thereby indicating that evidence supporting clinical use of rTMS for ASD is weak [51]. Evidence for the utility of TMS in other neurodevelopmental disorders like ADHD and intellectual disabilities remain similarly scant [52].

Migraine headache, which is a disease of the young but also seen in a significant portion of the elderly population, has been shown to benefit from TMS intervention [53]. TMS has FDA approval for the treatment of migraine in adults [48]. Later onset migraine or migraine persisting to advanced age is not only an exception to the rule, but also a cause for concern because of associated medical and psychiatric complexities [54]. Many studies estimate the prevalence of migraine in the elderly at 10% or less, although the relatively decreased inclusion of older adults in most studies of migraine makes such estimation unreliable [53, 55, 56]. Some of the complexities associated with migraine in older age include the higher prevalence of comorbidities which usually leads to the exclusion of many medications that are efficacious in younger patients. Also, the presence of comorbidities complicates the work up and accurate diagnosis of migraine in this age group. An RCT that compared preventive and acute treatment single pulse TMS with sham protocols among patients with migraine headaches over a period of 4 months found a significant reduction in not only headache days, but also in abortive medication use and headache consequences compared to placebo effect [57].

TMS has been used as a treatment in epileptic disorders. Results from RCTs of TMS on epilepsy are conflicting [58, 59]. While some studies have indicated that TMS can reduce epileptiform discharges, others have found no benefits compared to sham, or have found rebound or increased seizures activities following treatment for the disorder. TMS has shown efficacy in the treatment of refractory seizures, cortical mapping of brain seizure activities, and in studying the effects of antiepileptic medications in patients with seizures [60, 61].

Theoretically, some forms of TMS are believed to be able to modulate neuronal transmission and thereby alleviate some symptoms of multiple sclerosis, but there is no clear research evidence indicating therapeutic benefits in this disorder [62, 63]. Although TMS has not been approved for the treatment of any motor disorder, active research in this area has produced promising results. For instance, rTMS's influence on neuronal activities such as dopamine release, synaptic modulation, and neurogenesis in various areas of the brain including the basal ganglia indicates that high-frequency TMS may improve symptoms of Parkinson disease, especially when combined with other forms of treatment [64, 65]. Updated guidelines by the International Federation of Clinical Neurophysiology show that there is some evidence for reduction of depression and/or motor symptoms of Parkinson's disease by high-frequency rTMS applied to both prefrontal cortices [29]. Evidence on the benefits of TMS in other motor disorders such as dystonia, Huntington's disease, Tourette's syndrome remain scant and inconclusive. Modulation of cortical excitability through actions on neurotrophic factors like brain-derived neurotropic factor and long-term potentiation of neuronal activities by high-frequency rTMS has been shown to be a possible mechanism of the beneficial effects of TMS on Alzheimer's disease [66]. Though multiple meta-analyses have shown that prolonged TMS on multiple sites in the brain improved cognition in Alzheimer's disease, clinical use is still not recommended based on absence of long-term observational studies of the long-term effects of TMS in Alzheimer's disease. Studying the benefits of TMS in various neurodegenerative disorders offers a unique opportunity to understand the safety of TMS in the geriatric population.

There is evidence, albeit limited, suggesting that rTMS to the cortex contralateral to pain area leads to improvements in pain and quality of life in patients with neuropathic pain [67, 68]. Multiple RCTs have demonstrated these effects, although stimulation patterns vary substantially in these studies as well as the duration of pain reduction. While there is still no uniform protocol recommended for the use of TMS in patients with neuropathic pain due to paucity of research, current studies indicate that longer duration of stimulation and multiple sessions are likely to produce significant and longer-lasting pain reduction [67, 68]. Also, there is limited literature and no RCTs on the efficacy and safety of TMS specifically for geriatric neuropathic pain. The effects of TMS on tinnitus have been reported in multiple studies. These studies have suggested that application of TMS using standard protocol as in other disorders produced only modest improvements in symptoms [69, 70]. Individualized protocols, on the other hand, have been shown to provide better improvement for tinnitus.

7 Smoking Cessation in Older Adults

Smoking cessation is associated with enormous benefits such as a reduction in all-cause mortality associated with tobacco use and increased life expectancy, even at advanced age [71]. For instance, a study done exclusively among people aged 70 years and older found that quitting smoking at that age compared to not quitting was associated with a significant reduction in risk of death [72]. Very few studies have examined the prevalence of smoking or the effectiveness of smoking cessation strategies in older adults [72]. According to the U.S. Centers for Disease Control and Prevention (CDC), smoking prevalence in older adults is not too different from that of the general population [73]. While the CDC estimated it to be 8.5% in people aged 65 years and older in 2021, a recent study published in JAMA estimated a slightly increased prevalence of 9.4% in 2022 compared to 8.7% in 2011 [74].

Meanwhile, the same study found a significant reduction in smoking from 19.2% to 4.9% among people aged 18–24 within the same span of years indicating that while smoking significantly reduced in younger populations, it witnessed an increase in the older population [74]. This observation is related to a variety of factors including a relatively less aggressive interventions for smoking cessation among older adults compared to younger adults, beliefs among caregivers that older adults are less likely to quit at that age, or that quitting at that age will provide little to no benefit to the patients [75].

While smoking is less common in old age than in the younger population, older people are more likely to suffer adverse outcomes, are less likely to attempt to quit, and are less likely to be successful in their attempt to quit, but they are just as likely to benefit from smoking cessation as younger people [74–77]. Furthermore, age-related changes such as increased body fat and hepatic metabolism changes have implications for pharmacological interventions for smoking cessation. There are limited studies on the use of pharmacological treatment for smoking cessation in older adults. A study of existing literature on the use of pharmacotherapy for smoking cessation in older adults found that nicotine replacement therapy (NRT) had more evidence in this population and evidence remains scant for the use of other modalities such as bupropion or varenicline [78].

Deep TMS was approved for smoking cessation in 2020 in people aged 18 years and older, based on the result of a multi-center randomized control trial among 262 people aged 22 and 70 years old [79, 80]. Again, elderly adults were excluded in the landmark RCT that led to the approval of TMS for smoking cessation, adding to the lack of scientific evidence for the safety of TMS in the geriatric population. Currently, there are no studies demonstrating the efficacy and safety of repetitive TMS for smoking cessation in older adults [81]. There is evidence to suggest that older people may require special considerations when treated for smoking cessation using TMS [82]. Therefore, the lack of scientific study into the application of TMS in the geriatric population also precludes the evaluation of the effect of advanced age on the efficacy and safety of TMS when used to treat smoking cessation in older adults. However, extant evidence indicates that TMS is as safe and effective as currently approved medications for smoking cessation in adults including bupropion and varenicline, although efficacy comparison is difficult for various reasons [80].

8 Conclusion

Transcranial magnetic stimulation, particularly as adjunctive treatment in multiple neuropsychiatric disorders has proven to be non-invasive, efficacious, and associated with minimal adverse events compared to pharmacological interventions and has gained FDA approval for several disorders. The use of TMS has also been shown to be associated with minimal adverse effects in older adults despite the scarcity of studies in this population. It is important to consider the inclusion of the geriatric population in ongoing research to demonstrate the safety and efficacy of this intervention in this population as the USA and the world face a significant increase in the geriatric population in the coming years. We cannot rely on large naturalistic studies to determine treatment outcomes in older populations. Not having data on older adults may delay treatment and worsen outcomes for this already vulnerable population. This exclusion risks reducing equity for life-saving treatments. Potentially withholding a nonpharmacological treatment such as TMS from older patients when polypharmacy and deteriorating organ function are major clinical considerations may increase the likelihood of harm by comparison. Availability of evidence-based, non-invasive interventions to treat many common neuropsychiatric disorders in the elderly population will not only improve the quality of life of these patients but also reduce the rates of morbidity and mortality associated with the disorders occurring in older ages.

References

1. Hamlin D, Garman J. A brief history of transcranial magnetic stimulation. Am J Psych Res J. https://doi.org/10.1176/APPI.AJP-RJ.2023.180303. 2023;18(3):8–10.
2. Cohen SL, Bikson M, Badran BW, George MS. A visual and narrative timeline of US FDA milestones for transcranial magnetic stimulation (TMS) devices. Brain Stimul. 2022;15(1):73–5. https://doi.org/10.1016/j.brs.2021.11.010.
3. Yam P. Brain stimulation poised to move from last resort to frontline treatment. Proc Natl Acad Sci USA. 2024;121(7):e2401731121. https://doi.org/10.1073/PNAS.2401731121.
4. Shahin H, Safavi R, McDeavitt K, Livingston R. Role of electroconvulsive therapy and transcranial magnetic stimulation in treatment-resistant depression. Psychiatr Ann. 2024;54(6):e164–9. https://doi.org/10.3928/00485713-20240610-01.
5. Sabesan P, Lankappa S, Khalifa N, Krishnan V, Gandhi R, Palaniyappan L. Transcranial magnetic stimulation for geriatric depression: promises and pitfalls. World J Psych. 2015;5(2):170. https://doi.org/10.5498/WJP.V5.I2.170.
6. Carmi L, Tendler A, Bystritsky A, et al. Efficacy and safety of deep transcranial magnetic stimulation for obsessive-compulsive disorder: a prospective multicenter randomized double-blind placebo-controlled trial. Am J Psychiatry. 2019;176(11):931–8. https://doi.org/10.1176/appi.ajp.2019.18101180.
7. Singh S, Bajorek B. Defining 'elderly' in clinical practice guidelines for pharmacotherapy. Pharm Pract. 2014;12(4):489. https://doi.org/10.4321/S1886-36552014000400007.
8. Ismail Z, Ahmad WIW, Hamjah SH, Astina IK. The impact of population ageing: a review. Iran J Public Health. 2021;50(12):2451. https://doi.org/10.18502/IJPH.V50I12.7927.
9. Bureau UC. 2023. Accessed November 12, 2024. https://www.census.gov/programs-surveys/acs/news/updates/2023.html
10. da Costa Lane Valiengo L, Stella F, Forlenza OV. Mood disorders in the elderly: prevalence, functional impact, and management challenges. Neuropsychiatr Dis Treat. 2016;12:2105. https://doi.org/10.2147/NDT.S94643.
11. Cummings JL. Neuropsychiatry in the elderly. In: Contemporary neuropsychiatry. Tokyo: Springer; 2001. p. 107–15. https://doi.org/10.1007/978-4-431-67897-7_16.
12. Leading causes of death, and disability - PAHO/WHO | Pan American Health Organization. Accessed November 12, 2024. https://www.paho.org/en/enlace/leading-causes-death-and-disability
13. Profile of Older Americans | ACL Administration for Community Living. Accessed November 12, 2024. https://acl.gov/aging-and-disability-in-america/data-and-research/profile-older-americans

14. Mézière A, Blachier M, Thomas S, et al. Neuropsychiatric symptoms in elderly inpatients: a multicenter cross-sectional study. Dement Geriatr Cogn Dis Extra. 2013;3(1):123. https://doi.org/10.1159/000350805.
15. Lyketsos CG, Lopez O, Jones B, Fitzpatrick AL, Breitner J, Dekosky S. Prevalence of neuropsychiatric symptoms in dementia and mild cognitive impairment: results from the cardiovascular health study. JAMA. 2002;288(12):1475–83. https://doi.org/10.1001/JAMA.288.12.1475.
16. Voisin T, Sourdet S, Cantet C, et al. Descriptive analysis of hospitalizations of patients with Alzheimer's disease: a two-year prospective study of 686 patients from the REAL.FR study. J Nutr Health Aging. 2009;13(10):890–2. https://doi.org/10.1007/S12603-009-0247-Y.
17. Stern Y, Tang MX, Albert MS, et al. Predicting time to nursing home care and death in individuals with Alzheimer disease. JAMA. 1997;277(10):806–12. https://doi.org/10.1001/JAMA.1997.03540340040030.
18. Public Health and Aging: Trends in Aging — United States and Worldwide. Accessed November 12, 2024. https://www.cdc.gov/mmwr/preview/mmwrhtml/mm5206a2.htm
19. Tedeschini E, Levkovitz Y, Iovieno N, Ameral VE, Nelson JC, Papakostas GI. Efficacy of antidepressants for late-life depression: a meta-analysis and meta-regression of placebo-controlled randomized trials. J Clin Psychiatry. 2011;72(12):5967. https://doi.org/10.4088/JCP.10R06531.
20. Guerra M, Prina AM, Ferri CP, et al. A comparative cross-cultural study of the prevalence of late life depression in low and middle income countries. J Affect Disord. 2016;190:362–8. https://doi.org/10.1016/J.JAD.2015.09.004.
21. Cappon D, den Boer T, Jordan C, Yu W, Metzger E, Pascual-Leone A. Transcranial magnetic stimulation (TMS) for geriatric depression. Ageing Res Rev. 2022;74:101531. https://doi.org/10.1016/J.ARR.2021.101531.
22. Iriarte IG, George MS. Transcranial magnetic stimulation (TMS) in the elderly. Curr Psychiatry Rep. 2018;20(1):6. https://doi.org/10.1007/S11920-018-0866-2/METRICS.
23. BrainsWay Receives Expanded FDA Labeling to Treat Late Life Depression. Accessed November 12, 2024. https://www.brainsway.com/news_events/brainsway-receives-expanded-fda-labeling-to-treat-late-life-depression/
24. Tikka SK, Siddiqui MA, Garg S, Pattojoshi A, Gautam M. Clinical practice guidelines for the therapeutic use of repetitive transcranial magnetic stimulation in neuropsychiatric disorders. Indian J Psychiatry. 2023;65(2):270–88. https://doi.org/10.4103/indianjpsychiatry.indianjpsychiatry_492_22.
25. Nardone R, Bergmann J, Christova M, et al. Effect of transcranial brain stimulation for the treatment of Alzheimer disease: a review. Int J Alzheimers Dis. 2011;2012:687909. https://doi.org/10.1155/2012/687909.
26. Cheng J, Fairchild JK, McNerney MW, et al. Repetitive transcranial magnetic stimulation as a treatment for veterans with cognitive impairment and multiple comorbidities. 2021;85(4):1593–600. https://doi.org/10.3233/JAD-210349.
27. Ahmed MA, Darwish ES, Khedr EM, El Serogy YM, Ali AM. Effects of low versus high frequencies of repetitive transcranial magnetic stimulation on cognitive function and cortical excitability in Alzheimer's dementia. J Neurol. 2012;259(1):83–92. https://doi.org/10.1007/S00415-011-6128-4/METRICS.
28. Jung YH, Jang H, Park S, et al. Effectiveness of personalized hippocampal network–targeted stimulation in Alzheimer disease: a randomized clinical trial. JAMA Netw Open. 2024;7(5):e249220. https://doi.org/10.1001/JAMANETWORKOPEN.2024.9220.
29. Lefaucheur JP, Aleman A, Baeken C, et al. Evidence-based guidelines on the therapeutic use of repetitive transcranial magnetic stimulation (rTMS): an update (2014–2018). Clin Neurophysiol. 2020;131(2):474–528. https://doi.org/10.1016/J.CLINPH.2019.11.002.
30. Jo K, Jhoo JH, Mun YJ, et al. The effect of cognitive intervention on cognitive improvement in patients with dementia. Dement Neurocogn Disord. 2018;17(1):23. https://doi.org/10.12779/DND.2018.17.1.23.

31. Pagali SR, Kumar R, LeMahieu AM, et al. Efficacy and safety of transcranial magnetic stimulation on cognition in mild cognitive impairment, Alzheimer's disease, Alzheimer's disease-related dementias, and other cognitive disorders: a systematic review and meta-analysis. Int Psychogeriatr. 2024;36:880. https://doi.org/10.1017/S1041610224000085.
32. Drumond Marra HL, Myczkowski ML, Maia Memória C, et al. Transcranial magnetic stimulation to address mild cognitive impairment in the elderly: a randomized controlled study. Behav Neurol. 2015;2015(1):287843. https://doi.org/10.1155/2015/287843.
33. Padala PR, Padala KP, Lensing SY, et al. Repetitive transcranial magnetic stimulation for apathy in mild cognitive impairment: a double-blind, randomized, sham-controlled, cross-over pilot study. Psychiatry Res. 2018;261:312–8. https://doi.org/10.1016/J.PSYCHRES.2017.12.063.
34. Licher S, Ahmad S, Karamujić-Čomić H, et al. Genetic predisposition, modifiable-risk-factor profile and long-term dementia risk in the general population. Nat Med. 2019;25(9):1364–9. https://doi.org/10.1038/s41591-019-0547-7.
35. Moore BA, Barnett JE. Depression in neurological disorders: comorbidity or another clinical manifestation? In: Case studies in clinical psychological science: bridging the gap from science to practice; 2018. p. 1–7. https://doi.org/10.1093/MED/9780190603342.003.0007.
36. Conroy SK, Brownlowe KB, McAllister TW. Depression comorbid with stroke, traumatic brain injury, Parkinson's disease, and multiple sclerosis: diagnosis and treatment. Focus (Am Psychiatr Publ). 2020;18(2):150–61. https://doi.org/10.1176/APPI.FOCUS.20200004.
37. Li K, Mo D, Yu Q, Feng R, Li Y. Effect of repetitive transcranial magnetic stimulation on post-stroke comorbid cognitive impairment and depression: a randomized controlled trial. J Alzheimers Dis. 2024;101(1):337–52. https://doi.org/10.3233/JAD-240505.
38. Jiang S, Zhan C, He P, et al. Neuronavigated repetitive transcranial magnetic stimulation improves depression, anxiety and motor symptoms in Parkinson's disease. Heliyon. 2023;9(8):e18364. https://doi.org/10.1016/J.HELIYON.2023.E18364.
39. Wetherell JL, Lenze EJ, Stanley MA. Evidence-based treatment of geriatric anxiety disorders. Psychiatr Clin N Am. 2005;28(4):871–96. https://doi.org/10.1016/J.PSC.2005.09.006.
40. Kohn R, Westlake RJ, Rasmussen SA, Marsland RT, Norman WH. Clinical features of obsessive-compulsive disorder in elderly patients. Am J Geriatr Psychiatry. 1997;5(3):211–5. https://doi.org/10.1097/00019442-199700530-00004.
41. Klenfeldt IF, Karlsson B, Sigström R, et al. Prevalence of obsessive-compulsive disorder in relation to depression and cognition in an elderly population. Am J Geriatr Psychiatry. 2014;22(3):301–8. https://doi.org/10.1016/J.JAGP.2012.09.004.
42. Cath DC, Nizar K, Boomsma D, Mathews CA. Age-specific prevalence of hoarding and obsessive compulsive disorder: a population-based study. Am J Geriatr Psychiatry. 2017;25(3):245–55. https://doi.org/10.1016/J.JAGP.2016.11.006.
43. Dell'Osso B, Benatti B, Rodriguez CI, et al. Obsessive-compulsive disorder in the elderly: a report from the International College of Obsessive-Compulsive Spectrum Disorders (ICOCS). Eur Psychiatry. 2017;45:36–40. https://doi.org/10.1016/J.EURPSY.2017.06.008.
44. Pulular A, Levy R, Stewart R. Obsessive and compulsive symptoms in a National Sample of older people: prevalence, comorbidity, and associations with cognitive function. Am J Geriatr Psychiatry. 2013;21(3):263–71. https://doi.org/10.1016/J.JAGP.2012.11.011.
45. Jazi AN, Asghar-Ali ALIA. Obsessive-compulsive disorder in older adults: a comprehensive literature review. J Psychiatr Pract. 2020;26(3):175–84. https://doi.org/10.1097/PRA.0000000000000463.
46. Steuber ER, McGuire JF. A meta-analysis of transcranial magnetic stimulation in obsessive-compulsive disorder. Biol Psychiatry Cogn Neurosci Neuroimag. 2023;8(11):1145–55. https://doi.org/10.1016/J.BPSC.2023.06.003.
47. Perera MPN, Mallawaarachchi S, Miljevic A, Bailey NW, Herring SE, Fitzgerald PB. Repetitive transcranial magnetic stimulation for obsessive-compulsive disorder: a meta-analysis of randomized, sham-controlled trials. Biol Psychiatry Cogn Neurosci Neuroimag. 2021;6(10):947–60. https://doi.org/10.1016/J.BPSC.2021.03.010.

48. Vaishnavi S. Transcranial magnetic stimulation for developmental neuropsychiatric disorders with inflammation. Dev Neurosci. 2023;45(6):342. https://doi.org/10.1159/000535103.
49. Enticott PG, Fitzgibbon BM, Kennedy HA, et al. A double-blind, randomized trial of deep repetitive transcranial magnetic stimulation (rTMS) for autism Spectrum disorder. Brain Stimul. 2014;7(2):206–11. https://doi.org/10.1016/J.BRS.2013.10.004.
50. Gómez L, Vidal B, Maragoto C, et al. Non-invasive brain stimulation for children with Autism spectrum disorders: a short-term outcome study. Behav Sci. 2017;7:63. https://doi.org/10.3390/BS7030063.
51. Smith JR, DiSalvo M, Green A, et al. Treatment response of transcranial magnetic stimulation in intellectually capable youth and young adults with Autism spectrum disorder: a systematic review and meta-analysis. Neuropsychol Rev. 2022;33(4):834–55. https://doi.org/10.1007/S11065-022-09564-1.
52. Memon AM. Transcranial magnetic stimulation in treatment of adolescent attention deficit/hyperactivity disorder: a narrative review of literature. Innov Clin Neurosci. 2021;18(1–3):43. Accessed November 12, 2024 https://pmc.ncbi.nlm.nih.gov/articles/PMC8195561/
53. Haan J, Hollander J, Ferrari MD. Migraine in the elderly: a review. Cephalalgia. 2007;27(2):97–106. https://doi.org/10.1111/J.1468-2982.2006.01250.X.
54. Do TP, Remmers A, Schytz HW, et al. Red and orange flags for secondary headaches in clinical practice: SNNOOP10 list. Neurology. 2019;92(3):134–44. https://doi.org/10.1212/WNL.0000000000006697.
55. Prencipe M, Casini AR, Ferretti C, et al. Prevalence of headache in an elderly population: attack frequency, disability, and use of medication. J Neurol Neurosurg Psychiatry. 2001;70(3):377–81. https://doi.org/10.1136/JNNP.70.3.377.
56. Wijeratne T, Tang HM, Crewther D, Crewther S. Prevalence of migraine in the elderly: a narrated review. Neuroepidemiology. 2019;52(1–2):104–10. https://doi.org/10.1159/000494758.
57. Starling AJ, Tepper SJ, Marmura MJ, et al. A multicenter, prospective, single arm, open label, observational study of sTMS for migraine prevention (ESPOUSE study). Cephalalgia. 2018;38(6):1038–48. https://doi.org/10.1177/0333102418762525.
58. Walton D, Spencer DC, Nevitt SJ, Michael BD. Transcranial magnetic stimulation for the treatment of epilepsy. Cochrane Database Syst Rev. 2021;2021(4) https://doi.org/10.1002/14651858.CD011025.pub3.
59. Seynaeve L, Devroye A, Dupont P, Van Paesschen W. Randomized crossover sham-controlled clinical trial of targeted low-frequency transcranial magnetic stimulation comparing a figure-8 and a round coil to treat refractory neocortical epilepsy. Epilepsia. 2016;57(1):141–50. https://doi.org/10.1111/EPI.13247.
60. Hsu WY, Cheng CH, Lin MW, Shih YH, Liao KK, Lin YY. Antiepileptic effects of low frequency repetitive transcranial magnetic stimulation: a meta-analysis. Epilepsy Res. 2011;96(3):231–40. https://doi.org/10.1016/J.EPLEPSYRES.2011.06.002.
61. Amassian VE, Cracco RQ, Maccabee PJ, Cracco JB, Rudell A, Eberle L. Suppression of visual perception by magnetic coil stimulation of human occipital cortex. Electroencephalogr Clin Neurophysiol. 1989;74(6):458–62. https://doi.org/10.1016/0168-5597(89)90036-1.
62. Hausmann A, Weis C, Marksteiner J, Hinterhuber H, Humpel C. Chronic repetitive transcranial magnetic stimulation enhances c-fos in the parietal cortex and hippocampus. Mol Brain Res. 2000;76(2):355–62. https://doi.org/10.1016/S0169-328X(00)00024-3.
63. Liu M, Fan S, Xu Y, Cui L. Non-invasive brain stimulation for fatigue in multiple sclerosis patients: a systematic review and meta-analysis. Mult Scler Relat Disord. 2019;36 https://doi.org/10.1016/j.msard.2019.08.017.
64. Chervyakov AV, Chernyavsky AY, Sinitsyn DO, Piradov MA. Possible mechanisms underlying the therapeutic effects of transcranial magnetic stimulation. Front Hum Neurosci. 2015;9(June):130417. https://doi.org/10.3389/FNHUM.2015.00303/BIBTEX.
65. Arias-Carrión O. Basic mechanisms of rTMS: implications in Parkinson's disease. Int Arch Med. 2008;1(1):2. https://doi.org/10.1186/1755-7682-1-2.

66. Somaa FA, de Graaf TA, Sack AT. Transcranial Magnetic Stimulation in the Treatment of Neurological Diseases. Front Neurol. 2022 May 20;13:793253. https://doi.org/10.3389/fneur.2022.793253. PMID: 35669870; PMCID: PMC9163300.
67. Khedr EM, Kotb HI, Mostafa MG, et al. Repetitive transcranial magnetic stimulation in neuropathic pain secondary to malignancy: a randomized clinical trial. Eur J Pain. 2015;19(4):519–27. https://doi.org/10.1002/EJP.576.
68. André-Obadia N, Mertens P, Gueguen A, Peyron R, Garcia-Larrea L. Pain relief by rTMS: differential effect of current flow but no specific action on pain subtypes. Neurology. 2008;71(11):833–40. https://doi.org/10.1212/01.WNL.0000325481.61471.F0.
69. Cacace AT, Hu J, Romero S, Xuan Y, Burkard RF, Tyler RS. Glutamate is down-regulated and tinnitus loudness-levels decreased following rTMS over auditory cortex of the left hemisphere: a prospective randomized single-blinded sham-controlled cross-over study. Hear Res. 2018;358:59–73. https://doi.org/10.1016/J.HEARES.2017.10.017.
70. James GA, Thostenson JD, Brown G, et al. Neural activity during attentional conflict predicts reduction in tinnitus perception following rTMS. Brain Stimul. 2017;10(5):934–43. https://doi.org/10.1016/J.BRS.2017.05.009.
71. Kwon DM, Santiago-Torres M, Mull KE, Sullivan BM, Bricker JB. Older adults who smoke: Do they engage with and benefit from web-based smoking cessation interventions? Prev Med (Baltim). 2022;161:107118. https://doi.org/10.1016/J.YPMED.2022.107118.
72. Nash SH, Liao LM, Harris TB, Freedman ND. Cigarette smoking and mortality in adults aged 70 years and older: results from the NIH-AARP cohort. Am J Prev Med. 2017;52(3):276–83. https://doi.org/10.1016/J.AMEPRE.2016.09.036.
73. Cornelius ME, Loretan CG, Jamal A, et al. Tobacco product use among adults—United States, 2021. MMWR Morb Mortal Wkly Rep. 2023;72(18):475–83. https://doi.org/10.15585/MMWR.MM7218A1.
74. Meza R, Cao P, Jeon J, Warner KE, Levy DT. Trends in US adult smoking prevalence, 2011 to 2022. JAMA Heal Forum. 2023;4(12):e234213. https://doi.org/10.1001/JAMAHEALTHFORUM.2023.4213.
75. Henley SJ, Asman K, Momin B, et al. Smoking cessation behaviors among older U.S. adults. Prev Med Rep. 2019;16:100978. https://doi.org/10.1016/J.PMEDR.2019.100978.
76. Hunt LJ, Covinsky KE, Cenzer I, et al. The epidemiology of smoking in older adults: a National Cohort Study. J Gen Intern Med. 2023;38(7):1697–704. https://doi.org/10.1007/S11606-022-07980-W/FIGURES/3.
77. Cigarette Smoking Among Adults — United States. 2007. Accessed November 12, 2024. https://www.cdc.gov/mmwr/preview/mmwrhtml/mm5745a2.htm
78. Cawkwell PB, Blaum C, Sherman SE. Pharmacological smoking cessation therapies in older adults: a review of the evidence. Drugs Aging. 2015;32(6):443–51. https://doi.org/10.1007/S40266-015-0274-9.
79. Harmelech T, Hanlon CA, Tendler A. Transcranial magnetic stimulation as a tool to promote smoking cessation and decrease drug and alcohol use. Brain Sci. 2023;13(7):1072. https://doi.org/10.3390/BRAINSCI13071072/S1.
80. Zangen A, Moshe H, Martinez D, et al. Repetitive transcranial magnetic stimulation for smoking cessation: a pivotal multicenter double-blind randomized controlled trial. World Psychiatry. 2021;20(3):397–404. https://doi.org/10.1002/WPS.20905.
81. Ibrahim C, Tang VM, Blumberger DM, Le FB. Repetitive transcranial magnetic stimulation for smoking cessation. Can Med Assoc J. 2024;196(6):E187. https://doi.org/10.1503/CMAJ.230806.
82. Gersner R, Barnea-Ygael N, Tendler A. Moderators of the response to deep TMS for smoking addiction. Front Psych. 2023;13:1079138. https://doi.org/10.3389/FPSYT.2022.1079138.

The Ethics of TMS in the Treatment of Suicidal Ideation

Alyssa Bernanke, Claudia Roldan, and Brandon Hamm

1 Introduction

1.1 Epidemiology of Suicide

Suicide is a public health crisis that affects people of all ages and identities. The Centers for Disease Control and Prevention (CDC) reported suicide in the top nine leading causes of death among age groups 10–64 and the second leading cause of death among ages 10–14 and 25–34 [1]. Suicide attempt rates increased by approximately 36% between 2000 and 2022, with a record high of over 49,000 deaths by suicides [2]. In 2022, 13.2 million Americans reported seriously thinking about suicide, 3.8 million reported planning an attempt, and 1.6 million attempted suicide [1]. In 2020 alone, there were more than 1.3 million emergency department visits for suicidal ideation, attempt, or self-harm behavior (both suicidal and non-suicidal self-injury) [3].

Suicide rates vary by factors such as gender, race/ethnicity, and sexual orientation. Ethnicities with the highest suicide rates in the USA include non-Hispanic American Indian/Alaska Native people, followed by non-Hispanic white people [4]. The completed suicide rates for men are approximately four times higher than women, though women notably are more likely to attempt suicide [5, 6]. Older adults aged 55–64 have recently shown a concerning increase in suicide [2]. Populations at increased risk include young people who identify as non-heterosexual, veterans, people who live in rural areas, and those who work in certain professions

A. Bernanke (✉) · B. Hamm
Department of Psychiatry and Behavioral Sciences, Northwestern University, Chicago, IL, USA
e-mail: alyssa.bernanke@northwestern.edu; brandon.hamm2@nm.org

C. Roldan
Feinberg School of Medicine, Northwestern University, Chicago, IL, USA
e-mail: Claudia.roldan@northwestern.edu

V. Dubljević, J. R. Young (eds.), *TMS and Neuroethics*, Advances in Neuroethics,
https://doi.org/10.1007/978-3-031-92401-9_13

such as mining and construction [4]. Other risk factors include chronic pain [7] and mental health conditions such as major depression, dysthymia, generalized anxiety disorder, substance use disorder [8], and social isolation [9].

Protective factors include social support [10] and access to mental health treatment [11]. Despite the importance of mental health treatment in addressing suicidal thoughts and behaviors, one study found that the increases in suicide attempts, ideation, and planning among young adults in the USA have not been accompanied by a corresponding rise in mental health care in the USA. In fact, mental health care access has slightly declined among uninsured suicidal young adults [12].

Suicidal ideation exists on a spectrum of intensity, beginning with a general desire to die that lacks any concrete method, plan, intention, or action, also known as passive death wish. At the other end of the spectrum is active suicidal ideation, which involves a detailed plan and a determined intent to act on suicidal thoughts [13]. A suicide attempt is defined as intentional self-harm with the purpose of ending one's life [14]. Suicidal ideation is a significant risk factor for progression to suicide attempt. A cross-national study using retrospective age of onset data found that the risk of progression is most significant in the first year of suicidal ideation (OR = 117.4–123.1). This highly concerning statistic suggests that individuals who experience SI are *over 100 times more likely* to plan or attempt suicide within the first year compared to those who do not have SI. Those who had a suicide plan had a 7.5-fold higher risk than those who did not [15]. Among people with a depressive disorder, the severity of hopelessness predicted both suicidal ideation and attempts [16].

1.2 Transcranial Magnetic Stimulation in the Treatment of Suicidality

Suicidality is a broad term that encompasses ideation, behaviors, and intent [17]. There are several treatment options for active suicidal ideation. Inpatient psychiatric hospitalization (both voluntary and involuntary) is often used to acutely mitigate risk for individuals with active suicidal ideation. Involuntary hospitalization is ethically complex: It involves restricting the rights of the individual and cannot guarantee confidentiality when the risk of harm to self is credible and concerning. Hospitalized individuals are monitored closely for suicidal behaviors and treated with a variety of interventions.

Here, it may be worth mentioning that suicide risk is difficult to predict, and patients may live for years or decades without ever advancing to an attempt despite ongoing suicidal ideation. However, in itself, suicidal ideation is a form of suffering for which there is great need for alleviation. Chronic and acute suicidal ideation also represents different levels of risk and therefore warrant different interventions [13, 18]. Pharmacotherapy such as lithium [19], clozapine [20], or antidepressant therapy [21] may have antisuicidal benefits. Other interventions such as electroconvulsive therapy (ECT) [22] and ketamine [21] in adults can also be used to treat acute suicidal ideation. Psychotherapy, including cognitive behavioral therapy and dialectical behavioral therapy, as well as brief interventions, demonstrate good evidence in reducing suicidal ideation and some interventions may also reduce suicide attempts [23]. Pharmacotherapy may also be beneficial in reducing suicidal ideation but may take many weeks to take effect [24]. Despite these myriad options for the treatment of suicidal ideation, suicide rates remain devastatingly high, and further interventions are needed. Brain stimulation therapy such as transcranial magnetic stimulation (TMS) has, therefore, emerged as a potential therapeutic modality for acutely reducing suicidality and treating other mood disorders.

ECT is shown to reduce suicidality acutely and reduces the risk of death by suicide within 1 year of discharge for patients with severe depression [22, 25, 26]. ECT is a more cognitively impactful (acutely) and stigmatized (unfortunately) procedure than TMS. Thus patients may prefer TMS due to the perception that it is safer and less intimidating. The importance of treating depression to reduce the risk of suicide cannot be understated: One systematic review found that major depression had an attributable risk of 67% for suicide attempts in older adults [27]. A meta-analysis also showed severe depression and suicidal ideation predicted both suicide attempts and completed suicide [16]. Concerningly, among adolescents, around a third who have suicide ideation will progress to attempt, and depressive disorder is one predictor of suicide attempt [28–30].

Treatment of suicidality using TMS is an area of active study. One study showed bilateral TMS resolved suicidal ideation in 40.4% of patients, which was a significant reduction compared to 18.8% of sham controls [31]. Of note regarding this study, left unilateral repetitive TMS (rTMS; the FDA-approved protocol clinically available to patients) demonstrated reduction in suicidal ideation by 26.8%, which did not significantly separate from sham TMS. A meta-analysis of 415 patients in ten randomized controlled trials (RCTs) found that rTMS significantly reduced suicidal ideation, although this effect was only significant with non-treatment-resistant depression [32]. Another meta-analysis examining 16 studies found a significant decrease in suicidal ideation scores for rTMS in uncontrolled trials and a trend toward significance in controlled trials [33]. One exploratory study in adolescents with treatment-resistant depression found improvements in suicidal ideation over the course of treatment that was likely mediated by depressive symptom relief [34].

The amplitude of TMS's role in treating suicidality is unclear, and more large-scale studies are needed to establish appropriate treatment parameters, including whether bilateral dorsolateral prefrontal cortex rTMS is superior to left unilateral in reducing suicidal ideation.

2 Ethics Considerations of Suicidal Patients in Research Settings

2.1 Managing Risk in the Research Setting

Research ethics regarding suicidal patients is inherently delicate. Participant safety is a significant concern of any research study that includes suicidal patients.

Although suicide is a comparatively infrequent event in psychiatric health (of the >13 million Americans who seriously consider suicide, less than 0.4% will complete suicide), it represents the worst possible outcome, and risk must be carefully navigated. Virtue ethics, as it is employed in psychiatry, suggests the TMS clinician and researcher must employ sedulousness and veracity in patient interactions. In other words, they must be curious and diligent in eliciting symptoms, and they must be honest and transparent in the informed consent process. This includes discussing the risk of continued suicidal ideation after treatment and managing acute, active suicidal ideation should it arise during treatment.

Studies that assess suicidality as a primary or secondary endpoint often use validated tools such as the Columbia Suicide Severity Rating Scale (C-SSRS) [35] or the Beck Scale for Suicidal Ideation (BSI) [36]. These suicide assessment scales (SAS) are used as outcome measures but also as indicators that a more thorough risk evaluation must be performed. SAS are often used at specific timepoints in a study, allowing researchers to assess change over time. SAS may also be used to indicate when a patient's risk is more acute. Thus, researchers may find themselves at odds with the priority of participant safety: Participants who drop out of the study due to needing a higher level of care may dilute or bias results. Intention-to-treat analyses can help reduce this type of bias, but researchers should use the virtue of veracity when reporting results and indicate the reasons subjects withdraw from treatment. That being said, patient safety is of the highest importance, and clinicians should not hesitate to introduce higher levels of care or add additional treatments when warranted.

Research on the suicidal patient must undergo strict internal review board (IRB) approval and oversight. This is necessary to define the roles of the researchers when a participant requires additional evaluation [37]. However, there is a lack of standardization across IRBs. As Hom et al. state in their publication on the ethics in suicide-related research, "although IRBs must abide by certain federal guidelines and fulfill the overarching purposes of ensuring the protection of human subjects and ethical conduct of research, how these principles are applied are largely up to the discretion of individual panels." [37] This lack of standardization can provide researchers with the flexibility to create studies adaptable to their population of interest. However, it can also be confusing for the researcher, who may lack clear and consistent guidelines on how to manage emergencies that arise in studies.

Management of severe, acute suicidality will typically involve de-escalation of modifiable risk factors, often beginning with a psychiatric hospitalization to maintain safety. However, hospitalization may "pause" access to what otherwise might be effective interventions. Determining whether psychiatric hospitalization is required to maintain safety or if it is more appropriate to tolerate some risk to maintain treatment can be challenging. These decisions must be considered on an individual basis. It is here that casuistry may be a useful means of guiding the clinician. Relevant, paradigmatic cases may be incorporated into the training of the researcher or provider to help inform their clinical decision-making.

Other interventions, such as safety planning, can also be employed. There is an ample evidence base to support the use of safety planning interventions to reduce suicide risk [38–40]. Thus, trials that include suicidal patients may benefit from incorporating safety plans to reduce the need for escalation to a higher level of care.

2.2 Ethical Considerations in RCTs

Given that various psychological and social circumstances may contribute to suicidal ideation or suicide attempts, an individual's suicidal ideation and risk may not uniformly respond to TMS treatment. Studies on suicidality and TMS typically include participants who have a known psychiatric diagnosis of major depressive disorder (MDD), hence confining derived knowledge to unipolar depression contexts. Common exclusion criteria are bipolar depression, history of psychosis (whether associated with psychotic disorder or with MDD), history of substance use, and, surprisingly, history of suicide attempt in the prior 6 months [41, 42]. As a result of the imposed homogeneity of the patient populations within these studies, extrapolation to other disease processes and psychiatric comorbidities is challenging.

For studies in which suicidality is a primary endpoint, patient safety is critically important. This leads to unique ethical considerations. When conducted in the outpatient setting, these studies often exclude those participants who have had a suicide attempt within the past 6 months [41, 43]. Pan et al. performed a double-blinded, sham-controlled trial on treatment-naïve patients with Hamilton Depression Rating Scale (HAMD)-24 scores indicative of severe depression and Beck's Suicide Index indicating at least moderate risk [44]. The trial used rTMS with neuronavigation to identify the left dorsolateral prefrontal cortex, using a total dose of 42,000 pulses over a 7-day period. Participants in both arms were also treated with escitalopram, however, it is unlikely that the medication would take effect during the short duration of the study [45]. Notably, this study demonstrated a rapid and significant reduction in suicidal ideation as measured by SAS. A meta-analysis found that the standardized mean difference (a statistical method used to compare studies that assess the same outcome but may use different measures) for this study was −1.94, more than double the next closest study [46]. Meta-analysis suggests that neuronavigation techniques may have contributed to improved outcomes. It is also possible that the study used more acute patients as evidenced by their selection of patients who are severely depressed and moderate risk of suicide. The ethical considerations regarding this study are complicated: The study used participants who were higher risk and treatment-naive. It may be that those who are most at risk are those who can attain the greatest benefit. Exclusion of the most severely ill may dilute results, a concern that may be particularly pronounced in studies that are limited by the use of assessment tools with lower sensitivity to accurately reflect a change in state.

The use of sham-TMS arms raises a particular ethical question in research in suicidality contexts, as the risk is inherently higher. In one trial, a participant in the sham arm attempted suicide during the course of the study [47]. While the double-blind randomized controlled trial remains the gold standard for medical research, the presence of a control arm in studies on suicidality raises the ethical concern of withholding a potentially lifesaving treatment from individuals in need. Some studies, such as Pan et al., address this by using an active (rather than passive) sham control. Pan et al. used escitalopram, a current first-line treatment for depression, in both groups [44]. Use of an active control group can reduce risk by providing a treatment to the control group.

2.3 Extrapolation and Durability of Treatment

The question of who will benefit most from a TMS intervention for suicidality is of critical importance. As described above, the majority of studies are constricted to uncomplicated major depression, and thus whether suicidal ideation is being indirectly targeted by alleviation of the primary mood disorder is not yet clear. However, suicide is devastatingly common in several disorders that are excluded from most studies, such as bipolar and psychotic disorders. A handful of studies assess the role of TMS in other pathologies associated with suicidality, such as borderline personality disorder (BPD) [48–50]. These studies are limited by small cohorts of participants, but demonstrated efficacy in reducing symptoms based on a validated scale of BPD symptoms. Two of these studies reported statistically significant improvements in suicidal ideation specifically [49, 50]. The third did not comment on suicidality. That being said, extrapolation of current studies to other pathologies must be undertaken with caution. The psychopathologies underlying suicidality are varied and complex, and suicidality in major depressive disorder may be too dissimilar in psychogenesis, neuroconnectivity influence, and phenomenological manifestation to extrapolate TMS efficacy findings to suicidality in bipolar disorder or schizophrenia. Thus, further studies are warranted to assess whether and which protocols may be effective in lysing suicidal ideation and de-escalating suicidality for these disorders.

Durability is an important consideration in assessing interventions for the treatment of suicidality. Studies on TMS for suicidality employ a wide range of post-intervention follow-up: Some have none, while others follow participants for several months. Longer-term follow-up is more costly and introduces logistical complexity; however, when a treatment for suicidality is effective, it is important to consider whether the risk of suicide will increase in the event of treatment wearing off.

When a study treatment is effective for a participant in the acute setting, this can raise considerations of justice and beneficence when the treatment intervention has not yet reached mainstream accessibility. If these treatments have not yet been granted FDA approval, they are unlikely to be approved by insurance and may be financially prohibitive. When a study targeting suicidality concludes, participants may lose access to a treatment that was effective for them, and discontinuation may result in a return of suicidal thoughts and behaviors as their symptoms return. The precarious nature of such phenomena ought to be communicated during informed consent for such studies. Maintenance protocols may be beneficial in reducing the risk of relapse and should be considered when designing a clinical trial.

Few studies look at treatment durability. A study by Hines et al. performed a double-blind trial of 120 Veterans who received active TMS (n = 59) or sham TMS (n = 61) in a modified protocol, three times per day for 3 consecutive days. The Veterans were followed for 6 months and the study found that the antisuicidal benefits were sustained at the 6-month mark for both active and sham groups [47]. A high rate of placebo response in this trial hinders the ability to clearly attribute reduction in suicidality to TMS, although the TMS group did experience a more rapid decline in suicidal ideation than the sham control, which is a promising result that warrants further study. However, no clear conclusion about the durability of treatment can be drawn from this study.

The reasons for the high rate of response in placebo arms in studies on suicidality are likely multifactorial. Two primary considerations are perhaps the most relevant. First, studies on psychiatric illness are plagued by high placebo rates across modalities, including rTMS and pharmacotherapy studies [51, 52]. These results may be due to the length of the study, with shorter studies being more susceptible to placebo effects, as well as rater bias or unconscious desires of the participant [51]. Additionally, acute suicidal ideation fluctuates and naturally decreases over time [53]. Thus, the natural evolution of suicidal thoughts over time may influence study results.

Second, as Hines et al. describe in their study, TMS "is an active, technologic, ritualistic procedure that creates an implicit psychological association with reassurance and healing, consistent with other forms of cognitive conditioning." [47] Thus, the very act of TMS, which provides structure, a caring environment, and access to mental health workers, may in itself have a therapeutic, antisuicidal, and antidepressive benefit. This does not mean that TMS is ineffective in the treatment of suicidal ideation. From a clinical perspective, placebo effects that reduce patient suffering and improve safety should not be discounted, though are expected to have less reliable durability. Placebo response durability for depression may vary based on baseline illness, severity and chronicity, age and comorbidity, trial visit and treatment frequency, and outcome measure [54]. In addition, the Hines study demonstrated a more rapid improvement in the treatment arm compared to the control arm, despite their convergence at the end of the study and a lack of difference in durability between active and sham arms, which is a critical point. For the acutely suicidal patient, a matter of a few hours of active suicidal ideation may represent the difference between life and death.

3 Ethical Considerations for the Treatment of SI Using TMS in the Clinical Setting

3.1 Risk Management in the Outpatient Setting

TMS is becoming more widely available and utilized in outpatient settings. Data show some promise in the treatment of suicidality, and some clinicians recommend TMS to their patients for this complication. Ethical conflict arises if patient safety is not the foremost consideration when making recommendations for the treatment of suicidality.

TMS clinicians must navigate how suicidality is monitored and managed during high-frequency treatment. The role of the psychiatrist in the outpatient setting for TMS is typically confined to a preliminary interview to assess the patient's needs and appropriateness for treatment, establish the motor threshold, create the treatment guidelines and plan, and follow-up evaluation with course completion. Once these parameters are in place, the sessions themselves are typically run by a technician, with variable further physician oversight. This raises important safety considerations for the suicidal patient. A typical outpatient psychiatric evaluation will comprehensively elicit and evaluate suicidal thoughts, though such assessments are weeks to months apart. Though a TMS course may fall within a typical reassessment interval, a patient's presence in the psychiatrist's office several times per week should prompt careful consideration about the high frequency of suicide risk assessment during a 35-session TMS course, and what level of supervision is appropriate. TMS technicians should also be trained in the evaluation and management of suicidality, but norms regarding technician scope within this aspect are not standardized.

Many practices utilize depression screening tools that can be easily provided to patients to assess for suicidality. Commonly used scales are the Patient Health Questionnaire 9 (PHQ9) [55], Inventory of Depressive Symptomatology (IDS) [56], and the Hamilton Depression Rating Scale (HDRS, also known as HAM-D) [57]. Although primarily scales for depression, all of these scales contain a specific question about suicidal ideation. These scales and more risk-specific SAS are quick, efficient, and can be helpful in screening for suicidal ideation. If positive, SAS can alert the provider to do a more thorough risk assessment for the patient. Effective use of SAS requires practices that have procedures in place for the assessment and management of patients when they screen positive for suicidal ideation. If necessary, psychiatric hospitalization may be required to maintain patient safety. Although there are emerging TMS technologies and accelerated protocols that may more rapidly achieve results, TMS as is currently typically available in the community may take several weeks to achieve optimal effect [42, 43, 58, 59]. Until these newer technologies and protocols are readily accessible, standard rTMS requiring many weeks of treatment remains a suboptimal and ethically fraught treatment for the acutely suicidal patient but may be reasonably employed for low-risk or nascently suicidal patients.

3.2 Best Use and Misuse

Current FDA-approved indications for the use of TMS in the clinical setting are narrow. FDA approval is a slow and costly process [60, 61]. Off-label use is unlikely to be covered by insurance. For example, there are mixed reports regarding the effectiveness of TMS in bipolar depression, though some studies suggest there is some benefit [62–65]. However, the treatment of bipolar depression is not an FDA-approved indication, limiting access for these individuals.

There are no studies evaluating the effects of TMS on suicidality in schizophrenia, despite a lifetime suicide risk of approximately 10% [66]. There is a huge unmet need for these patients. Some studies suggest improvements in negative symptoms of schizophrenia, which may improve overall mood outcomes and reduce suicide risk, but further studies that specifically monitor suicidal ideation and attempts are warranted [67]. Some studies support the use of TMS in BPD, as described previously [48–50]. PTSD is another pathology commonly associated with suicidal ideation and elevated suicide risk [68–70]. Again, there are few studies and no clear consensus on benefit; most studies do not address suicidality as a primary endpoint [71, 72].

However, patients who are struggling with suicidality may be motivated to try alternative therapies in the hopes of alleviating their suffering. As TMS gains further popularity and accessibility, it may attract a broader patient population, highlighting the need for critical and evidence-driven assessment of the nuanced role of TMS as an intervention for the spectrum of suicidality. This presents a practical dilemma for TMS clinicians assessing whether TMS may be an appropriate therapy to try in light of limited or mixed data for a particular disorder. Extrapolation to other pathologies must be done with great care, open communication of risks and benefits of treatment, alternative options or adjunctive treatments, and continued reassessment of the efficacy throughout treatment.

Suicidal patients are particularly vulnerable because of their intense suffering and desperation for relief from symptoms. There are an increasing number of independent subspecialists offering TMS to patients in the community who may focus more exclusively on TMS management and not offer many other treatment approaches within their practices. Conflict of interest and consideration of nonmaleficence arise here if patients desperate and vulnerable in the setting of suicidal ideation are exploited by clinicians who recommend TMS for inappropriate indications, or who offer unrealistic expectations of TMS benefits.

3.3 Equity

As discussed in Part I of this chapter, there are known differences in suicide rates among ethnicities in the USA [4]. Many studies do not include ethnicity in their sample descriptions [41, 73, 74]. Additionally, there are deficits in resources, training, and opportunities to engage in equity-oriented research [75]. This lack of diversity limits the generalizability of a treatment and, importantly, does not allow for a thorough evidence base to support its use in these populations.

The FDA indications for TMS are narrow, and insurance does not cover off-label use, even if research supports the benefits of treatment. TMS has high out-of-pocket costs, further limiting care to those who are most vulnerable. Lower socioeconomic status is associated with an increased risk of depression, suicidal ideation, and attempted or completed suicides [76, 77]. Those who suffer from depressive disorders and suicidal ideation may also have lower occupational functioning that worsens their economic status. A study of the Wisconsin workforce demonstrated that each additional point on the Patient Health Questionnaire-9 (PHQ-9) resulting in an additional 1.65% productivity loss [78]. In 2002, it was estimated that depression costs $44 billion per year in lost productivity in the USA alone [79]. Thus, those who are most at risk and most in need of treatment may not be able to receive employment-facilitated insurance-covered care.

We are once again faced with the question of who is appropriate for treatment. When suicidal ideation is the result of poor occupational functioning, food insecurity, personal debt, or other concerns associated with low socioeconomic status, this likely reduces the benefits of TMS, although, to our knowledge, there are no studies at this time that directly address this question [76, 80, 81].

Newer technologies with more efficient protocols may allow for patients to be treated more rapidly, and as TMS becomes a more popular treatment, some authors propose that this may naturally reduce cost [82, 83]. As it currently stands, however, there exists a great need to expand care and reduce out-of-pocket costs to high-risk populations.

3.4 TMS in the Inpatient Setting

Suicidality can be one of the most challenging phenomena to manage on an inpatient psychiatric unit, particularly when it is related to a treatment-resistant major depressive episode. While decompensated psychosis can often respond rapidly to antipsychotic medication intervention, for example, therapies used to target episodes of MDD may take weeks to reach full effect. Further, many individuals with treatment-resistant depression, require more than one medication trial to find the most effective pharmacological treatment [84]. While pharmacotherapy for depression may take weeks to months for optimal efficacy, inpatient psychiatric facilities are much more suited for much shorter duration admissions. Importantly, the period after hospitalization is particularly high risk in individuals with suicidality: A large meta-analysis indicated that during the 3 months post-discharge, individuals had a suicide rate that is 44 times higher than the population rate [85]. Although the population studied is at much higher risk than the general population, there remains an urgent need for treatments that could reduce suicide for these patients.

As described in Sect. 2, studies suggest brief interventions in the inpatient unit may be useful in the treatment of suicidal ideation. In particular, intermittent theta burst stimulation (iTBS) is capable of producing a comparable effect to rTMS in far less time [59]. TMS is less stigmatized and has a more desirable side effect profile than ECT. It utilizes minimal space, does not involve anesthesiologists or specialized nursing staff, and requires far less training to perform adequately compared to ECT. Although TMS has a narrower set of indications (there are no data supporting its use in the treatment of catatonia and minimal data for positive symptoms of psychosis, for example), accelerated TMS, like the Stanford Accelerated Intelligent Neuromodulation Therapy (SAINT) protocol, could deliver timely de-escalation of depression and suicidal ideation to enable more prompt transition to an outpatient level of care. Indeed, some studies use accelerated protocols in the inpatient setting to good effect [42, 86, 87]. At this time, accelerated TMS does not have wide insurance coverage, but an argument can still be made that if it were accessible during inpatient hospitalizations, it could be a viable option to expand treatment in underfunded or standalone facilities. If accelerated TMS could substantially shorten length of stay for psychiatric hospitalizations involving severe depression and high suicide risk, it might, in turn increase bed availability, increasing access to a crisis resource in scarcity.

3.5 Future Directions

The future role of TMS in the treatment of suicidal ideation is promising, though important questions remain that may help determine the optimal populations and treatment conditions. For example, few studies compare the efficacy of TMS compared to its interventional psychiatry counterparts such as ECT and ketamine, thus it is as yet unclear how TMS efficacy compares to these other treatments [88, 89]. TMS is also unlikely to work as quickly as ketamine to relieve symptoms, as ketamine can achieve a reduction of suicidal thoughts in a matter of hours [21, 90]. However, rapid protocols are being developed and show early promise. Future studies that replicate the rapid effect of the SAINT trial, create protocols specific to the treatment of suicide, and refine these protocols to optimize their effects could have substantial impact.

Future RCTs should also seek to expand diversity and inclusivity, thus broadening their impact and generalizability. Particular attention should be paid to the concept of equity, ensuring that there is adequate representation and consideration of the unique needs of different populations.

Further study to clarify the role of maintenance treatment may help reduce relapse of symptoms and improve outcomes. Additionally, functional neuroimaging studies may help to clarify the neurobiological effects of TMS on suicidal ideation. From an ethical standpoint, TMS has the potential to increase patient autonomy by targeting emerging suicidal ideation before it escalates to requiring a higher level of care. The ethics of involuntary hospitalization for suicidality are complex and beyond the scope of this chapter; suffice to say that the ethics of TMS for the treatment of suicidality are arguably less fraught and, if effective, TMS would represent a more acceptable alternative for suicide treatment for some patients. TMS also presents an alternate choice to ketamine and ECT to those for whom these treatments are intimidating, inaccessible, or medically contraindicated.

TMS is thought to induce neuroplasticity [91–93]. This alone may account for some of the therapeutic benefit of the treatment, as new data support increased neuroplasticity as a primary mechanism by which traditional antidepressants, as well as ketamine, improve depression [94–96]. It may also account for the evidence that supports a combination of psychotherapy and medication is more effective than either therapy alone. Future studies may benefit from incorporating psychotherapeutic techniques, particularly those that target suicidality such as dialectical behavioral therapy, collaborative assessment and management for suicidality, or cognitive behavioral therapy for suicide, in conjunction with treatment in order to enhance learning and improve outcomes.

One may argue that the combined use of psychotherapy and TMS to promote behavioral change could raise concerns about autonomy. Although dissimilar in many ways, the debate around interventional psychiatric treatments has gained recent attention with the rise in popularity of psychedelic-assisted psychotherapy. Individuals undergoing these treatments are put into a vulnerable state, and some have transformational experiences that can have lasting effects on their metaphysical beliefs [97]. Some express concern that individuals in this vulnerable state may be subject to manipulation by the therapist, although, unlike its counterparts in the psychedelics world, there is little concern TMS would produce the altered state of consciousness necessary for this to occur. That being said, the risks of TMS must be clearly stated in the informed consent process [98]. Studies that utilize TMS for suicidality would likewise benefit from continuously modifying and updating the informed consent process to reflect the participant's experience.

To some extent, societal values appear to weigh in the direction of reducing suicide, even at the expense of autonomy [99, 100]. Inpatient psychiatric hospitalization, even involuntarily, is considered the ethical standard of care over allowing an individual the autonomy to end their own life. Consider, even, the societal and physician backlash against physician-assisted death for the terminally ill in the USA [101]. Internationally (i.e., Canada and Belgium), the use of physician-assisted death for psychiatric disorders raises fierce debate [102]. Even in countries where assisted death is acceptable for intolerable severe mental illness, any treatment that stands to reduce premature death from psychiatric illness would likely be preferable.

Suicidality is a complex phenomenon, and it is unlikely that any one modality will be suitable for every patient. While current studies on the use of TMS show some promise in the treatment of suicidality, the imposed homogeneity of the research context may not generalize to the community. Future studies should clarify within which psychopathologies suicidality respond best to TMS, whether incorporation of concomitant psychotherapy could further de-escalate suicide risk and suicidal ideation, and look for ways to expand treatment to those who are underprivileged and under-resourced.

References

1. Centers for Disease Control and Prevention. Risk and Protective Factors for suicide. [cited 2024 May 1 2024]; Available from: https://www.cdc.gov/suicide/factors/index.html
2. Curtin SC, et al. Provisional estimates of suicide by demographic characteristics: United States, 2022. 2023.
3. Centers for Disease Control and Prevention, N.C.f.I.P.a.C. Web-based Injury Statistics Query and Reporting System (WISQARS) [online]). [cited 2024 July 3]; Available from: https://wisqars.cdc.gov

4. Centers for Disease Control and Prevention. Suicide data and statistics. 2024 [10/3/2024]; Available from: https://www.cdc.gov/suicide/facts/data.html#:~:text=Some%20groups%20have%20disproportionately%20high,and%20non%2DHispanic%20White%20people
5. Bommersbach TJ, et al. Why are women more likely to attempt suicide than men? Analysis of lifetime suicide attempts among US adults in a nationally representative sample. J Affect Disord. 2022;311:157–64.
6. Miranda-Mendizabal A, et al. Gender differences in suicidal behavior in adolescents and young adults: systematic review and meta-analysis of longitudinal studies. Int J Public Health. 2019;64(2):265–83.
7. Tang NK, Crane C. Suicidality in chronic pain: a review of the prevalence, risk factors and psychological links. Psychol Med. 2006;36(5):575–86.
8. Bernal M, et al. Risk factors for suicidality in Europe: results from the ESEMED study. J Affect Disord. 2007;101(1):27–34.
9. Shaw RJ, et al. Living alone, loneliness and lack of emotional support as predictors of suicide and self-harm: a nine-year follow up of the UK biobank cohort. J Affect Disord. 2021;279:316–23.
10. Hu F-H, et al. Effects of social support on suicide-related behaviors in patients with severe mental illness: a systematic review and meta-analysis. J Affect Disord. 2023;328:324–33.
11. McClellan C, Ali MM, Mutter R. Impact of mental health treatment on suicide attempts. J Behav Health Serv Res. 2021;48(1):4–14.
12. Han B, et al. National Trends in the prevalence of suicidal ideation and behavior among young adults and receipt of mental health care among suicidal young adults. J Am Acad Child Adolesc Psychiatry. 2018;57(1):20–27.e2.
13. Harmer B, et al. Suicidal ideation. Treasure Island, FL: StatPearls; 2024.
14. Health, N.I.o.M. [cited 2024 September 10]; Available from: https://www.nimh.nih.gov/health/publications/suicide-faq
15. Nock MK, et al. Cross-national prevalence and risk factors for suicidal ideation, plans and attempts. Br J Psychiatry. 2008;192(2):98–105.
16. Riera-Serra P, et al. Clinical predictors of suicidal ideation, suicide attempts and suicide death in depressive disorder: a systematic review and meta-analysis. Eur Arch Psychiatry Clin Neurosci. 2023;
17. Nestor BA, Sutherland S. Theory of mind and suicidality: a meta-analysis. Arch Suicide Res. 2022;26(4):1666–87.
18. Klonsky ED, et al. The three-step theory of suicide: description, evidence, and some useful points of clarification. Prev Med. 2021;152(Pt 1):106549.
19. Cipriani A, et al. Lithium in the prevention of suicide in mood disorders: updated systematic review and meta-analysis. BMJ. 2013;346:f3646.
20. Modestin J, Dal Pian D, Agarwalla P. Clozapine diminishes suicidal behavior: a retrospective evaluation of clinical records. J Clin Psychiatry. 2005;66(4):534–8.
21. Mann JJ, Michel CA, Auerbach RP. Improving suicide prevention through evidence-based strategies: a systematic review. Am J Psychiatry. 2021;178(7):611–24.
22. Kaster TS, et al. Risk of suicide death following electroconvulsive therapy treatment for depression: a propensity score-weighted, retrospective cohort study in Canada. Lancet Psychiatry. 2022;9(6):435–46.
23. D'Anci KE, et al. Treatments for the prevention and management of suicide: a systematic review. Ann Intern Med. 2019;171(5):334–42.
24. Cheng Q, et al. Analysis of time-course, dose-effect, and influencing factors of antidepressants in the treatment of acute adult patients with major depression. Int J Neuropsychopharmacol. 2020;23(2):76–87.
25. Lin CH, Chen CC, Huang CJ. Electroconvulsive therapy versus fluoxetine in suicidal resolution for patients with major depressive disorder. J ECT. 2020;36(4):234–41.
26. Fink M, Kellner CH, McCall WV. The role of ECT in suicide prevention. J ECT. 2014;30(1):5–9.

27. Beghi M, et al. Suicidal behaviour in older age: a systematic review of risk factors associated to suicide attempts and completed suicides. Neurosci Biobehav Rev. 2021;127:193–211.
28. Nock MK, et al. Prevalence, correlates, and treatment of lifetime suicidal behavior among adolescents. JAMA Psychiatry. 2013;70(3):300.
29. Glenn CR, Nock MK. Improving the prediction of suicidal behavior in youth. Int J Behav Consult Ther. 2014;9(3):7.
30. Ong MS, et al. Risk factors for suicide attempt in children, adolescents, and young adults hospitalized for mental health disorders. Child Adolesc Ment Heal. 2021;26(2):134–42.
31. Weissman CR, et al. Bilateral repetitive transcranial magnetic stimulation decreases suicidal ideation in depression. J Clin Psychiatry. 2018;79(3)
32. Chen GW, et al. Efficacy and tolerability of repetitive transcranial magnetic stimulation on suicidal ideation: a systemic review and meta-analysis. Front Psych. 2022;13:884390.
33. Mehta S, et al. The effect of repetitive transcranial magnetic stimulation on suicidal ideation in treatment-resistant depression: a meta-analysis. J Clin Psychiatry. 2022;83(2)
34. Croarkin PE, et al. High-frequency repetitive TMS for suicidal ideation in adolescents with depression. J Affect Disord. 2018;239:282–90.
35. Posner K, et al. The Columbia-suicide severity rating scale: initial validity and internal consistency findings from three multisite studies with adolescents and adults. Am J Psychiatry. 2011;168(12):1266–77.
36. Beck RW, Morris JB, Beck AT. Cross-validation of the suicidal intent scale. Psychol Rep. 1974;34(2):445–6.
37. Hom MA, et al. Ethical issues and practical challenges in suicide research. Crisis. 2017;38(2):107–14.
38. Ferguson M, et al. The effectiveness of the safety planning intervention for adults experiencing suicide-related distress: a systematic review. Arch Suicide Res. 2022;26(3):1022–45.
39. Nuij C, et al. Safety planning-type interventions for suicide prevention: meta-analysis. Br J Psychiatry. 2021;219(2):419–26.
40. Abbott-Smith S, et al. Suicide prevention: what does the evidence show for the effectiveness of safety planning for children and young people? - a systematic scoping review. J Psychiatr Ment Health Nurs. 2023;30(5):899–910.
41. Desmyter S, et al. Accelerated intermittent theta burst stimulation for suicide risk in therapy-resistant depressed patients: a randomized, sham-controlled trial. Front Hum Neurosci. 2016;10:480.
42. George MS, et al. A two-site pilot randomized 3 day trial of high dose left prefrontal repetitive transcranial magnetic stimulation (rTMS) for suicidal inpatients. Brain Stimul. 2014;7(3):421–31.
43. Yesavage JA, et al. Effect of repetitive transcranial magnetic stimulation on treatment-resistant major depression in US veterans: a randomized clinical trial. JAMA Psychiatry. 2018;75(9):884–93.
44. Pan F, et al. Neuronavigation-guided rTMS for the treatment of depressive patients with suicidal ideation: a double-blind, randomized, sham-controlled trial. Clin Pharmacol Ther. 2020;108(4):826–32.
45. Landy K, Rosani A, Estevez R. Escitalopram. Treasure Island, FL: StatPearls; 2024.
46. Cui Y, et al. Efficacy of transcranial magnetic stimulation for reducing suicidal ideation in depression: a meta-analysis. Front Psych. 2021;12:764183.
47. Hines CE, et al. Repetitive transcranial magnetic stimulation promotes rapid psychiatric stabilization in acutely suicidal military service members. J ECT. 2022;38(2):103–9.
48. Cailhol L, et al. Borderline personality disorder and rTMS: a pilot trial. Psychiatry Res. 2014;216(1):155–7.
49. Reyes-Lopez J, et al. Clinical improvement in patients with borderline personality disorder after treatment with repetitive transcranial magnetic stimulation: preliminary results. Braz J Psych. 2018;40(1):97–104.

50. Calderon-Moctezuma AR, et al. Improvement in borderline personality disorder symptomatology after repetitive transcranial magnetic stimulation of the dorsomedial prefrontal cortex: preliminary results. Braz J Psych. 2020;43(1):65–9.
51. Stolk P, et al. Meta-analysis of placebo rates in major depressive disorder trials. Ann Pharmacother. 2003;37(12):1891–9.
52. Razza LB, et al. A systematic review and meta-analysis on placebo response to repetitive transcranial magnetic stimulation for depression trials. Prog Neuro-Psychopharmacol Biol Psychiatry. 2018;81:105–13.
53. Bryan CJ, et al. Nonlinear change processes and the emergence of suicidal behavior: a conceptual model based on the fluid vulnerability theory of suicide. New Ideas Psychol. 2020;57:100758.
54. Papakostas GI, Ostergaard SD, Iovieno N. The nature of placebo response in clinical studies of major depressive disorder. J Clin Psychiatry. 2015;76(4):456–66.
55. Kroenke K, Spitzer RL, Williams JB. The PHQ-9: validity of a brief depression severity measure. J Gen Intern Med. 2001;16(9):606–13.
56. Beck AT, et al. An inventory for measuring depression. Arch Gen Psychiatry. 1961;4:561–71.
57. Hamilton M. A rating scale for depression. J Neurol Neurosurg Psychiatry. 1960;23(1):56–62.
58. O'Reardon JP, et al. Efficacy and safety of transcranial magnetic stimulation in the acute treatment of major depression: a multisite randomized controlled trial. Biol Psychiatry. 2007;62(11):1208–16.
59. Williams NR, et al. High-dose spaced theta-burst TMS as a rapid-acting antidepressant in highly refractory depression. Brain. 2018;141(3):e18.
60. Wimmer NJ, et al. Assessing the cost burden of United States FDA-mandated post-approval studies for medical devices. J Health Care Finance. 2016;
61. Wouters OJ, McKee M, Luyten J. Estimated Research and Development investment needed to bring a new medicine to market, 2009-2018. JAMA. 2020;323(9):844–53.
62. Dolberg OT, et al. Transcranial magnetic stimulation in patients with bipolar depression: a double blind, controlled study. Bipolar Disord. 2002;4(Suppl 1):94–5.
63. Nahas Z, et al. Left prefrontal transcranial magnetic stimulation (TMS) treatment of depression in bipolar affective disorder: a pilot study of acute safety and efficacy. Bipolar Disord. 2003;5(1):40–7.
64. Cohen RB, et al. Clinical predictors associated with duration of repetitive transcranial magnetic stimulation treatment for remission in bipolar depression: a naturalistic study. J Nerv Ment Dis. 2010;198(9):679–81.
65. Fitzgerald PB, et al. A negative double-blind controlled trial of sequential bilateral rTMS in the treatment of bipolar depression. J Affect Disord. 2016;198:158–62.
66. Siris SG. Suicide and schizophrenia. J Psychopharmacol. 2001;15(2):127–35.
67. Lorentzen R, et al. The efficacy of transcranial magnetic stimulation (TMS) for negative symptoms in schizophrenia: a systematic review and meta-analysis. Schizophrenia (Heidelb). 2022;8(1):35.
68. Bryan CJ. Treating PTSD within the context of heightened suicide risk. Curr Psychiatry Rep. 2016;18(8):73.
69. Chou PH, Ito M, Horikoshi M. Associations between PTSD symptoms and suicide risk: a comparison of 4-factor and 7-factor models. J Psychiatr Res. 2020;129:47–52.
70. Stanley IH, et al. PTSD symptoms among trauma-exposed adults admitted to inpatient psychiatry for suicide-related concerns. J Psychiatr Res. 2021;133:60–6.
71. Philip NS, et al. Theta-burst transcranial magnetic stimulation for posttraumatic stress disorder. Am J Psychiatry. 2019;176(11):939–48.
72. Philip NS, et al. Repetitive transcranial magnetic stimulation for depression and posttraumatic stress disorder in veterans with mild traumatic brain injury. Neuromodulation. 2023;26(4):878–84.
73. Wilkening J, Witteler F, Goya-Maldonado R. Suicidality and relief of depressive symptoms with intermittent theta burst stimulation in a sham-controlled randomized clinical trial. Acta Psychiatr Scand. 2022;146(6):540–56.

74. Cole EJ, et al. Stanford accelerated intelligent neuromodulation therapy for treatment-resistant depression. Am J Psychiatry. 2020;177(8):716–26.
75. Baumann AA, et al. Assessing researchers' capabilities, opportunities, and motivation to conduct equity-oriented dissemination and implementation research, an exploratory cross-sectional study. BMC Health Serv Res. 2022;22(1):731.
76. Kim JL, et al. Effect of socioeconomic status on the linkage between suicidal ideation and suicide attempts. Suicide Life Threat Behav. 2016;46(5):588–97.
77. Madigan A, Daly M. Socioeconomic status and depressive symptoms and suicidality: the role of subjective social status. J Affect Disord. 2023;326:36–43.
78. Beck A, et al. Severity of depression and magnitude of productivity loss. Ann Fam Med. 2011;9(4):305–11.
79. Stewart WF, et al. Cost of lost productive work time among US workers with depression. JAMA. 2003;289(23):3135–44.
80. van Heeringen K, Mann JJ. The neurobiology of suicide. Lancet Psychiatry. 2014;1(1):63–72.
81. Cohen A, et al. Social inequalities in response to antidepressant treatment in older adults. Arch Gen Psychiatry. 2006;63(1):50–6.
82. Zengin G, et al. The efficacy and safety of transcranial magnetic stimulation in treatment-resistant bipolar depression. Psychiatr Danub. 2022;34(2):236–44.
83. Carnell BL, et al. How effective is repetitive transcranial magnetic stimulation for bipolar depression? J Affect Disord. 2017;209:270–2.
84. Gaynes BN, et al. Primary versus specialty care outcomes for depressed outpatients managed with measurement-based care: results from STAR*D. J Gen Intern Med. 2008;23(5):551–60.
85. Chung DT, et al. Suicide rates after discharge from psychiatric facilities: a systematic review and meta-analysis. JAMA Psychiatry. 2017;74(7):694–702.
86. Badawi A, et al. Repetitive transcranial magnetic stimulation: influence on stress and early responsiveness outcomes for depression, anxiety, and stress. Psychiatry Q. 2022;93(1):385–91.
87. Barnes R, et al. Equal remission rates and reduced length of hospital stay with twice-daily repetitive transcranial magnetic stimulation (rTMS) for major depression - a large naturalistic retrospective cohort association study. Prog Neuro-Psychopharmacol Biol Psychiatry. 2023;127:110820.
88. Rhee TG, et al. Efficacy and safety of ketamine vs electroconvulsive therapy among patients with major depressive episode: a systematic review and meta-analysis. JAMA Psychiatry. 2022;79(12):1162–72.
89. Ronnqvist I, Nilsson FK, Nordenskjold A. Electroconvulsive therapy and the risk of suicide in hospitalized patients with major depressive disorder. JAMA Netw Open. 2021;4(7):e2116589.
90. Abbar M, et al. Ketamine for the acute treatment of severe suicidal ideation: double blind, randomised placebo controlled trial. BMJ. 2022;376:e067194.
91. King ES, Tang AD. Intrinsic plasticity mechanisms of repetitive transcranial magnetic stimulation. Neuroscientist. 2022:10738584221118262.
92. Ramachandra V, Giese MA, Benali A. The effects of low-intensity repetitive transcranial magnetic stimulation on white matter plasticity and depression. Biol Psychiatry Glob Open Sci. 2022;2(2):92–4.
93. Jannati A, et al. Assessing the mechanisms of brain plasticity by transcranial magnetic stimulation. Neuropsychopharmacology. 2023;48(1):191–208.
94. D'Sa C, Duman RS. Antidepressants and neuroplasticity. Bipolar Disord. 2002;4(3):183–94.
95. Vargas MV, et al. Psychedelics promote neuroplasticity through the activation of intracellular 5-HT2A receptors. Science. 2023;379(6633):700–6.
96. Aleksandrova LR, Phillips AG. Neuroplasticity as a convergent mechanism of ketamine and classical psychedelics. Trends Pharmacol Sci. 2021;42(11):929–42.
97. Timmermann C, et al. Psychedelics alter metaphysical beliefs. Sci Rep. 2021;11(1):22166.
98. Jacobs E. Transformative experience and informed consent to psychedelic-assisted psychotherapy. Front Psychol. 2023;14:1108333.

99. Emanuel EJ, et al. Attitudes and practices of euthanasia and physician-assisted suicide in the United States, Canada, and Europe. JAMA. 2016;316(1):79–90.
100. Gale C, Barak Y. Euthanasia, medically assisted dying or assisted suicide: time for psychiatrists to say no. Australas Psychiatry. 2020;28(2):160–3.
101. Snyder Sulmasy L, et al. Ethics and the legalization of physician-assisted suicide: an American College of Physicians Position Paper. Ann Intern Med. 2017;167(8):576–8.
102. Grassi L, et al. Debating euthanasia and physician-assisted death in people with psychiatric disorders. Curr Psychiatry Rep. 2022;24(6):325–35.

Transcranial Magnetic Stimulation for Substance Use Disorders: Ethical Considerations

Dhvani D. Mehta, Melissa Hiebert, Victor M. Tang, Heather B. Ward, and Tony P. George

1 Introduction

The high prevalence of substance use disorders (SUDs) is a significant public health concern. SUDs contribute to approximately half a million deaths annually [1, 2], making them a major contributor to global morbidity and mortality. Although various therapeutic interventions have been substantiated, including pharmacological and psychosocial therapies, relapse rates continue to be high [3, 4]. Thus, the development of novel and effective therapeutic interventions for treatment-resistant SUDs is imperative. Clinical and preclinical research has linked substance misuse with mesolimbic hypodopaminergic function [5], as well as dysfunction in prefrontal areas responsible for cognitive control (e.g., dorsolateral prefrontal cortex (DLPFC) [6] and subcortical reward system areas (e.g., nucleus accumbens) [7]. Consequently, neuromodulation (NM) techniques may be encouraging in the treatment of addictions due to their ability to modulate SUD-related neural circuitry and functional abnormalities [8–10]. Repetitive transcranial magnetic stimulation (rTMS)

Supported in part by grants from NIDA (R21-DA-043949) and the Canadian Institutes of Health Research (CIHR; PJT-190053) to Dr. George.

D. D. Mehta · V. M. Tang · T. P. George (✉)
Addictions Division, CAMH, Toronto, ON, Canada

Department of Psychiatry, Temerty Faculty of Medicine, University of Toronto, Toronto, ON, Canada
e-mail: tony.george@camh.ca

M. Hiebert
Clinical Research Division, CAMH, Toronto, ON, Canada

H. B. Ward
Department of Psychiatry and Behavioral Sciences, Vanderbilt University Medical Center, Nashville, TN, USA

V. Dubljević, J. R. Young (eds.), *TMS and Neuroethics*, Advances in Neuroethics,
https://doi.org/10.1007/978-3-031-92401-9_14

is a non-invasive neuromodulation (NM) intervention with considerable efficacy in recent years for the treatment of neurological and psychiatric disorders [11–13].

rTMS utilizes alternating magnetic fields to modulate electrical currents in localized cortical tissue [14]. Typically, low-frequency (LF; ≤1 Hz) stimulation produces local inhibitory effects while high-frequency (HF; ≥5 Hz) stimulation is excitatory [15, 16]. Numerous studies have demonstrated enduring functional and structural neuroplastic changes in target regions [17, 18], and increased DA release in the mesolimbic system [19–22], thus making it a promising brain-based approach for SUDs. Accordingly, evidence is accumulating for the efficacy of rTMS in reducing craving and consumption across several SUDs [8, 9, 13], which could help address the high disease burden of SUDs for which evidence-based treatments are urgently needed. Amongst NM modalities, it is the most commonly studied, and has the best evidence supporting its efficacy in treating neuropsychiatric disorders.

Ultimately, the advent of rTMS as a therapeutic intervention for SUDs represents a potential advance. However, it also raises ethical considerations that must be carefully considered [23–25]. This includes crucial issues such as informed consent, risk mitigation, the balance between experimental innovation and evidence-based practice, and equitable access to treatment. These are important for the ethical deployment of rTMS, given the vulnerability of people with SUDs. Informed consent is highly relevant to TMS clinical practice, and it is crucial that patients are fully aware of the experimental nature of the treatment, its potential risks and benefits, and alternative therapies [24]. Risk mitigation is another critical ethical concern. rTMS, while generally considered safe, carries potential risks such as seizures, headaches, and mood disturbances. These risks must be carefully managed through thorough pre-treatment screening, continuous monitoring, and the implementation of robust safety protocols. Balancing the drive for innovation with the principles of evidence-based practice is essential. While rTMS holds significant promise, its use must be grounded in rigorous scientific research. This involves conducting well-designed studies to establish efficacy and safety as well as ensuring that clinical practice is informed by the latest evidence and healthcare providers are well informed of the evolving scientific landscape. Equitable access to rTMS treatment is another fundamental ethical issue. The high cost and technical demands of rTMS can limit its availability, potentially exacerbating existing healthcare disparities. Ethical practice demands efforts to promote accessibility and affordability, ensuring that all patients, regardless of socioeconomic status or geographic location, can benefit from advanced treatment options.

This chapter addresses the ethics relevant to rTMS and SUDs, examining issues of accessibility, guidelines for patient inclusion, and treatment strategies to improve clinical outcomes and reduce burden. Moreover, we evaluate the balance between potential benefits and risks, and consider ethical obligations to maximize well-being and minimize harm. By examining these ethical dimensions, we aim to provide a comprehensive framework for clinicians, researchers, and policymakers to responsibly integrate rTMS into SUD treatment protocols, ensuring that this promising technology is utilized in a manner that is both effective and ethically sound.

2 Benefits and Risks

2.1 Efficacy

Several randomized controlled trials (RCTs) have demonstrated the efficacy of rTMS for SUD treatment (Table 1). In a recent systematic review and meta-analysis of NM for SUDs, Mehta et al. [13] found that rTMS reduced subjective and cue-induced substance craving and/or consumption when treating tobacco, compared to sham treatment. While effects of rTMS on alcohol use disorder (AUD) were less consistent than that of other substances, subsequent meta-analyses found that multi-session rTMS produced significantly greater reductions in alcohol craving and consumption compared to sham [13, 26]. Investigations of cannabis are limited, but preliminary studies in cannabis use disorder (CUD) [27] and comorbid CUD-schizophrenia [28] populations have observed greater reductions in cannabis use outcomes, including cannabis use frequency [27], and cannabis consumption and craving [28] following 20 session of HF rTMS versus sham. This highlights the potential efficacy of this treatment for CUD and the need for further research.

It is important to note that the application of rTMS in SUD treatment has primarily focused on substances such as alcohol and tobacco, leaving a notable gap in research and clinical implementation for other addictive substances such as cannabis, opioids, and stimulants. This gap may stem from several factors, including the complex neurobiology of different substances, variations in treatment response, and logistical challenges in conducting controlled trials across diverse addiction populations. Furthermore, regulatory approval and funding constraints may hinder the expansion of rTMS applications to encompass a broader range of SUDs. Addressing this gap requires concerted efforts from researchers and clinicians to expand the scope of rTMS research and ensure equitable access to innovative SUD treatments. Given that there are several evidence-based treatments available for AUD, TUD,

Table 1 A summary of substance use treatment outcomes from randomized controlled trials investigating the use of rTMS for SUDs

Repetitive transcranial magnetic stimulation (rTMS) [Total N = 2478; 52 Studies]		
Substance use disorder	Population	Studies with positive outcomes (Effect Size—Active vs. Control)
Alcohol	N = 607; 16 studies	7/16 (**44%**) *Hedge's g = −1.01, 95% CI [−1.62, −0.40]*
Tobacco	N = 781; 16 studies	14/16 (**88%**) *Hedge's g = −1.36, 95% CI [−2.09, −0.63]*
Cannabis	N = 105; 3 studies	2/3 (**66%**) *Hedge's g = −0.16, 95% CI [−0.64, 0.32]*
Cocaine	N = 227; 6 studies	3/6 (**50%**) *Hedge's g = −0.73, 95% CI [−1.57, 0.11]*
Methamphetamine	N = 519; 8 studies	7/8 (**88%**) *Hedge's g = −1.45, 95% CI [−3.22, 0.32]*
Opioids	N = 239; 4 studies	3/4 (**75%**) *Hedge's g = −0.99, 95% CI [−2.25, 0.27]*

and opioid use disorder [29], research efforts should shift to SUDs with a lack of evidence-based biological treatments, such as cannabis and stimulants.

Additionally, studies in the field of rTMS for addictions have predominantly compared active rTMS to sham stimulation or no intervention. While this is import to establish feasibility, tolerability, and efficacy of rTMS for SUDs, it is imperative that efficacious rTMS protocols be compared with currently available evidence-based treatments for SUDs, such as behavioral interventions and pharmacotherapies.

2.2 Safety

One of the notable advantages of rTMS is the favorable side effect profile compared to pharmacotherapy and invasive NM procedures like deep brain stimulation (DBS) [30]. Studies have reported minimal and transient adverse effects, such as mild headache or scalp discomfort during treatment sessions, which are generally well-tolerated by patients [12]. Moreover, the non-invasive nature of rTMS minimizes the risk of systemic side effects associated with medication-based treatments, making it an attractive option for people seeking alternative approaches to addiction management, or who are treatment-resistant to conventional addiction treatments.

Nonetheless, there are concerns regarding severe side effects (e.g., seizures), though this can be mitigated by maintaining strict inclusion/exclusion criteria [30]. Furthermore, some individuals may experience discomfort or intolerance to rTMS treatment, especially during initial sessions or with higher intensity stimulation protocols ($\geq$ 5 Hz). Importantly, when treating SUDs with rTMS, acute intoxication of substances, particularly stimulants, and/or withdrawal from substances such as alcohol, raise safety concerns as they may increase the risk of seizure [30, 31]. These risks underscore the importance of thorough patient screening, careful monitoring during treatment, and adherence to established safety protocols. Moreover, personalized adjustment of stimulation parameters and gradual acclimatization may enhance patient comfort and treatment adherence.

Another key safety consideration is the need for detailed reporting of adverse events in clinical studies. While rTMS is well-tolerated by people with addiction, as it is by patients with major depressive disorder (MDD), the published literature often lacks comprehensive data on the tolerability and long-term safety of rTMS for SUDs. Addressing this gap through rigorous reporting and long-term follow-up studies is vital to establishing a robust safety profile for rTMS in this context and the potential for relapse prevention.

Ethical Implications Participants should be fully informed about the potential risks and benefits of rTMS for SUDs, including the experimental nature [24]. Informed consent must be obtained without coercion, considering the vulnerability of SUD populations. Comprehensive pre-treatment assessments are necessary to identify any underlying conditions that might heighten the risk of adverse effects. Moreover, continuous monitoring before, during, and after rTMS sessions is crucial to address any negative reactions.

3 Populations

3.1 Inclusion/Exclusion Guidelines

When devising inclusion and exclusion criteria for SUD treatment with rTMS, adherence to established clinical guidelines and empirical evidence is crucial [13]. Patients should ideally have a clinically diagnosed SUD, confirmed by standardized diagnostic criteria such as the DSM-5. Priority should be given to candidates demonstrating motivation for treatment engagement and a sincere commitment to overcoming addictive behaviors, as well as those demonstrating resistance to conventional interventions or experiencing frequent relapses. Moreover, consideration should extend to patients presenting with certain comorbid psychiatric conditions such as MDD, due to the established efficacy of rTMS for these disorders [32]. While the majority of studies have studied patients 18–65 years [13], some studies may consider older adults on a case-by-case basis. Younger patients (under 18) are typically excluded due to limited research on safety and efficacy in this group, as well as the ethical constraints on obtaining informed consent.

Conversely, exclusion criteria should target individuals at heightened risk for adverse outcomes, or those unlikely to derive substantive benefits from rTMS. This includes patients with unstable psychiatric disorders, such as active psychosis, severe depression or mania, as the presence of these conditions may impede treatment response. In cases of pregnancy, published reports in MDD and other indications have not noted adverse events or increased perinatal complications [33–35]. However, safety data is still extremely limited and there are no such findings in people with SUDs. As such, pregnancy should be an exclusion due to potential fetal risks associated with rTMS exposure and the potential ramifications of rTMS-induced seizures [30]. A rigorous pre-treatment assessment process is imperative to ensure the absence of contraindications to rTMS and identify risk factors associated with seizure induction during rTMS [30, 31]. This includes a personal/family history of epilepsy/seizure, past stroke or head injury with neurologic damage, and the presence of metallic implants. Acute intoxication of stimulants (e.g., cocaine, methamphetamine) could also increase risk of seizure, and should be carefully monitored prior to rTMS administration. Lastly, neurological disorders and medical disorders associated with lowered seizure threshold should be considered. This includes metabolic abnormalities (e.g., hypoglycemia, hyperglycemia, renal failure/uremia, liver failure), sleep deprivation, and electrolyte imbalance (e.g., with anorexia nervosa). Notably, withdrawal from substances, such as alcohol, may also increase seizure risk, in which case detoxification may be required prior to inclusion.

Concomitant medications may increase cortical excitability or lower seizure threshold, such as amphetamines [36, 37], and thus may increase rTMS-related seizure risk. However, empirical evidence shows that rTMS in combination with these medications does not heighten the risk for seizure induction, and rates remain low [30, 38]. Consequently, there are currently no medication contraindications for rTMS. Nonetheless, patients should be on a stable medication regimen (2–4 weeks) prior to receiving rTMS. Fluctuations in medication levels can affect neural excitability and potentially increase the risk of adverse effects. Additionally, ensuring medication stability minimizes variables that could confound treatment outcomes, allowing for a clearer assessment of effectiveness and safety of rTMS. It also protects the patient from unnecessary risks. Patients with SUDs often have co-occurring psychiatric or medical conditions that require ongoing pharmacological management. Sudden changes in their medication regimen can lead to destabilization, exacerbating symptoms or leading to withdrawal effects, which could interfere with rTMS treatment response.

Furthermore, patients who are unable or unwilling to adhere to treatment requirements, including session attendance and compliance with safety protocols, may not be suitable candidates for rTMS, or require modifications to their treatment course which may compromise efficacy. Importantly, limitations in access to rTMS facilities or financial constraints (e.g., insurance) necessitate careful consideration, as these factors may impede treatment feasibility and equitable patient access.

3.2 Psychiatric Comorbidities

Comorbid psychiatric disorders are common among people with SUDs, and can exacerbate the severity of addiction and hinder recovery efforts [39–41]. rTMS has been shown to be effective in treating MDD and anxiety, with several studies indicating that it can alleviate symptoms and improve overall functioning [42]. Despite the exceedingly high rates of SUDs in psychiatric disorders, few studies have tested rTMS interventions in populations with comorbid psychiatric disorders. Notably, three rTMS studies examining TUD participants with co-occurring SCZ observed significant reductions in tobacco craving and consumption [43–45], signifying promise in the use of rTMS in treating co-occurring SUD and psychiatric disorders, and warranting further research in clinical trials to determine safety and efficacy. Overall, the high prevalence of co-occurring psychiatric disorders amongst people with SUDs signifies a greater urgency to develop effective treatments for this population, and supports the inclusion of people with co-occurring disorders to access rTMS.

Nonetheless, certain psychiatric conditions may pose risks, and a rigorous risk-benefit analysis tailored to each patient should be performed by clinicians. For example, people with bipolar disorder may experience treatment-emergent mania in response to rTMS, though this is rare [46]. It is crucial to stabilize severe psychiatric symptoms prior to initiating rTMS, and continuously monitor for adverse effects to ensure patient safety. It is also important to consider that the current evidence base for efficacy and safety in approved conditions, such as MDD, are based upon randomized controlled trials that systematically excluded people with comorbid SUDs. Thus, there remains a great deal of uncertainty about whether rTMS has the same effect on people with MDD with a comorbid SUD. These are important details to include in informed consent discussions such patients receiving rTMS.

3.3 Equitable Access

There are prominent barriers to access that impede on the widespread adoption of rTMS for addiction treatment. Accessibility of rTMS for SUDs is significantly more limited compared to its use for other psychiatric disorders, such as MDD and obsessive-compulsive disorder (OCD). rTMS received FDA approval for MDD in 2008 and for OCD in 2018 [47], supported by substantial clinical evidence demonstrating its efficacy and safety. Consequently, established treatments for MDD and OCD benefit from robust clinical infrastructure, widespread clinician training, and extensive public and professional awareness, all of which enhance their accessibility. In contrast, rTMS for addiction is still emerging, and while studies show promise, the field has yet to establish standardized protocols and determine long-term efficacy. For SUDs, a deep TMS coil (H4) that modulates both the bilateral PFC and insula has been cleared by the FDA as efficacious for smoking cessation, representing the first and only rTMS indication for addictions [48]. As a result, there is limited availability of trained providers and fewer specialized treatment centers. Ultimately, achieving regulatory approval is integral in bridging this gap and allowing insurance reimbursement, a process that depends on increasing investment in research and rigorous clinical trials. Insurance coverage often does not always follow the scientific evidence closely, and even treatments that have consensus agreement amongst healthcare providers does not mean that insurance companies will provide reimbursement [49].

The underrepresentation of females and minorities in rTMS trials for SUDs also poses significant concerns on equitable access, leading to a limited understanding of how different groups respond to treatment [50]. A large number of studies in addiction are not balanced for sex, focusing on males. Thus, sex-related differences in treatment outcomes are unclear. This is particularly relevant as there is strong reason to believe that sex/gender is a critical moderator of rTMS treatment efficacy, with data suggesting larger TMS-related neuroplastic responses in women during periods of high estradiol [51], with implications for understanding menstrual cycle effects in treatment response. Similarly, research studies often lack diversity in participant demographics, with samples consisting predominantly of Caucasians. This is unfortunate, given that racial/ethnic minorities suffer more severe consequences from SUDs, including higher incarceration rates, higher alcohol-related problems, and higher drug overdose mortality [52]. This underrepresentation hinders the generalizability of study findings and may perpetuate disparities in healthcare access and outcomes. Additionally, without sufficient representation from diverse groups, there is a risk of overlooking important nuances in SUD pathology and treatment response, further impeding progress in developing effective interventions. Addressing this issue requires purposeful efforts to recruit and retain diverse participants in clinical trials, ensuring that research findings accurately reflect the diverse populations impacted by SUDs. Investigation of demographic predictors of rTMS treatment response is essential, as it allows for the development of personalized treatment approaches and optimization of therapeutic protocols.

4 Study Designs and Treatment Parameters

High- Versus Low-Frequency rTMS Variability in treatment efficacy of rTMS for SUDs can be attributed to differences in stimulation parameters. In particular, frequency-dependent rTMS effects on local and distant regional brain activity may have implications for clinical therapeutics in SUDs [53, 54], and should be evaluated from both a clinical and ethical perspective. Due to the hypothesized role of dopaminergic dysfunction in addiction neurocircuitry [13], most rTMS studies have used HF stimulation (e.g., ≥5 Hz) regardless of SUD. A select few studies [44–47] have compared the effects of LF (1 Hz) and HF (10 Hz) stimulation and found that 10-Hz rTMS significantly reduced substance craving and/or consumption, suggesting that HF rTMS stimulation parameters may have greater therapeutic potential in comparison to LF stimulation. Nonetheless, there was significant heterogeneity on the SUD of interest and the stimulation region of interest across these studies. It should also be noted that in depression, meta-analyses have shown equivalent efficacy between LF and HF rTMS [48]. Given the complexities surrounding hyperactive ventral striatal activity and hypoactive prefrontal control in addiction, further research is required to understand the effects of LF vs HF stimulation on these regions and the effects on reducing problematic substance use. LF stimulation is also more tolerable and safer than HF and has the potential to be developed into an at-home coil [49]. LF rTMS offers a potentially safer option for patients at higher risk for seizures, while HF rTMS may provide greater efficacy but carries higher seizure risks. Both approaches require thorough patient education, transparent communication about the benefits and risks, and a commitment to ongoing research to validate their use in SUD treatment.

Accelerated rTMS There has been emerging interest in the use of accelerated protocols (e.g., more than one session daily) due to their potential for rapid and robust therapeutic effects. Studies in depression have found that accelerated protocols are safe and well-tolerated, presenting a similar side effect profile to daily rTMS [55–57]. However, the adverse effects on cognition and mood are unclear and require further investigation. Moreover, an accelerated schedule of rTMS sessions may impose a significant burden on participants, requiring them to commit to multiple sessions within a short timeframe. Ultimately, the ethical administration of this treatment hinges on the availability of robust scientific evidence to support its efficacy across different addiction types and patient populations. Martinotti et al. [58] conducted the only randomized sham-controlled addictions study to adopt an accelerated approach, and reported unfavorable cocaine use outcomes following twice daily stimulation. Nonetheless, Steele et al. [59] have found that three iTBS sessions/day delivered over 10 days was tolerable and produced reductions in cocaine consumption. Overall, the efficacy of accelerated rTMS paradigms for SUDs is not well established and the durability of therapeutic effects remain unclear. However, these accelerated schedules can impose significant burdens on participants. The increased frequency and intensity of sessions may lead to heightened physical and mental fatigue, discomfort, and potential side effects such as headaches or scalp discomfort. Additionally, the demanding schedule can interfere with daily routines, work, and personal responsibilities, potentially causing logistical and financial challenges. Providing comprehensive support, such as flexible scheduling, transportation assistance, and continuous monitoring of participant well-being, is crucial to mitigate these challenges and ensure that the accelerated rTMS protocols are both effective and manageable for patients.

Cue-Provocation Inclusion of cue-reactivity procedures with NM has emerged as an important methodological refinement, as it may enhance rTMS effects. It has been postulated that cue-exposure prior to rTMS may activate craving-related neurocircuitry making it more amenable to modulatory effects, and subsequent stimulation could then disrupt drug-related memory consolidation [43]. Accordingly, studies of rTMS in addiction has shown that activation of craving through the presentation of drug cues prior to stimulation may be important in optimizing treatment effects [60, 61]. Moreover, ethical considerations are present around the potential distress or discomfort induced by cue exposure, particularly in vulnerable people undergoing addiction treatment, and potential contributions to subsequent relapse to substance use. Importantly, the risk of death from relapse varies by substance; a single relapse on opioids is significantly more dangerous than relapse on tobacco. While cue exposure may be essential for therapeutic insight and neural modulation, safeguarding the well-being and autonomy of participants remains of greatest importance. Striking a balance between scientific advancement and the protection of individuals' psychological and emotional well-being is imperative. The frontal cortex, which is central in addictions neurobiology and craving-related mechanisms in particular, has a primary function in cognitive control and executive function as well [62]. As such, the relevant target neurocircuitry may also be activated through the use of cognition-centered provocation. Cognitive activation strategies have also been proposed; the advantage of cognitive activation paired with rTMS is that there is less risk of a negative clinical effect when compared with craving activation through the presentation of drug cues, which for patients may be a destabilizing and an aversive procedure that may be difficult to tolerate.

Substance Dependence Detoxification preceding rTMS treatment for SUDs may enhance the efficacy of rTMS by stabilizing patients' physiological state and facilitating better engagement in therapy. Accordingly, several studies in rTMS for SUDs have required participants to detoxify prior to beginning treatment in order to mitigate risks associated with acute withdrawal symptoms [63, 64]. However, no studies have compared treatment outcomes between recently detoxified and non-detoxified participants. Beyond efficacy, detoxification may be necessary and/or preferred for substances that increase the risk of seizure. For instance, acute intoxication of substances (e.g., cocaine, methamphetamine, and amphetamine) or acute withdrawal from a substance (e.g., alcohol, benzodiazepines, and barbiturates) can lower seizure threshold. As such, detoxification paradigms may be ideal when treating these SUDs (e.g., AUD, cocaine use disorder) to mitigate risk. Detoxification does pose challenges that must be considered. Patients may experience discomfort and distress during the process, potentially affecting their willingness to continue treatment. Additionally, detoxification often requires medical oversight in an inpatient facility. Many studies of rTMS for SUDs were conducted in rehabilitation facilities, which are often private requiring insurance or cash payment, and are inaccessible to many without the financial means. Furthermore, the time, resources, and potential for residual withdrawal symptoms further complicate the decision-making process.

Cognition When utilizing rTMS in addiction treatment, its impact on cognition is a critical consideration. While rTMS primarily targets neural circuits implicated in addictive behaviors, it also impacts cognitive functions, such as attention, memory, and executive control. Understanding the cognitive effects of rTMS is essential for optimizing treatment outcomes and minimizing potential adverse effects on cognition. Research suggests that rTMS can modulate cognitive processes through its influence on cortical excitability and network connectivity. Few systematic reviews have examined the effects of rTMS on cognitive functioning in people with SUDs. While there are some reports of positive rTMS-related effects on cognitive functioning [57, 58], particularly with HF stimulation, there has been considerable heterogeneity across studies and the specific effects of rTMS on cognition in the context of SUD treatment remain relatively understudied. This underscores the importance of utilizing the ethical principles of beneficence and non-maleficence to mitigate any potential cognitive side effects of rTMS. Longitudinal studies are needed to assess the long-term cognitive outcomes of rTMS treatment of addictions.

Combination With Other Treatments Combining rTMS with behavioral and pharmacotherapies offers a comprehensive treatment approach that addresses the neurobiological, physiological, and behavioral components of addiction [65]. Behavioral therapies, such as cognitive-behavioral therapy (CBT), offers strategies to manage triggers, develop healthier coping skills, and alter maladaptive learning patterns that contribute to addictive behaviors. Moreover, rTMS may create more receptive brain states, potentially improving the effectiveness of psychotherapeutic interventions. According to Wilkinson et al. [66], rTMS can create a state of enhanced neural plasticity which can be harnessed by cognitive and behavioral interventions to improve treatment outcomes and relapse rates. Pharmacotherapies, which include medications to manage withdrawal symptoms and reduce the risk of relapse, complement this combination by addressing the physiological aspects of addiction. rTMS, when combined with pharmacotherapy, may potentiate effects by simultaneously targeting reward pathways, managing withdrawal symptoms, and stabilizing neural circuits, leading to more sustained recovery. Together, this comprehensive approach ensures that all aspects of addiction are addressed, providing a robust framework for rehabilitation by tailoring interventions to individual needs. This may be particularly advantageous in cases of psychiatric comorbidities, such as psychosis, depression or anxiety. Integrated treatment approaches that target both conditions simultaneously may be more effective and ethically appropriate than treating each condition in isolation. While some rTMS studies in addiction have implemented adjunctive psychotherapeutic and/or pharmacological intervention [43], few have attempted to discern the clinical impact of each therapy during treatment, and augmentation of treatment outcomes.

Neuroimaging Given the complexity and heterogeneity of SUDs, a personalized medicine approach, which tailors treatment to an individual's unique characteristics, holds significant promise. Neuroimaging techniques such as functional magnetic resonance imaging (fMRI), diffusion tensor imaging (DTI), and positron emission tomography (PET) may play an essential role in this approach by providing insights into addiction-related structural and functional neural alterations. Incorporating neuroimaging into rTMS treatment facilitates the identification of biomarkers that predict individual responses, which may help to determine which patients are most likely to benefit. Neuroimaging also allows for the customization of rTMS protocols, adjusting stimulation parameters based on the patient's unique neural profile. This has been investigated significantly in studies for depression [67], wherein rTMS protocols have increasingly used subgenual anterior cingulate cortex (sgACC) functional connectivity to individualize treatment targets, based on its relevance to TMS treatment mechanisms and depression pathology [68]. Regular imaging assessments during treatment can monitor progress and guide adjustments, ensuring that rTMS remains effective over time. Lastly, as traditional targeting systems do not consider inter-individual differences in brain morphology, neuronavigation-guided approaches may achieve greater precision with personalized targets, limiting inter-individual variability. This precise targeting may also reduce the risk of stimulating non-relevant brain areas, which can lead to unwanted side effects and less effective treatment. rTMS studies in depression have demonstrated the benefits of such an approach and found that clinical outcomes were significantly improved when patients were stimulated closer to fMRI-personalized targets [69]. Selected rTMS studies in addiction have integrated MRI-neuronavigation [70–76], though the efficacy of this approach in comparison to non-personalized targeting has not yet been determined.

5 Conclusions

In discussing ethics of rTMS for SUD treatment, it is essential to balance the promise of this technology with the ethical imperatives that govern medical practice. rTMS offers a novel, non-invasive approach to addressing addiction, a condition that has profound personal, social, and economic implications. However, the ethical deployment of rTMS necessitates a careful consideration of several critical factors. At the forefront is rigorous validation of the efficacy and safety of rTMS for SUDs through clinical trials. The current body of research is still evolving, and optimal treatment parameters remain to be determined. Continuous evaluation of the effectiveness of rTMS, alongside a meticulous assessment of adverse effects and a conscious inclusion of historically marginalized populations (e.g., women, comorbid psychiatric disorders, and racialized minorities), is essential to establish a robust evidence base that can guide ethical clinical practice and improve accessibility. In addition to safety monitoring, an emphasis on cognitive outcomes is needed as it is notably underrepresented in the current research landscape. Future research should adopt a multimodal approach, combining rTMS with behavioral interventions and pharmacotherapies. This may be particularly valuable in comorbid psychiatric conditions, which constitutes a significant percentage of the SUD population. Finally, personalized medicine approaches using neuroimaging may enhance rTMS effectiveness, while fostering a patient-centered approach to care. By prioritizing rigorous scientific validation, equitable access, and a patient-centered approach to care, we can harness the benefits and minimize risks of rTMS, while upholding the highest ethical standards.

References

1. Globalchange.gov. Key substance use and mental health indicators in the United States: results from the 2021 National Survey on drug use and health. Substance Abuse and Mental Health Services Administration; 2021.
2. UNODC. World drug report 2020. United Nations Office on Drugs and Crime; 2020.
3. NIDA. Drugs, brains, and behavior: the science of addiction. National Institute on Drug Abuse, National Institutes of Health, U.S. Department of Health and Human Services; 2007.
4. Kabisa E, et al. Determinants and prevalence of relapse among patients with substance use disorders: case of Icyizere Psychotherapeutic Centre. Subst Abuse Treat Prev Policy. 2021;16:13.
5. Nutt DJ, Lingford-Hughes A, Erritzoe D, Stokes PRA. The dopamine theory of addiction: 40 years of highs and lows. Nat Rev Neurosci. 2015;16:305–12.
6. Goldstein RZ, Volkow ND. Dysfunction of the prefrontal cortex in addiction: neuroimaging findings and clinical implications. Nat Rev Neurosci. 2011;12:652–69.
7. Everitt BJ, Belin D, Economidou D, Pelloux Y, Dalley JW, Robbins TW. Neural mechanisms underlying the vulnerability to develop compulsive drug-seeking habits and addiction. Philos Trans R Soc B Biol Sci. 2008;363:3125–35.
8. Salling M, Martinez D. Brain stimulation in addiction. Neuropsychopharmacol. 2016;41:2798–809.
9. Coles AS, Kozak K, George TP. A review of brain stimulation methods to treat substance use disorders. Am J Addict. 2018;27:71–91.
10. Steele VR. A circuit-based approach to treating substance use disorders with noninvasive brain stimulation. Biol Psychiatry. 2021;89:944–6.
11. George MS, Taylor JJ, Short EB. The expanding evidence base for rTMS treatment of depression. Curr Opin Psychiatry. 2013;26:13–8.
12. Slotema CW, Blom JD, Hoek HW, Sommer IEC. Should we expand the toolbox of psychiatric treatment methods to include repetitive transcranial magnetic stimulation (rTMS)? A meta-analysis of the efficacy of rTMS in psychiatric disorders. J Clin Psychiatry. 2010;71:873–84.
13. Mehta DD, Praecht A, Ward HB, Sanches M, Sorkhou M, Tang VM, Steele VR, Hanlon CA, George TP. A systematic review and meta-analysis of neuromodulation therapies for substance use disorders. Neuropsychopharmacology. 2024;49:649–80.

14. Burt T, Lisanby SH, Sackeim HA. Neuropsychiatric applications of transcranial magnetic stimulation: a meta analysis. Int J Neuropsychopharmacol. 2002;5:73–103.
15. Speer AM, Kimbrell TA, Wassermann EM, Repella J, Willis MW, Herscovitch P, et al. Opposite effects of high and low frequency rTMS on regional brain activity in depressed patients. Biol Psychiatry. 2000;48:1133–41.
16. Fitzgerald PB, Fountain S, Daskalakis ZJ. A comprehensive review of the effects of rTMS on motor cortical excitability and inhibition. Clin Neurophysiol. 2006;117:2584–96.
17. Tang AD, Bennett W, Bindoff AD, Bolland S, Collins J, Langley RC, et al. Subthreshold repetitive transcranial magnetic stimulation drives structural synaptic plasticity in the young and aged motor cortex. Brain Stimul Basic Transl Clin Res Neuromodulation. 2021;14:1498–507.
18. Jin Y, Bai X, Jiang B, Guo Z, Mu Q. Repetitive transcranial magnetic stimulation induces quantified functional and structural changes in subcortical stroke: a combined arterial spin labeling perfusion and diffusion tensor imaging study. Front Hum Neurosci. 2022;16:829688.
19. Keck ME, Welt T, Müller MB, Erhardt A, Ohl F, Toschi N, et al. Repetitive transcranial magnetic stimulation increases the release of dopamine in the mesolimbic and mesostriatal system. Neuropharmacology. 2002;43:101–9.
20. Strafella AP, Paus T, Fraraccio M, Dagher A. Striatal dopamine release induced by repetitive transcranial magnetic stimulation of the human motor cortex. Brain. 2003;126:2609–15.
21. Strafella AP, Paus T, Barrett J, Dagher A. Repetitive transcranial magnetic stimulation of the human prefrontal cortex induces dopamine release in the caudate nucleus. J Neurosci. 2001;21:RC157.
22. Cho SS, Strafella AP. rTMS of the left dorsolateral prefrontal cortex modulates dopamine release in the ipsilateral anterior cingulate cortex and orbitofrontal cortex. PLoS One. 2009;4:e6725.
23. Zuk P, Torgerson L, Sierra-Mercado D, Lázaro-Muñoz G. Neuroethics of neuromodulation: an update. Curr Opin Biomed Eng. 2018;8:45–50.
24. Boyle Q, Illes J. Neuromodulation and opioid use disorder: ethical opportunities for Canada. Can J Neurol Sci. 2023;50:s26–33.
25. Mattioli F, Maglianella V, D'Antonio S, Trimarco E, Caligiore D. Non-invasive brain stimulation for patients and healthy subjects: current challenges and future perspectives. J Neurol Sci. 2024:456.
26. Sorkhou M, Stogios N, Sayrafizadeh N, Hahn MK, Agarwal SM, George TP. Non-invasive neuromodulation of dorsolateral prefrontal cortex to reduce craving in alcohol use disorder: a meta-analysis. Drug Alcohol Depend Rep. 2022;4:100076.
27. Sahlem GL, Kim B, Baker NL, Wong BL, Caruso MA, Campbell LA, et al. A preliminary randomized controlled trial of repetitive transcranial magnetic stimulation applied to the left dorsolateral prefrontal cortex in treatment seeking participants with cannabis use disorder. Drug Alcohol Depend. 2024;254:111035.
28. Bidzinski KK, Lowe DJE, Sanches M, Sorkhou M, Boileau I, Kiang M, et al. Investigating repetitive transcranial magnetic stimulation on cannabis use and cognition in people with schizophrenia. NPJ Schizophr. 2022;8:2.
29. Berglund M, Thelander S, Salaspuro M, Franck J, Andréasson S, Öjehagen A. Treatment of alcohol abuse: an evidence-based review. Alcohol Clin Exp Res. 2003;27:1645–56.
30. Rossi S, Antal A, Bestmann S, Bikson M, Brewer C, Brockmöller J, et al. Safety and recommendations for TMS use in healthy subjects and patient populations, with updates on training, ethical and regulatory issues: expert guidelines. Clin Neurophysiol. 2021;132:269–306.
31. McClintock SM, Reti IM, Carpenter LL, McDonald WM, Dubin M, Taylor SF, et al. Consensus recommendations for the clinical application of repetitive transcranial magnetic stimulation (rTMS) in the treatment of depression. J Clin Psychiatry. 2018;79:16cs10905.
32. Blyth SH, Zabik NL, Krosche A, Prisciandaro JJ, Ward HB. rTMS for co-occurring psychiatric and substance use disorders: narrative review and future directions. Curr Addict Rep. 2024;11:342–51.
33. Ferrão YA, da Silva R, de MF. Repetitive transcranial magnetic stimulation for the treatment of major depression during pregnancy. Braz J Psychiatry. 2018;40:227–8.

34. Hızlı Sayar G, Ozten E, Tufan E, Cerit C, Kağan G, Dilbaz N, et al. Transcranial magnetic stimulation during pregnancy. Arch Womens Ment Health. 2014;17:311–5.
35. Nahas Z, Bohning DE, Molloy MA, Oustz JA, Risch SC, George MS. Safety and feasibility of repetitive transcranial magnetic stimulation in the treatment of anxious depression in pregnancy: a case report. J Clin Psychiatry. 1999;60:50–2.
36. Ziemann U, Tam A, Bütefisch C, Cohen LG. Dual modulating effects of amphetamine on neuronal excitability and stimulation-induced plasticity in human motor cortex. Clin Neurophysiol Off J Int Fed Clin Neurophysiol. 2002;113:1308–15.
37. Buchmann J, Dueck A, Gierow W, Zamorski H, Heinicke S, Heinrich H, et al. Modulation of motorcortical excitability by methylphenidate in adult voluntary test persons performing a go/nogo task. J Neural Transm Vienna Austria. 1996;2010(117):249–58.
38. Lerner AJ, Wassermann EM, Tamir DI. Seizures from transcranial magnetic stimulation 2012–2016: results of a survey of active laboratories and clinics. Clin Neurophysiol Off J Int Fed Clin Neurophysiol. 2019;130:1409–16.
39. Swendsen JD, Merikangas KR. The comorbidity of depression and substance use disorders. Clin Psychol Rev. 2000;20:173–89.
40. Starzer MSK, Nordentoft M, Hjorthøj C. Rates and predictors of conversion to schizophrenia or bipolar disorder following substance-induced psychosis. Am J Psychiatry. 2018;175:343–50.
41. Kandel DB, Johnson JG, Bird HR, Weissman MM, Goodman SH, Lahey BB, et al. Psychiatric comorbidity among adolescents with substance use disorders: findings from the MECA study. J Am Acad Child Adolesc Psychiatry. 1999;38:693–9.
42. Somani A, Kar SK. Efficacy of repetitive transcranial magnetic stimulation in treatment-resistant depression: the evidence thus far. Gen Psychiatry. 2019;32:e100074.
43. Wing V, Bacher I, Wu B, Daskalakis Z, George T. High frequency repetitive transcranial magnetic stimulation reduces tobacco craving in schizophrenia. Schizophr Res. 2012;139:264–6.
44. Prikryl R, Ustohal L, Kucerova H, Kasparek T, Jarkovsky J, Hublova V, et al. Repetitive transcranial magnetic stimulation reduces cigarette consumption in schizophrenia patients. Prog Neuro-Psychopharmacol Biol Psychiatry. 2014;49:30–5.
45. Moeller SJ, Gil R, Weinstein JJ, Baumvoll T, Wengler K, Fallon N, et al. Deep rTMS of the insula and prefrontal cortex in smokers with schizophrenia: proof-of-concept study. Schizophrenia. 2022;8:1–9.
46. Xia G, Gajwani P, Muzina DJ, Kemp DE, Gao K, Ganocy SJ, et al. Treatment-emergent mania in unipolar and bipolar depression: focus on repetitive transcranial magnetic stimulation. Int J Neuropsychopharmacol. 2008;11:119–30.
47. Cotovio G, Ventura F, Rodrigues da Silva D, Pereira P, Oliveira-Maia AJ. Regulatory clearance and approval of therapeutic protocols of transcranial magnetic stimulation for psychiatric disorders. Brain Sci. 2023;13:1029.
48. Zangen A, Moshe H, Martinez D, Barnea-Ygael N, Vapnik T, Bystritsky A, et al. Repetitive transcranial magnetic stimulation for smoking cessation: a pivotal multicenter double-blind randomized controlled trial. World Psych Off J World Psych Assoc WPA. 2021;20:397–404.
49. Weissman CR, Bermudes RA, Voigt J, Liston C, Williams N, Blumberger DM, et al. Repetitive transcranial magnetic stimulation for treatment-resistant depression: mismatch of evidence and insurance coverage policies in the United States. J Clin Psychiatry. 2023;84:46813.
50. Peebles IS, Phillips TO, Hamilton RH. Toward more diverse, inclusive, and equitable neuromodulation. Brain Stimulat. 2023;16:737–41.
51. Hanlon CA, McCalley DM. Sex/gender as a factor that influences transcranial magnetic stimulation treatment outcome: three potential biological explanations. Front Psych. 2022;13:869070.
52. Acevedo A, Garnick DW, Lee MT, Horgan CM, Ritter G, Panas L, et al. Racial/ethnic differences in substance abuse treatment initiation and engagement. J Ethn Subst Abus. 2012;11:1–21.
53. Martin-Rodriguez JF, Ruiz-Veguilla M, Alvarez de Toledo P, Aizpurua-Olaizola O, Zarandona I, Canal-Rivero M, et al. Impaired motor cortical plasticity associated with cannabis use disorder in young adults. Addict Biol. 2021;26:e12912.

54. Nardone R, Bergmann J, Christova M, Lochner P, Tezzon F, Golaszewski S, et al. Non-invasive brain stimulation in the functional evaluation of alcohol effects and in the treatment of alcohol craving: a review. Neurosci Res. 2012;74:169–76.
55. Fitzgerald PB, Hoy KE, Elliot D, Susan McQueen RN, Wambeek LE, Daskalakis ZJ. Accelerated repetitive transcranial magnetic stimulation in the treatment of depression. Neuropsychopharmacology. 2018;43:1565–72.
56. Chen L, Hudaib A-R, Hoy KE, Fitzgerald PB. Efficacy, efficiency and safety of high-frequency repetitive transcranial magnetic stimulation applied more than once a day in depression: a systematic review. J Affect Disord. 2020;277:986–96.
57. Cole EJ, Stimpson KH, Bentzley BS, Gulser M, Cherian K, Tischler C, et al. Stanford accelerated intelligent neuromodulation therapy for treatment-resistant depression. Am J Psychiatry. 2020;177:716–26.
58. Martinotti G, Pettorruso M, Montemitro C, Spagnolo PA, Acuti Martellucci C, Di Carlo F, et al. Repetitive transcranial magnetic stimulation in treatment-seeking subjects with cocaine use disorder: a randomized, double-blind, sham-controlled trial. Prog Neuro-Psychopharmacol Biol Psychiatry. 2022;116:110513.
59. Steele VR, Maxwell AM, Ross TJ, Stein EA, Salmeron BJ. Accelerated intermittent theta-burst stimulation as a treatment for cocaine use disorder: a proof-of-concept study. Front Neurosci. 2019;13:1147.
60. Dinur-Klein L, Dannon P, Hadar A, Rosenberg O, Roth Y, Kotler M, et al. Smoking cessation induced by deep repetitive transcranial magnetic stimulation of the prefrontal and insular cortices: a prospective, randomized controlled trial. Biol Psych. 2014;76:742–9.
61. Amiaz R, Levy D, Vainiger D, Grunhaus L, Zangen A. Repeated high-frequency transcranial magnetic stimulation over the dorsolateral prefrontal cortex reduces cigarette craving and consumption. Addiction. 2009;104:653–60.
62. Friedman NP, Robbins TW. The role of prefrontal cortex in cognitive control and executive function. Neuropsychopharmacology. 2022;47:72–89.
63. Mishra B, Nizamie S, Das B, Praharaj S. Efficacy of repetitive transcranial magnetic stimulation in alcohol dependence: a sham-controlled study. Addiction. 2010;105:49–55.
64. Hoven M, Schluter RS, Schellekens AF, van Holst RJ, Goudriaan AE. Effects of 10 add-on HF-rTMS treatment sessions on alcohol use and craving among detoxified inpatients with alcohol use disorder: a randomized sham-controlled clinical trial. Addiction. 2023;118:71–85.
65. Ekhtiari H, Tavakoli H, Addolorato G, Baeken C, Bonci A, Campanella S, et al. Transcranial electrical and magnetic stimulation (tES and TMS) for addiction medicine: a consensus paper on the present state of the science and the road ahead. Neurosci Biobehav Rev. 2019;104:118–40.
66. Wilkinson ST, Holtzheimer PE, Gao S, Kirwin DS, Price RB. Leveraging neuroplasticity to enhance adaptive learning: the potential for synergistic somatic-behavioral treatment combinations to improve clinical outcomes in depression. Biol Psychiatry. 2019;85:454–65.
67. Lu F, Cui Q, Zou Y, Guo Y, Luo W, Yu Y, et al. Effects of rTMS intervention on functional neuroimaging activities in adolescents with major depressive disorder measured using resting-state fMRI. Bioeng Basel Switz. 2023;10:1374.
68. Elbau IG, Lynch CJ, Downar J, Vila-Rodriguez F, Power JD, Solomonov N, et al. Functional connectivity mapping for rTMS target selection in depression. Am J Psychiatry. 2023;180:230–40.
69. Cash RFH, Cocchi L, Lv J, Fitzgerald PB, Zalesky A. Functional Magnetic Resonance Imaging–Guided Personalization of Transcranial Magnetic Stimulation Treatment for Depression. JAMA Psychiatry. 2021;78:337–9.
70. Herremans S, Baeken C, Vanderbruggen N, Vanderhasselt M, Zeeuws D, Santermans L, et al. No influence of one right-sided prefrontal HF-rTMS session on alcohol craving in recently detoxified alcohol-dependent patients: results of a naturalistic study. Drug Alcohol Depend. 2012;120:209–13.
71. Herremans S, Vanderhasselt M, De Raedt R, Baeken C. Reduced intra-individual reaction time variability during a go-NoGo task in detoxified alcohol-dependent patients after one right-sided dorsolateral prefrontal HF-rTMS session. Alcohol Alcohol. 2013;48:552–7.

72. Herremans S, Van Schuerbeek P, De Raedt R, Matthys F, Buyl R, De Mey J, et al. The impact of accelerated right prefrontal high-frequency repetitive transcranial magnetic stimulation (rTMS) on cue-reactivity: an fMRI study on craving in recently detoxified alcohol-dependent patients. PLoS One. 2015;10
73. Jansen JM, van den Heuvel OA, van der Werf YD, de Wit SJ, Veltman DJ, van den Brink W, et al. The effect of high-frequency repetitive transcranial magnetic stimulation on emotion processing, reappraisal, and craving in alcohol use disorder patients and healthy controls: a functional magnetic resonance imaging study. Front Psychiatry. 2019:10.
74. Pripfl J, Tomova L, Riecansky I, Lamm C. Transcranial magnetic stimulation of the left dorsolateral prefrontal cortex decreases cue-induced nicotine craving and EEG delta power. Brain Stimulat. 2014;7:226–33.
75. Trojak B, Meille V, Achab S, Lalanne L, Poquet H, Ponavoy E, et al. Transcranial magnetic stimulation combined with nicotine replacement therapy for smoking cessation: a randomized controlled trial. Brain Stimulat. 2015;8:1168–74.
76. Terraneo A, Leggio L, Saladini M, Ermani M, Bonci A, Gallimberti L. Transcranial magnetic stimulation of dorsolateral prefrontal cortex reduces cocaine use: a pilot study. Eur Neuropsychopharmacol. 2016;26:37–44.

Regulatory Perspectives on Transcranial Magnetic Stimulation

Michael Pflanzer and Veljko Dubljević

1 Introduction

Paradigm shifts represent an innovation or development that leaves a lasting impression on humanity and the world—for better or worse. Consider the invention of the Gutenberg Press or the development of the atomic bomb. These developments are of such importance that history is often framed in the context of such technologies (e.g., the Atomic Age). Similarly, the twenty-first-century digital age has ushered in many transformational innovations, including the internet, artificial intelligence (AI), genetic engineering, and quantum computing. Among these advancements, the first quarter of the twenty-first century has been particularly notable for the development of neurological devices, most prominently brain–computer interfaces such as Neuralink and Synchron.[1] Despite still undergoing clinical trials [33, 38], these technologies have captured the public's attention and have been extensively covered by mainstream media (e.g., [7, 8, 34]). In contrast, some neurological devices, despite their great potential, seem to receive far less of the scrutiny and attention they warrant.

Transcranial magnetic stimulation (TMS)[2] is a prime example. Despite not receiving the same level of public scrutiny, the technology's range and scope have

[1] Synchron use a minimally invasive approach by implanting the device via blood vessels, offering less risk but with potentially lower signal fidelity compared to Neuralink's direct brain interface, which is more invasive but promises greater precision [39, 42].

[2] In this chapter, aside from the introduction of Barker's initial TMS device, we use TDS (TDS regulation, TDS technology, etc.) as an umbrella term that includes related technology (e.g., rTMS, tDCS, iTBS).

M. Pflanzer
North Carolina State University, Raleigh, NC, USA
e-mail: mspflanz@ncsu.edu

V. Dubljević (✉)
Department of Philosophy and Religious Studies, North Carolina State University, Raleigh, NC, USA
e-mail: veljko_dubljevic@ncsu.edu

V. Dubljević, J. R. Young (eds.), *TMS and Neuroethics*, Advances in Neuroethics,
https://doi.org/10.1007/978-3-031-92401-9_15

expanded significantly since its first approval by the Food and Drug Administration (FDA) in 2008 [59]. TMS was initially developed in the early 1980s at the University of Sheffield in England by Anthony Barker and colleagues as a method for mapping neural pathways and studying brain function [1]. TMS is a non-invasive procedure that uses magnetic fields to stimulate neurons in the brain [24]. A TMS device consists of an electromagnetic coil placed against the scalp. When activated, the device generates a magnetic pulse that passes through the skull and induces an electric current in targeted brain regions. The region of interest determines the specific shape and type of coil, and neuronal activity is moderated by the specific current used to either excite or inhibit brain regions, on the frequency and pattern of the pulses [29]. In the late 1980s and early 1990s, researchers began examining the therapeutic application of TMS technology for psychiatric conditions. These repetitive TMS (rTMS) devices promise non-invasive and relatively pain-free intervention options for patients who are medication-resistant or cannot tolerate pharmacological intervention due to comorbid conditions [25]. These procedures, known for their safety and tolerability, are typically performed on an outpatient basis, with sessions lasting about 20–40 min over several weeks [46, 47, 68].

Initially approved for the treatment of depression, rTMS has since been approved by the FDA for the treatment of a variety of applications, including migraines, smoking cessation [2], obsessive-compulsive disorder [62], and comorbid anxiety disorder [3]. Most recently, the FDA extended the approval of rTMS for adolescents with treatment-resistant depression [63]. Recently, researchers have developed a novel form of rTMS, intermittent theta-burst stimulation (iTBS), which promises a 90% reduction in treatment time [23]. Reduced clinical visit duration and lower self-reported experiences of discomfort [69] arguably means that the therapy could be offered to more patients who would enjoy the benefits of rTMS sooner and with fewer side effects. And while it is yet unclear as to whether TMS technologies represent a neurological paradigm shift, their potential to revolutionize psychiatric treatments is both inspiring and intriguing. As with many emerging technologies, this potential is often reflected in media portrayals that tend to amplify their promise. Research suggests that the way TMS is represented in the media does not always align with the measured tone of academic discourse [48].

An FDA-proposed ban on electrical stimulation devices for individuals with aggressive or self-injurious behaviors [64] also prompts concern as it indicates its authority to draw a line when it is revealed that certain applications of medical devices are inappropriate for certain conditions [65]. Furthermore, the recent claims by do-it-yourself (DIY) and self-improvement communities that TMS devices can be built and used at home for therapeutic and enhancement purposes also lead to questions regarding governmental roles in regulating both DIY and commercial TMS devices for cognitive enhancement purposes. Additionally, while we are unaware of existing Direct-to-Consumer (DTC) rTMS technology, it is good practice to proactively consider the risk factors of such applications as the technology develops.

This chapter reflects upon the history of regulatory approvals for rTMS and the need to consider the future roles of regulators of brain stimulation technologies. We will first review the regulatory history of rTMS before discussing how the increased range of clinical, off-label, and non-clinical applications renews the discussion of whether the scope of rTMS regulation should be expanded. This chapter concludes by discussing the hurdles of rTMS regulation and providing suggestions for the effective regulation of rTMS technology.

2 Current Range of Approval

Under the 1906 Pure Food and Drug Act and subsequent consumer protection laws, the FDA provides regulatory oversight for medical treatments and devices. Such oversight is currently only explicitly relevant to medical devices used for specific therapeutic applications that have been clinically demonstrated to significantly improve medical condition(s) while only causing "acceptable" or minimal risks of adverse effects.[3]

The FDA's regulatory scope expands as more clinical applications for TMS devices are approved. The first FDA approval for rTMS was awarded to Neuronetics in 2008, whose NeuroStar TMS Therapy system was approved for the treatment of adults with medication-resistant Major Depressive Disorder (MDD) [59].

In 2018, the FDA approved the first iTBS devices for treatment-resistant depression [32]. iTBS offers the advantages of significantly shorter treatment durations and lower stimulation intensities compared to traditional rTMS while effectively enhancing cortical excitability and functional brain connectivity [11]. This makes it a versatile and efficient therapeutic option for various neurological and psychiatric conditions with a lower risk of adverse effects [16, 23]. In the same year, rTMS received clearance from the US regulatory agency for the treatment of obsessive-compulsive disorder [62], and the approval for smoking cessation arrived only 2 years later, in 2020 [2].

Recent clinical studies demonstrate the therapeutic benefits of rTMS for adolescent MDD [36, 49], and in 2024, the FDA expanded its approval for NeuroStar rTMS Therapy to adolescents aged 15–21 with medication-resistant MDD [40]. There are several other indications of rTMS that are undergoing clinical trials. The National Institute for Health and Care Research (NIHR, the UK analog of the FDA) is also currently investigating the appropriateness of rTMS for adolescent MDD, as indicated by an ongoing call for participants for a feasibility study [37].

rTMS is also being considered for treating the symptoms of autism. These studies often show clinical improvement in autism symptomology but often have small sample sizes (e.g., [41, 74]), and research indicates that nearly one in four research participants experience one or more adverse effects from TMS treatment [22, 26, 30].[4] Research has examined the therapeutic value of rTMS for other conditions

[3] In Sect. 4.2, we suggest that the FDA should clarify whether enhancement applications of medical technology fall within its regulatory scope when such devices cannot be classified as "low risk."

[4] The most common adverse effects were headache and irritability which were reported by 16% and 14% of participants respectively. However, it should be noted that adverse effects may be under-reported (see [22, 30]).

including comorbid anxiety disorder, schizophrenia, and auditory hallucinations [11]. The FDA has not publicly reported approved trials for any of these conditions.

3 Current Regulatory Scope

The FDA regulates medical devices under the Federal Food, Drug, and Cosmetic Act (FD&C Act), ensuring that devices are safe and effective for their intended uses as specified by the manufacturer. While clinicians can use their medical expertise to prescribe pharmaceuticals and medical devices for unapproved purposes (known as "off-label usage"), the FDA regulates how manufacturers market and promote their devices [72]. Manufacturers are prohibited from marketing or promoting devices for off-label uses, with enforcement actions for violations including fines, warning letters, and other penalties.

Additionally, the FDA can modify approved device labeling if off-label use generates significant risk [58]. Conversely, manufacturers are encouraged to conduct clinical studies using Investigational Device Exemptions (IDEs) when off-label use results in significant clinical benefits [66]. The FDA reviews the clinical evidence from these studies and can approve new labeling for additional applications of medical devices.

The FDA does not always publicly report approved trials. Instead, disclosing such trials generally lies within the discretion of the device manufacturer or sponsor. Given the increased off-label usage of rTMS technology, it is reasonable to assume that numerous clinical trials are underway, even if their existence is not yet public (e.g., [28]). The FDA assesses the risks and benefits of these applications as part of its regulatory role in the USA.

Conformité Européenne (CE) marking is an analog in Europe to the U.S. FDA. CE marking indicates that a device complies with European Union legislation. Most rTMS devices with FDA approval have also been cleared or given the CE mark in Europe; however, the protocols associated with CE approval are less clear, typically provided by the device manufacturer rather than the regulatory agency [11]. Sometimes devices receive CE marking before they receive FDA approval. BrainsWay reports, for example, that their device has been CE-marked for many applications which have not received FDA clearance.[5]

The regulatory scope remains limited when medical device usage extends beyond the control of clinical practitioners and device manufacturers [50]. This underscores the need for transparency and robust data sharing to inform and advance clinical practice. Transcranial direct current stimulation (tDCS), for example, is a medical technology that precedes rTMS and similarly works by activating neural networks for the relief of clinical symptoms. Until recently, some of the major benefits of

[5] BrainsWay claims that its Deep TMS device has been CE-marked in Europe to safely treat Alzheimer's disease, autism, bipolar disorder, chronic pain, multiple sclerosis, Parkinson's disease, post-stroke rehabilitation, post-traumatic stress disorder, and the negative symptoms of schizophrenia [4]. None of these applications currently have FDA approval.

tDCS included its relative portability and affordability, which enabled providers to prescribe the devices to patients for at-home care [9]. While these advantages are no less salient than before, the growing favor of rTMS devices over tDCS devices translates to the DIY community's growing interest in developing and implementing at-home rTMS devices. Nevertheless, some of the same concerns raised toward tDCS such as safety, autonomy, privacy, and informed consent [12] also apply to TMS. Although there is significant scholarship regarding the potential to regulate at-home tDCS devices [17, 18, 67], there is scarce discussion of at-home TMS devices and how and whether regulatory agencies might expand their regulatory scope to provide oversight to DIY and DTC applications of the technology.

4 The Need to Expand the Regulatory Scope

4.1 At-Home TMS Devices

Media sensationalization of the benefits of TMS [48] could be driving interest in at-home applications of TMS. Recently, DIY enthusiasts have been developing and sharing plans for homemade "TMS devices" and notes about their non-clinical experimentation. Instructables.com is a website that hosts hundreds of thousands of DIY "Instructables" or blueprints for creating projects using a wide variety of different skills. It claims to be the largest online DIY community and currently.

There are at least two Instructables for building quasi-TMS devices. The first is a blueprint for a "real working 'Brain Ray'" powered by capacitors harvested from disposable cameras. While the DIY inventor (identified as "Quicksilv3rflash") stops short of calling this device a TMS device (despite frequent references to clinical TMS devices), they claim it "works by inducing electrical currents in neuronal tissue through a large electromagnetic pulse" [54]. The other Instructable is for a single- or paired-pulse stimulation "TMS machine" and not an rTMS machine capable of generating the needed high frequency pulse rate indicated for clinical applications [44]. However, some early experiments showcase the enhancement potential of even limited TMS machines demonstrating that limited TMS can inhibit the activity of specific brain regions in neurotypical participants, resulting in temporary savant-like abilities [51, 52]. While some might rightfully suggest that such research is dated and has not yet been replicated, the belief that at-home TMS devices *might* be useful for enhancement purposes could prove inspirational for DIY enthusiasts.

Both Instructables authors cite factors the cost and accessibility of rTMS treatment as motivations for their designs. Notably, they also disclose the dangers of handling the components necessary (which include high-voltage transformers and microwave capacitors) to build their devices and the subsequent operation of the completed at-home TMS devices.

Quicksilv3rflash's explicit goal extends beyond simple innovation, revealing an alarming agenda for using nootropic drugs and TMS techniques for personality modification, with the ultimate goal of "rewrit[ing] the personalities of myself and the volunteers in order to convert us to blissed-out altruist super-soldiers to save the world" [44]. While it is unclear whether such expressions are sincere or satirical, they highlight, at the very least, a *belief* that this potential exists, underlining the necessity for regulatory oversight. That being said, outlandish intentions (such as the long-term goal of "human/AI symbiosis") seem to be the driving force in more recent neurotechnological inventions [45]. This is underscored by Neuralink CEO Elon Musk's recent claim that "cybernetic[6] superpowers are likely in the future" [35]. Nevertheless, Quicksilv3rflash's cavalier attitude toward safety and ethical considerations, including their admission of exploiting regulatory gaps, raises serious concerns about the unchecked proliferation of DIY medical technologies.

These concerns include safety risks, lack of supervision, and efficacy. Together, these factors generate legal and ethical issues that support increasing the regulatory scope of agencies such as the FDA to regulate DIY and DTC devices and implement stricter penalties for improper off-label usage of the technology [19].

First, there are significant safety risks of TMS usage that amateurs and enthusiasts working with DTC and homemade TMS devices must consider. Even if the plans for DIY devices are well-articulated and result in functional TMS devices, many variables, such as the quality of the craftsmanship or the materials used in their construction, can result in inferior final products. Furthermore, these devices would then be used and experimented with by people who lack proficiency in operating the devices and may not understand the risks. DIY TMS devices presumably lack the precision and safety measures of clinically approved devices. Clinical TMS is carefully calibrated and administered by professionals to ensure accurate targeting and safety (e.g. [43]). In contrast, at-home setups can pose significant risks due to improper coil placement, incorrect dosage, and lack of safety protocols.

Evidence suggests, for example, that TMS has been associated with seizures, primarily in epileptic patients [30]. Furthermore, *clinical* TMS has been associated with serious adverse effects, including psychiatric events [21]. Importantly, at-home TMS devices are likely to lack the sophistication of the clinical TMS devices produced by companies with regulatory approval. And while some of the adverse effects of clinical TMS could be associated with the clinical diagnoses that precipitated the need for such therapy,[7] there still exists a clear risk of injury when devices are mishandled and the as-of-yet-unknown risks of improper coil placement. While it is tempting to suggest that preexisting conditions trigger these vulnerabilities, at-home applications of TMS will not have the same pre-screening evaluations as rTMS in clinical settings. If the clinical administration of TMS has the potential for side effects and adverse events (e.g., inducing seizures), then we must consider that at-home administration of DIY devices is likely to see such risks amplified.

[6] Dictionary.com defines cybernetics as "the study of automatic control and communication functions in both living organisms and mechanical and electronic systems, involving the application of statistical mechanics to feedback" [15].

[7] For example, [21], where a patient with schizophrenia experienced a psychological event after receiving clinical TMS.

Another major issue with at-home TMS is the absence of professional oversight. In clinical settings, TMS treatments are individualized and closely monitored by healthcare providers who adjust parameters based on patient response and safety guidelines. DIY and DTC devices would not offer this level of supervision, leading to potential misuse and inconsistent treatment outcomes. Furthermore, while clinically administered treatment often does not enjoy the same degree of oversight as clinical trials, the lack of oversight deprives the scientific community of valuable data regarding dosage, effectiveness, and range of applications of at-home TMS technology. Clinicians and researchers use the data they collect from trials and experiments and share them with the community, contributing to a greater understanding of technology's therapeutic and enhancement applications [31]. While reports of successful at-home experimentation with TMS technology may circulate in DIY and other social communities, the usefulness of these data is questionable without the rigorous controls that approved studies and trials enjoy.

The efficacy of personal TMS devices is questionable compared to professional treatments. Clinical rTMS is backed by extensive research and FDA approval for treating conditions like MDD. DIY devices often use lower currents, which produce weaker magnetic fields and lack the technological sophistication of clinical counterparts, resulting in less effective treatment and slower, less meaningful results.

Additionally, ethical issues arise from the potential harm users may experience without the benefit of professional medical advice and oversight. Amateur inventors will likely fill research gaps with hearsay and anecdotal evidence from forums, blogs, and other internet sources [70]. Another naive experimenter might assume that if 15 min of therapy is beneficial, 60 min would be even better, not realizing the potential dangers of overstimulation. In a 2016 Open Letter, Wurzman and colleagues warned of the significant risks of DIY tDCS technology. These risks include difficulties isolating the intended brain region, enhancement in some areas to the detriment of others, changes that may last longer than anticipated, small changes having an unpredictably large impact, individual differences causing different impacts using the same parameters, and the fact that the risk/benefit ratio is different for enhancement vs. therapeutic applications [73]. These risks are all also relevant to DIY TMS, and if we consider that the voltage of a DIY TMS kit is roughly 100 times that of the DIY tDCS devices [44, 55], then it is safe to assume that those with reservations about DIY tDCS will also be critical of DIY TMS. While DIY disclaimers are important, they are not enough. Notably, the worst possible effect of DIY tDCS is temporary respiratory paralysis [5], while severe adverse events and death are possible with DIY TMS. These factors underscore the critical need for stringent regulations that fill a regulatory scope gap and prevent potentially dangerous and unethical applications of TMS. While the authors of this chapter acknowledge that at-home TMS is not yet a widespread phenomenon, we still urge regulators to consider proactive rather than reactive regulation of new technology.

4.2 TMS for Enhancement Purposes

Although the ethics of TMS for enhancement purposes is beyond the scope of this paper,[8] we shall briefly identify potential ethical situations created by enhancement applications of TMS technology.

We first examine the safety component and consider whether DIY and DTC technology fall within the scope of the FDA. DTC brain stimulation devices, including those using rTMS, are typically classified as medical devices and do not qualify as low-risk general wellness devices under FDA policy, even when marketed for enhancement rather than therapeutic purposes. According to the FDA's General Wellness Policy for Low-Risk Devices, a device may avoid premarket review only if it is "intended for only general wellness use" and presents low risk to the safety of users and other persons [61]. However, rTMS devices directly interact with brain function and carry risks including seizures, cognitive changes, and unintended psychological effects, making them ineligible for this low-risk category. For example, in its Class II Special Controls Guidance for rTMS Systems, the FDA highlights the need for safety protocols, including clear risk labeling and strict operational parameters, due to the "potential for serious adverse events" [60]. Given these risks, the FDA classifies rTMS devices as Class II or higher, requiring oversight through either the 510(k) premarket notification process or premarket approval (PMA) for Class III devices. Therefore, it is arguable that enhancement-oriented rTMS devices marketed directly to consumers, without medical supervision, are subject to FDA regulation under these stringent pathways due to their significant safety implications. We strongly urge the FDA and other regulatory bodies to clarify their position on moderate- and high-risk devices, including DTC devices or other rTMS devices marketed for enhancement purposes, and DIY TMS devices.

To facilitate proactive regulation, it is essential to identify the regulatory hurdles and knowledge gaps regarding the safety and efficacy of clinical and at-home TMS technology. The next section of this paper will address regulatory hurdles and offer potential solutions for improving both the quality of regulation and the research which guides it.

5 Understanding the Regulatory Scope

5.1 Regulatory Hurdles

Ten years ago, researchers believed that TMS's cost and proficiency requirements would limit its penetration and complicate the development and application of effective regulation. Today, anyone with the capital and an internet connection can attempt to build an at-home TMS device and begin experimenting on themselves

[8] See Dubljević [18] or Dubljević et al. [19] for a comprehensive discussion of the ethics and regulation of enhancement applications of neurostimulation devices.

and others for therapeutic and enhancement purposes. What can a regulatory body whose scope is historically limited to therapeutic medical devices marketed by industry and prescribed by clinical experts do when the public is free to experiment with no clinical training or oversight?

The high costs and technical proficiency required to administer TMS once seemed like natural barriers limiting its widespread use outside clinical settings [6]. While there is little evidence of widespread marketing of at-home TMS devices, the rise of DIY neurostimulation, including tDCS, has raised concerns about safety and unsupervised use. Instructions for building DIY neurostimulation devices, including TMS, are increasingly available online, which presents challenges for regulatory bodies like the FDA. Traditionally, the FDA has focused on therapeutic medical devices marketed by established manufacturers and prescribed by clinical professionals. As neurostimulation technologies become more accessible and move beyond clinical oversight, regulatory agencies must consider how to manage the potential risks associated with unsupervised, non-clinical use of these powerful technologies.

The gatekeeper model suggests that medical practitioners should leverage medical expertise and act as gatekeepers for applying pharmaceutical interventions for approved and off-label usage, including enhancement. Some research suggests that this approach can be applied to medical devices with the caveat that such gatekeeping is best left to therapeutic rather than enhancement technology applications [17]. This distinction is critical because the FDA's regulatory authority has traditionally focused on therapeutic devices, leaving enhancement technologies in a regulatory gray area. The case of early commercial tDCS headsets, such as those marketed by companies like Foc.us for cognitive enhancement, highlights this issue [53]. These devices were marketed for non-therapeutic uses, evading FDA scrutiny, and raising questions about whether the FDA has jurisdiction to regulate devices intended for enhancement rather than treatment.

Dubljević proposed extending the Economic Disincentives Model (EDM) from stimulant drugs to stimulation devices [17]. The EDM is a regulatory approach aimed at reducing the use of stimulant drugs by making their consumption less attractive through financial and economic penalties. The EDM could provide a licensing pathway for citizens to purchase and use TMS (or, as argued by Dubljević, tDCS) in compliance with approved protocols. Key components of the EDM include incentives for compliance (e.g., subsidies, insurance premium discounts, avoiding the costs of clinical TMS), penalties for non-compliance (e.g., taxes, fines, and penalties), and communications (e.g., public education campaigns, restrictions on improper advertising). Dubljević notes, however, that this approach does not resolve the issue of non-commercial uses of the technology; it merely offers a safer alternative.

As such, it is difficult to say how successful the regulation of enhancement applications of technology would be, especially when it concerns those in the DIY community or those who buy an unregulated device for enhancement purposes. People are more likely to disobey legislation and policies when they find them unnecessarily constraining [17, 70]. The recent uptick in distrust in science will likely exacerbate this effect, particularly in an era where the public places a greater premium on healthcare autonomy. Recall, for example, the explicit desires of *Instructables* user Quicksilv3rflash to rebel against the "extortion" of medical gatekeepers. If DIY is a resistance to the "ivory tower" research model, then further regulation is unlikely to be successful. Instead, Wexler [70] suggests increasing communication to understand the motivations of the community. Rather than a prescriptive approach, regulators should attempt to discern the needs and motivations that inspire their interest in DIY TMS.

5.2 Strategies to Overcome Regulatory Hurdles

This section attempts to offer solutions to the regulatory hurdles by (1) increasing public trust in regulations by increasing transparency and promoting stakeholder feedback, (2) improving regulatory compliance by improving public access to approved TMS protocols and imposing penalties on individuals or organizations that disregard the regulations and protocols, and (3) encourages legal and policy reforms, improving the regulatory framework, and developing risk mitigation strategies.

5.2.1 Establish Clear Guidelines and Standards

One of the primary concerns with DIY TMS is the lack of oversight by trained medical professionals, which raises significant risks due to amateurs' probable ignorance of appropriate treatment protocols. To address these concerns, comprehensive protocols might be established for TMS use, detailing parameters such as intensity, frequency, duration, and target areas. Defining rigorous safety standards is crucial to prevent misuse or overuse, ensuring all devices meet stringent criteria before approval. Making these protocols (along with the safety risks of violating them) publicly available to DIY enthusiasts would increase the likelihood of laypeople following them. Additionally, implementing certification programs for both commercial and DIY TMS kits is a potential avenue for ensuring adherence to these established standards, thereby enhancing the safety and efficacy of TMS applications outside professional medical settings. However, it is important to note that DIY TMS is not yet popular enough to justify a widespread certification campaign.

Considering the rapid development of TMS for therapeutic and enhancement purposes, studies could explore the effectiveness of different regulatory models. Comparative analyses between jurisdictions—such as the FDA's risk-based model and the European Union's CE marking—could help determine the most effective frameworks for balancing innovation with public safety. Such research could inform an adaptable regulatory model tailored to neurostimulation technologies, which vary in both intended use and potential risk.

5.2.2 Increase Public and Professional Awareness

Increasing awareness among healthcare professionals and the public is vital to ensuring TMS's safe and effective use. Launching educational campaigns can help disseminate information about TMS's benefits, risks, and proper use, keeping the public well-informed. Additionally, developing training programs for healthcare professionals is essential to ensure they can properly administer and monitor TMS treatments, thereby enhancing patient safety and treatment efficacy. The success of the Tobacco-Free Campaign by the U.S. Department of Health and Human Services, for example, illustrates the potential benefits of public outreach campaigns [57]. A similar online outreach strategy could be adopted to explain the risks, benefits, and treatment protocols for rTMS therapy.

Understanding public opinion on enhancement uses of TMS, including ethical concerns and perceptions of safety, could be highly beneficial. Surveys and focus groups could examine attitudes toward DIY and commercially available enhancement devices, helping policymakers gage public support for regulation or awareness campaigns. Additionally, studies on public perception could provide insights into educational needs, helping regulators to craft messages that promote safe use and informed decision-making.

5.2.3 Encourage Ethical Research and Transparency

Promoting ethical research and transparency is fundamental to the responsible advancement of TMS technology. Developing and enforcing ethical guidelines ensures that TMS studies are conducted responsibly, safeguarding participant welfare and scientific integrity. Encouraging the sharing of data and findings within the scientific community enhances transparency and facilitates the validation and replication of research results, contributing to the overall body of knowledge.

There is a notable absence of extensive longitudinal research on the long-term effects of TMS. While many studies assess the immediate and short-term safety and efficacy of TMS, especially in therapeutic contexts like depression, research that tracks users over multiple years is limited. This gap in data makes it challenging to assess potential long-term risks or unintended consequences, such as cumulative effects on cognitive function, emotional regulation, or neuroplasticity changes over time.

Research funding priorities might also play a role. Despite TMS being used to treat hundreds of thousands of patients for FDA-approved conditions, compared to tDCS's estimated user base of under ten thousand, past neuroethics-specific funding has exclusively supported tDCS-related grants [20]. This discrepancy, despite the broader clinical impact of TMS, may partially explain the relative scarcity of ethics research centered on TMS, highlighting the need for funding decisions to consider real-world usage and impact.

The scarcity of long-term studies applies particularly to non-therapeutic or enhancement applications of TMS, which are less frequently the focus of formal clinical research. Many regulatory and ethical concerns surrounding TMS use in the enhancement domain are rooted in this lack of long-term data, as it leaves unanswered questions about safety, effectiveness, and broader neurological impacts. Addressing this research gap would be invaluable for informing policy decisions and developing guidelines that ensure the safe, responsible use of TMS over the long term.

Unfortunately, enforcing ethical guidelines on amateur researchers and DIYers would be difficult. For example, DIY members such as Quicksilv3rflash see themselves as the resistance to traditional academia and pride themselves on disregarding regulatory controls. Even if the DIY community is interested in collaborating and sharing research findings with clinical and academic researchers, the lack of scientific rigor (e.g., experimental controls, validated measures, etc.) limits the usefulness of their findings. On the other hand, anecdotal and other forms of field research data may have greater external validity, and such reports provide insights useful to clinical and academic researchers. Prior work on DIY users of tDCS has demonstrated that it is possible to reach and collect data from these otherwise "hidden" groups [27].

5.2.4 Promote Collaborative Efforts

Involving stakeholders in the regulatory process is crucial for developing well-rounded and effective regulations. Engaging patients, clinicians, and the public ensures that diverse perspectives are considered, enhancing the relevance and acceptance of regulatory measures. Creating feedback mechanisms allows for continuous input from users and professionals, facilitating the ongoing improvement and adaptation of regulations to meet evolving needs and challenges. A robust feedback loop also increases the likelihood that the needs of all stakeholders—even DIY enthusiasts—are considered, which in turn increases the likelihood that stakeholders will observe TMS regulations and approved protocols.

While communication and outreach are beneficial, true stakeholder collaboration is vital for TMS's effective regulation and advancement. Fostering collaboration among neuroscientists, ethicists, engineers, and regulatory experts can comprehensively address the interdisciplinary aspects of TMS regulation [71]. Additionally, encouraging public-private partnerships between government agencies, academic institutions, and private companies can drive research and development efforts while ensuring safety and compliance with regulatory standards.

If an infrastructure for educating and certifying the public was successful, TMS researchers might consider pushing for citizen science initiatives. The mission of Citizen Science, for example, is "to nurture collaboration between the federal government and the public to advance inclusive participation in scientific discovery and research" [10]. The Crowdsourcing and Citizen Science Act [56] authorizes citizen science organizations to provide financial awards and other prizes to individuals and groups in exchange for participating in scientific research. The benefits of such initiatives include increased public support for science, decreased federal spending, and a quicker pace of scientific and technological development [13, 14]. Citizen Science initiatives present an opportunity for TMS stakeholders to enjoy these advantages while also benefiting from the increased oversight of DIY TMS operation and adherence to established TMS protocols.

5.2.5 Implement Risk Mitigation Strategies

Implementing robust risk mitigation strategies is essential to identify and address potential hazards associated with TMS. Conducting thorough risk assessments for commercial and DIY TMS devices can help identify and mitigate risks early. Establishing monitoring and reporting systems for adverse events and long-term effects of TMS use ensures ongoing safety and helps build a comprehensive understanding of the technology's impact. These reporting mechanisms could be available to all those who wish to use them, not only academic and clinical researchers but also DIY researchers/enthusiasts/participants, even those who choose not to engage with citizen science initiatives.

5.2.6 Enhance Accessibility and Affordability

Enhancing the accessibility and affordability of TMS technologies involves providing subsidies or funding for research and development to support the creation of safe and effective TMS devices. Advocating for wide insurance coverage of TMS treatments can make these therapies more accessible and affordable for patients, thereby broadening the reach and impact of TMS applications. Furthermore, increasing the affordability of rTMS treatments undermines the financial motivations for developing DIY TMS devices. This, in turn, increases the likelihood that patients interested in TMS therapy receive such therapy under the supervision of a trained professional, thereby decreasing the likelihood of an adverse effect and increasing the overall safety of rTMS interventions.

5.2.7 Strengthen the Regulatory Framework

Strengthening the regulatory framework for TMS involves close collaboration with regulatory bodies such as the FDA, NIH, and other health organizations (e.g., NIMH, WHO) to create unified regulations. Establishing a system for periodic reviews and updates to these regulations is necessary to keep pace with new research findings and technological advancements, ensuring that the regulatory framework remains relevant and effective.

5.2.8 Legal and Policy Reforms

Legal and policy reforms are necessary to address the specific regulatory needs of TMS technologies. Advocating new legislation or amendments to existing laws can help create a regulatory environment that adequately addresses the challenges posed by TMS. Strengthening enforcement mechanisms, including imposing penalties for non-compliance, ensures that regulations are effectively upheld, maintaining high safety and ethical standards. As Dubljević [17] argues, extending the EDM to neurological devices would allow the public to purchase and become licensed to use TMS at home under approved protocols and increase the oversight and enforcement potential of regulatory agencies such as the FDA.

The necessity for legal and policy reforms for TMS can be contextualized by examining the approaches used for evaluating mental health apps, as highlighted by Wexler and Reiner [71]. They note that while the nonprofit organization Psyberguide provides consumer-oriented ratings for mental health apps, the American Psychiatric Association has developed a framework for clinical professionals to assess the safety and efficacy of these apps. These contrasting approaches underscore the absence of a unified evaluation system for DTC neurotechnology, including TMS devices.

Wexler and Reiner [71] propose a balanced approach of establishing an independent working group tasked with surveying the main domains of DTC neurotechnology and providing concise appraisals of potential harms and efficacy. This group's evaluations would not entail an exhaustive review of every product, but would rather highlight the evidence base, identify potential risks, and pinpoint gaps in current knowledge. Such an approach could be particularly beneficial for TMS regulation, offering a structured method to assess the technology's safety and efficacy without the resource burden of individual product evaluations.

To ensure broad dissemination, the working group would collaborate with outreach organizations and media outlets to reach key consumer groups [71]. This body would also serve as a resource for regulatory agencies such as the FDA and FTC, as well as for third-party organizations monitoring advertising claims, industry stakeholders, social and medical scientists, funding agencies, and the public. By drawing on diverse expertise, the working group would continuously survey and update the landscape of TMS and other DTC neurotechnology, anticipate future developments, and address ethical concerns.

Establishing such an independent body to monitor and evaluate TMS technology is imperative, particularly as government agencies and private enterprises continue to fund research into new brain modulation methods. As Wexler and Reiner [71] suggest, the current generation of DTC neurotechnology may represent only the beginning, highlighting the urgent need for comprehensive oversight to safeguard public health and ensure ethical practices.

6 Conclusion

The evolving landscape of TMS technology highlights the necessity for comprehensive regulatory frameworks that address both clinical and non-clinical applications. As TMS devices continue to expand their therapeutic potential, the increasing interest from DIY communities and the associated risks highlight critical gaps in current regulations. As Wexler and Reiner [71] suggest establishing an independent working group offers a balanced approach to evaluating and disseminating information about DTC neurotechnology. This strategy could ensure that consumers, healthcare professionals, and regulatory bodies are well-informed about TMS devices' safety, efficacy, and ethical implications.

Public outreach and educational campaigns are essential to enhancing awareness and understanding of TMS and promoting its safe and effective use. Similar strategies can be employed to communicate the benefits and risks associated with TMS therapy by drawing on the success of other public health initiatives. Furthermore, fostering interdisciplinary collaboration and stakeholder engagement is crucial for developing robust regulations that can adapt to future advancements in neurotechnology.

Strengthening the regulatory framework for TMS involves updating existing laws and implementing new policies that can address the unique challenges posed by this technology. The EDM presents a viable option for regulating enhancement uses of TMS, providing a safer alternative to unregulated DIY experimentation. By enhancing public trust, promoting ethical research, and encouraging transparent data sharing, the regulatory landscape can effectively keep pace with technological advancements while safeguarding public health.

As government agencies and private enterprises continue to invest in brain modulation research, the imperative to establish comprehensive oversight mechanisms for TMS and other neurotechnology becomes increasingly clear. Proactively developing regulatory frameworks will mitigate potential risks and support the ethical advancement of these promising technologies, ensuring their benefits are maximized for all stakeholders.

References

1. Barker AT, Jalinous R, Freeston IL. Non-invasive magnetic stimulation of human motor cortex. Lancet. 1985;1(8437):1106–7.
2. BrainsWay. Smoking addiction treatment with deep TMS. n.d.-a. https://www.brainsway.com/treatments/smoking-addiction/
3. BrainsWay. FDA clears BrainsWay deep TMS™ system for decreasing anxiety symptoms in depressed patients. n.d.-b. https://www.brainsway.com/news_events/fda-clears-brainsway-deep-tms-system-for-decreasing-anxiety-symptoms-in-depressed-patients/
4. BrainsWay. Our TMS treatments: a platform for mental health treatment. n.d.-c. https://www.brainsway.com/our-treatments/
5. Brunoni AR, et al. Clinical research with transcranial direct current stimulation (tDCS): challenges and future directions. Brain Stimul. 2011;5:175. https://doi.org/10.1016/j.brs.2011.03.002.
6. Burke MJ, Fried PJ, Pascual-Leone A. Transcranial magnetic stimulation: neurophysiological and clinical applications. In: D'Esposito M, Grafman JH, editors. Handbook of clinical neurology, vol. 163. Elsevier, 2019. p. 73–92. https://doi.org/10.1016/B978-0-12-804281-6.00005-7.
7. Capoot A. Neuralink rival Synchron's brain implant now lets people control apple's vision pro with their minds. CNBC. 2024, Jul 30. https://www.cnbc.com/2024/07/30/neuralink-rival-synchron-offers-thought-control-with-apple-vision-pro-.html
8. Chapell B. What to know about Elon Musk's Neuralink, which put an implant into a human brain. NPR. 2024, Jan 30. https://www.npr.org/2024/01/30/1227850900/elon-musk-neuralink-implant-clinical-trial
9. Charvet LE, Kasschau M, Datta A, Knotkova H, Stevens MC, Alonzo A, Loo C, Krull KR, Bikson M. Remotely-supervised transcranial direct current stimulation (tDCS) for clinical trials: guidelines for technology and protocols. Front. Syst. Neurosci. 2015;9:26. https://doi.org/10.3389/fnsys.2015.00026
10. Citizenscience.gov. About Citizenscience.gov. U.S. General Services Administration. n.d.. Retrieved online from https://www.citizenscience.gov/about/#
11. Cotovio G, Ventura F, Rodrigues da Silva D, Pereira P, Oliveira-Maia AJ. Regulatory clearance and approval of therapeutic protocols of transcranial magnetic stimulation for psychiatric disorders. Brain Sci. 2023;13(7):1029. https://doi.org/10.3390/brainsci13071029.
12. Day P, Twiddy J, Dubljević V. Present and emerging ethical issues with tDCS use: a summary and review. Neuroethics. 2022; https://doi.org/10.1007/s12152-022-09508-9.

13. de Sherbinin A, Bowser A, Tyng-Ruey T, Cooper C, Danielsen F, Edmunds R, Elias P, Faustman E, Hultquist C, Mondardini R, Popescu I, Shonowo A, Sivakumar K. The critical importance of citizen science data. Front. Clim. 2021;3 https://doi.org/10.3389/fclim.2021.650760.
14. Dick DM. Rethinking the way we do research: the benefits of community-engaged, citizen science approaches and nontraditional collaborators. Alcohol Clin Exp Res. 2017;41:1849–56. https://doi.org/10.1111/acer.13492.
15. Dictionary.com. Cybernetics. In Dictionary.com. n.d.. Retrieved 24 Oct 2024 from https://www.dictionary.com/browse/cybernetics
16. Ding Q, Zhang S, Chen S, et al. The effects of intermittent theta burst stimulation on functional brain network following stroke: an electroencephalography study. Front Neurosci. 2021;15:755709. https://doi.org/10.3389/fnins.2021.755709.
17. Dubljević V, Saigle V, Racine E. The rising tide of transcranial direct current stimulation (tDCS) in the media and academic literature. Neuron. 2014;82:731–6.
18. Dubljević V. Neurostimulation devices for cognitive enhancement: toward a comprehensive regulatory framework. Neuroethics. 2015;8(2):115–26.
19. Dubljević V, Coates McCall I, Illes J. Neuroenhancement at work: addressing the ethical, legal and social implications. In: Trempe-Martineau J, Racine E, editors. Organizational neuroethics. Cham: Springer; 2019. p. 87–103.
20. Dubljević V, Young JR, Appelbaum LG. Diversifying the bioethics funding landscape: the case of TMS. Am J Bioeth. 2021;22(1):28–30.
21. Garg H, Kumar S, Singh S, Kumar N, Verma R. New onset obsessive compulsive disorder following high frequency repetitive transcranial magnetic stimulation over left dorsolateral prefrontal cortex for treatment of negative symptoms in a patient with schizophrenia. Clin Psychopharmacol Neurosci. 2019;17(3):443–5. https://doi.org/10.9758/cpn.2019.17.3.443.
22. Giustiniani A, Vallesi A, Oliveri M, et al. A questionnaire to collect unintended effects of transcranial magnetic stimulation: a consensus based approach. Clin Neurophysiol. 2021; https://www.ifcn.info/docs/TMS-Questionnaire-Manuscript.pdf
23. Goodman MS, Vila-Rodriguez F, Barwick M, et al. A randomized sham-controlled trial of high-dosage accelerated intermittent theta burst rTMS in major depression: study protocol. BMC Psychiatry. 2024;24:28. https://doi.org/10.1186/s12888-023-05470-9.
24. Green RM, Pascual-Leone A, Wassermann EM. Ethical guidelines for rTMS research. IRB Ethic Hum Res. 1997;19:1–7.
25. Horvath JC, Perez JM, Forrow L, Fregni F, Pascual-Leone A. Transcranial magnetic stimulation: a historical evaluation and future prognosis of therapeutically relevant ethical concerns. J Med Ethics. 2011;37(3):137–43.
26. Huashuang Z, Yang L, Chensheng H, Jing X, Bo C, Dongming Z, Kangfu L, Shi-Bin W. Prevalence of adverse effects associated with transcranial magnetic stimulation for autism Spectrum disorder: a systematic review and meta-analysis. Front Psych. 2022;13:875591. https://doi.org/10.3389/fpsyt.2022.875591.
27. Jwa A. Early adopters of the magical thinking cap: a study on do-it-yourself (DIY) transcranial direct current stimulation (tDCS) user community. J Law Biosci. 2015;2(2):292–335. https://doi.org/10.1093/jlb/lsv017.
28. Kedzior KK, Gerkensmeier I, Schuchinsky M. Can deep transcranial magnetic stimulation (DTMS) be used to treat substance use disorders (SUD)? A systematic review. BMC Psych. 2018;18:137.
29. Kim TD, Hong G, Kim J, Yoon S. Cognitive enhancement in neurological and psychiatric disorders using transcranial magnetic stimulation (TMS): a review of modalities, potential mechanisms and future implications. Experim. Neurobiol. 2019;28(1):1–16. https://doi.org/10.5607/en.2019.28.1.1.
30. Lerner AJ, Wassermann EM, Tamir D. Seizures from transcranial magnetic stimulation 2012–2016: results of a survey. Clin Neurophysiol. 2019;130(8):1409–16.
31. Maizey L, Allen CPG, Dervinis M, Verbruggen F, Varnava A, Kozlov M, et al. Comparative incidence rates of mild adverse effects to transcranial magnetic stimulation. Clin Neurophysiol. 2013;124:536. https://doi.org/10.1016/j.clinph.2012.07.024.

32. Mendlowitz AB, Shanbour A, Downar J, Vila-Rodriguez F, Daskalakis ZJ, Isaranuwatchai W, Blumberger DM. Implementation of intermittent theta burst stimulation compared to conventional repetitive transcranial magnetic stimulation in patients with treatment resistant depression: a cost analysis. PLoS One. 2019;14(9):e0222546. https://doi.org/10.1371/journal.pone.0222546.
33. Mitchell P, Lee SCM, Yoo PE, et al. Assessment of safety of a fully implanted endovascular brain-computer Interface for severe paralysis in 4 patients: the Stentrode with thought-controlled digital SWITCH (SWITCH) study. JAMA Neurol. 2023;80(3):270–8. https://doi.org/10.1001/jamaneurol.2022.4847.
34. Mullin E. The race to put brain implants in people is heating p. Wired. 2023, Dec 23. https://www.wired.com/story/the-race-to-put-brain-implants-in-people-is-heating-up/
35. Musk E [@elonmusk]. I think cybernetic superpowers are likely in the future. [Post.] X. 2024, July 1. https://x.com/elonmusk/status/1811314207649931624
36. Narang P, Madigan K, Sarai S, Lippmann S. Is transcranial magnetic stimulation appropriate for treating adolescents with depression? Innov Clin Neurosci. 2019;16(9–10):33–5. PMID: 32082948; PMCID: PMC7009324
37. National Institute for Health and Care Research. 23/155 repetitive transcranial magnetic stimulation for depression in adolescents. National Institute for Health and Care Research. 2023. Retrieved online from https://www.nihr.ac.uk/documents/23155-repetitive-transcranial-magnetic-stimulation-for-depression-in-adolescents/32880
38. Neuralink. Neuralink's first-in-human clinical trial is open for recruitment. 2023. Retrieved online from https://neuralink.com/blog/first-clinical-trial-open-for-recruitment/
39. Neuralink. PRIME study progress update. 2024. Retrieved from https://neuralink.com/blog/prime-study-progress-update/
40. Neuronetics. NeuroStar® advanced therapy receives FDA clearance as a first-line add-on treatment for adolescents with depression [Press Release]. 2024. Retrieved online from https://ir.neuronetics.com/news-releases/news-release-details/neurostarr-advanced-therapy-receives-fda-clearance-first-line/
41. Oberman LM, Rotenberg A, Pascual-Leone A. Use of transcranial magnetic stimulation in autism spectrum disorders. J Autism Dev Disord. 2015;45(2):524–36. https://doi.org/10.1007/s10803-013-1960-2.
42. Oxley TJ, et al. Minimally invasive endovascular stent-electrode array for high-fidelity, chronic recordings of cortical neural activity. Nat Biotechnol. 2016;34(3):320–7. https://doi.org/10.1038/nbt.3428.
43. Perera T, George MS, Grammer G, Janicak PG, Pascual-Leone A, Wirecki TS. The clinical TMS Society consensus review and treatment recommendations for TMS therapy for major depressive disorder. Brain Stimul. 2016;9:336–46.
44. Quicksilv3rflash. Transcranial magnetic stimulation (TMS) device. Autodesk Instructables. n.d.. Retrieved online May 31, 2024 from https://www.instructables.com/Transcranial-Magnetic-Stimulation-TMS-Device/
45. Reed J, McFadden J. Neuralink: can Musk's brain technology change the world? BBC News 2024. Retrieved 11 July 2024 from https://www.bbc.com/news/health-68169082
46. Rossi S, Hallett M, Rossini PM, Pascual-Leone A, Safety of TMS Consensus Group. Safety, ethical considerations, and application guidelines for the use of transcranial magnetic stimulation in clinical practice and research. Clin Neurophysiol. 2009;120(12):2008–39.
47. Rossi S, Antal A, Bestmann S, et al. Safety and recommendations for TMS use in healthy subjects and patient populations, with updates on training, ethical and regulatory issues: expert guidelines. Clin Neurophysiol. 2021;132(1):269–306.
48. Scheper A, Rosenfeld C, Dubljević V. The public impact of academic and print media portrayals of TMS: shining a spotlight on discrepancies in the literature. BMC Med Ethic. 2022; https://doi.org/10.1186/2Fs12910-022-00760-5.
49. Sigrist C, Vöckel J, MacMaster FP, et al. Transcranial magnetic stimulation in the treatment of adolescent depression: a systematic review and meta-analysis of aggregated and individual-patient data from uncontrolled studies. Eur Child Adolesc Psychiatry. 2022;31:1501–25. https://doi.org/10.1007/s00787-022-02021-7.

50. Simon DA. Off-label preemption. Wisconsin Law Rev. 2024;1079 https://doi.org/10.59015/wlr.WCYP5227.
51. Snyder A, Bahramali H, Hawker T, Mitchell DJ. Savant-like numerosity skills in normal people by magnetic pulses. Perception. 2006;35:837–45.
52. Snyder A. Explaining and inducing savant skills: privileged access to lower level, less-processed information. Phil Trans Royal Soc Britain. 2009;364:1399–405.
53. Steenbergen L, Sellaro R, Hommel B, Lindenberger U, Kühn S, Colzato LS. "Unfocus" on foc.us: commercial tDCS headset impairs working memory. Exp Brain Res. 2016;234(3):637–43. https://doi.org/10.1007/s00221-015-4391-9.
54. TheHomebrewGuru. How to build a real working "brain ray" (transcranial magnetic stimulation demo). Instructables.com. n.d.. Retrieved online May 31, 2024 from https://www.instructables.com/How-to-Build-a-Real-Working-Brain-Ray-Transcran/
55. Tylerpridemilligan. DIY wearable TDCS device. Instructables.com. n.d.. Retrieved June 3, 2024 from https://www.instructables.com/DIY-Wearable-TDCS-Device/
56. United States Code. Crowdsourcing and Citizen Science Act, 15 U.S.C. § 3724. 2024. Retrieved from https://www.law.cornell.edu/uscode/text/15/3724
57. U.S. Department of Health and Human Services. Tobacco-free campaigns. BeTobaccoFree.gov. n.d.. https://betobaccofree.hhs.gov/tobacco-free-campaigns
58. U.S. Food and Drug Administration. Guidance on medical device patient labeling: final guidance for industry and FDA staff. U.S. Department of Health and Human Services. 2001. Retrieved from https://www.fda.gov/cdrh_docs/pdf/1128.pdf
59. U.S. Food and Drug Administration. 510(k) summary: NeuroStar TMS Therapy System (K083538). 2008. https://www.accessdata.fda.gov/cdrh_docs/pdf8/k083538.pdf
60. U.S. Food and Drug Administration. Class II special controls guidance document: repetitive transcranial magnetic stimulation (rTMS) systems. 2011. Retrieved 6 Nov 2024 from https://www.fda.gov/medical-devices/guidance-documents-medical-devices-and-radiation-emitting-products/repetitive-transcranial-magnetic-stimulation-rtms-systems-class-ii-special-controls-guidance
61. U.S. Food and Drug Administration. General wellness: Policy for low risk devices: Guidance for industry and Food and Drug Administration staff. U.S. Department of Health and Human Services. 2016. https://www.fda.gov/media/90652/download
62. U.S. Food and Drug Administration. FDA permits marketing of transcranial magnetic stimulation for treatment of obsessive-compulsive disorder. (2018, August 17). https://www.fda.gov/news-events/press-announcements/fda-permits-marketing-transcranial-magnetic-stimulation-treatment-obsessive-compulsive-disorder
63. U.S. Food and Drug Administration. 510(k) summary: NeuroStar Advanced Therapy System (K231926). 2024a. https://www.accessdata.fda.gov/cdrh_docs/pdf23/K231926.pdf
64. U.S. Food and Drug Administration. Proposed ban on electrical stimulation devices for self-injurious or aggressive behavior. Fed Regist. 2024b;89(59):20882–9. Retrieved from https://www.govinfo.gov/content/pkg/FR-2024-03-26/pdf/2024-06037.pdf
65. U.S. Food and Drug Administration. Medical device bans. n.d.-a. https://www.fda.gov/medical-devices/medical-device-safety/medical-device-bans
66. U.S. Food and Drug Administration. Investigational device exemption (IDE). n.d.-b. https://www.fda.gov/medical-devices/premarket-submissions-selecting-and-preparing-correct-submission/investigational-device-exemption-ide
67. Voarino N, Dubljević V, Racine E. tDCS for memory enhancement: a critical analysis of the speculative aspects of ethical issues. Front Hum Neurosci. 2017;10 https://doi.org/10.3389/fnhum.2016.00678.
68. Wassermann EM. Risk and safety of repetitive transcranial magnetic stimulation: report and suggested guidelines from the international workshop on the safety of repetitive transcranial magnetic stimulation, June 5–7, 1996. Electroencephalogr Clin Neurophysiol. 1998;108(1):1–16.

69. Wei K, Luo Y, Bai Y, Chen T, Bi B. Toward a neurocircuit-based sequential bilateral transcranial magnetic stimulation for treatment of obsessive-compulsive disorder. Brain Stimul. 2023;16(6):1585–7. https://www.brainstimjrnl.com/article/S1935-861X(23)01940-X/fulltext
70. Wexler A. The practices of do-it-yourself brain stimulation: implications for ethical considerations and regulatory proposals. J Med Ethics. 2016;42(4):211–5. http://www.jstor.org/stable/44014348
71. Wexler A, Reiner P. Oversight of direct-to-consumer neurotechnologies. Science. 2019;363:234–5. https://doi.org/10.1126/science.aav0223.
72. Wexler A, Nagappan A, Kopyto D, et al. Off-label promotion of transcranial magnetic stimulation on provider websites. Brain Stimul. 2021;14:723–4.
73. Wurzman R, Hamilton RH, Pascual-Leone A, Fox MD. An open letter concerning do-it-yourself users of transcranial direct current stimulation. Ann Neurol. 2016;80:1–4. https://doi.org/10.1002/ana.24689.
74. Yang Y, Song P, Wang Y. Assessing the impact of repetitive transcranial magnetic stimulation on effective connectivity in autism spectrum disorder: an initial exploration using TMS-EEG analysis. Heliyon. 2024;10(11):e31746. https://doi.org/10.1016/j.heliyon.2024.e31746.

MIX
Papier aus verantwortungsvollen Quellen
Paper from responsible sources
FSC® C105338
FSC
www.fsc.org

If you have any concerns about our products,
you can contact us on
ProductSafety@springernature.com

In case Publisher is established outside the EU,
the EU authorized representative is:
Springer Nature Customer Service Center GmbH
Europaplatz 3, 69115 Heidelberg, Germany

Printed by Libri Plureos GmbH
in Hamburg, Germany